RECENT DEVELOPMENTS IN SUPERCONDUCTIVITY RESEARCH

RECENT DEVELOPMENTS IN SUPERCONDUCTIVITY RESEARCH

BARRY P. MARTINS
EDITOR

Nova Science Publishers, Inc.
New York

NOTICE TO THE READER

LIBRARY OF CONGRESS CATALOGING-IN-PUBLICATION DATA
Recent developments in superconductivity research / Barry P. Martins, editor.
    p. cm.
Includes bibliographical references and index.
ISBN 13 978-1-60021-462-2
ISBN 10 1-60021-462-2
Superconductivity--Research. I. Martins, Barry P.
QC611.96.R43   2006
621.3'5--dc22   2006030548

Published by Nova Science Publishers, Inc.   ✦ New York

CONTENTS

PREFACE

Superconductivity is the ability of certain materials to conduct electrical current with no resistance and extremely low losses. High temperature superconductors, such as La2-xSrxCuOx (Tc=40K) and YBa2Cu3O7-x (Tc=90K), were discovered in 1987 and have been actively studied since. In spite of an intense, world-wide, research effort during this time, a complete understanding of the copper oxide (cuprate) materials is still lacking. Many fundamental questions are unanswered, particularly the mechanism by which high-Tc superconductivity occurs. More broadly, the cuprates are in a class of solids with strong electron-electron interactions. An understanding of such "strongly correlated" solids is perhaps the major unsolved problem of condensed matter physics with over ten thousand researchers working on this topic. High-Tc superconductors also have significant potential for applications in technologies ranging from electric power generation and transmission to digital electronics. This ability to carry large amounts of current can be applied to electric power devices such as motors and generators, and to electricity transmission in power lines. For example, superconductors can carry as much as 100 times the amount of electricity of ordinary copper or aluminum wires of the same size. Many universities, research institutes and companies are working to develop high-Tc superconductivity applications and considerable progress has been made. This new volume brings together new leading-edge research in the field.

To synthesize a new superconductor which has a critical temperature, T_c exceeding the room temperature, one needs to know what chemical components to start with. Chapter 1 presents analysis of experimental data which allow one to draw a conclusion about components and the structure of a potential room-temperature superconductor. The two essential components of a room-temperature superconductor are large organic molecules (polymers, tissues) and atoms/molecules which are magnetic in the intercalated state. This conclusion is fully based on experimental facts known today, and does not require any assumptions about the mechanism of room-temperature superconductivity. This, however, does not mean that to synthesize a room-temperature superconductor is an easy task.

The discovery of ceramic-based high temperature superconductors (HTS) during 1980's opened the possibility of applying the technology to electric power devices such as power transmission cable, high magnetic field magnets, SQUID, motor and magnetic levitation trains etc. Dissipation phenomena in high temperature superconductors are governed by the microstructure that develops during the preparation process. Therefore, superconductors prepared either in the form of single crystals, thin films or polycrystalline plays an important role for understanding superconductivity as well as for practical applications. Also various doping elements play an impressive role in any superconducting systems.

In superconductors where the dc electrical conductivity diverges to infinity below T_c, the thermal conduction is almost a unique measurement to study the transport properties below T_c. The figure of merit is the deciding factor for the quality of thermoelectric materials. In order to increase the whole figure of merit, it is of interest to replace the p-type leg of the Peltier junction by a thermoelectrically passive material with a figure of merit close to zero. This is why it is interesting to study the figure of merit of the ceramic superconductors. One of the important thermomagnetic transport quantities is the electrothermal conductivity and is shown to be one of the powerful probes of high-temperature superconductors. Potent applications for electrothermal conductivity of superconductors are actuators in MEMS technologies, superconducting bolometric detectors, electrothermal rockets etc.

Chapter 2 investigates the relationships among macroscopic physical properties and features at atomic level for the high-T_c superconducting (Bi(Pb)Sr(Ba)2223) material, which was prepared by a solid state reaction method. The samples were analyzed by dc electrical resistivity, ac susceptibility, thermal transport, electrothermal conductivity and thermoelectric properties all as a function of temperature (from room down to LN_2 temperature). Room temperature X-ray diffraction studies were also done. Samples are investigated for thermal transport properties, i.e. thermal conductivity, thermal diffusivity and heat capacity per unit volume, by an Advantageous Transient Plane Source method. Simultaneous measurement of thermal conductivity and thermal diffusivity makes it possible to estimate specific heat and the Debye temperature Θ_D, as well as to separate the electron and phonon contributions to thermal conductivity and diffusivity. Thermoelectric power (Seebeck Coefficient) and electrothermal conductivity are also measured. Using electrical resistivity, thermal conductivity and thermoelectric power, the figure of merit factor is estimated.

$(Bi,Pb)_2Sr_2Ca_2Cu_3O_x$ (BSCCO) high-temperature superconductor (HTS) tape has been widely studied for use in the insert coil of 1 GHz NMR magnets because of its high upper critical magnetic field at 4.2 K. This coil system is required to have a high magnetic field with excellent field stability and field homogeneity, in order to improve its spectral resolution and sensitivity. To obtain this high magnetic field, multiple superconducting coils are assembled vertically and thus require many joints to interconnect them. In addition, to maintain the field stability, the coil system must be operated in the persistent current mode, thereby necessitating the connection of the coils to a persistent current switch or a flux pump. Therefore, it is considered that the joint between the superconducting tapes is an important factor in designing and developing HTS magnets for industrial applications.

In Chapter 3, the authors joined Bi-2223/Ag multifilamentary superconductor tape and fabricated closed double-pancake coils using the resistive-joint and superconducting-joint methods. The critical current ratio (CCR) of the jointed tape and the decay characteristics, joint resistance, and n-value of the closed pancake coils were estimated by the standard four-point probe technique and field decay technique. In addition, the joint resistance of the closed coil was evaluated as a function of the critical current of coil, contact length, sweep time, and external magnetic flux density. It was observed that the measured value of the CCR was higher in the jointed tape made by the resistive-joint method than in that made by the superconducting-joint method. On the other hand, the joint resistance was measured to be 4 orders of magnitude smaller in the superconducting-joint coil, in which approximately 40% of the critical current was retained in the persistent current mode and the joint resistance was 0.18 nΩ. This better and longer retention of the magnetic field in the superconducting-joint coil is believed to be due to the direct connection between the superconducting cores. In addition, the measured results were compared with the numerical values calculated by the 4th order Runge-Kutta technique.

In chapter 4, STO-based buffer layers will be prepared using chemical solution deposition (CSD) on Ni (200) tapes. First, the results show that seed layer is necessary for fabricating (200)-oriented STO buffer layers; second, the effects of addition of Ba element into seed layers are studied, the results show that when the seed layer is $Sr_{0.7}Ba_{0.3}TiO_3$ one can obtain highly (200)-oriented STO buffer layers when the annealing temperature is 900℃, which can be attributed to the orientation improvement of seed layers induced by the appearance of BaF_2 liquid phase; third, the conductive buffer layers $Sr_{1-x}La_xTiO_3$ (LSTO) ($0\leq x\leq0.4$) are fabricated using CSD method, the results show that it is possible to fabricate high-quality conductive LSTO buffer layers for YBCO coated conductors. Finally, thick STO buffer layers are fabricated using multi-seeded technique, the results show that the orientation of thick STO buffer layers can be improved using the so-called multi-seeded technique.

Chapter 5 presents a novel route to prepare the superconductive sodium cobalt oxyhydrates using aqueous permanganate solution. Immersing γ-$Na_{0.7}CoO_2$ in $KMnO_4$ (aq.) leads to two distinct novel phases of $(Na,K)_x(H_2O)_yCoO_2$, depending on the concentration of the $KMnO_4$ (aq.) solution. These two phases differ in the c-axis but show little change from the parent compound in the a-axis. The phase transformation can be viewed as a topotactic process. The c ≈ 19.6 Å phase has bi-layers of water molecule inserted into the lattice, whereas the c ≈ 13.9 Å phase has mono-layers of water molecule inserted into the lattice essentially without changing the skeleton of the parent compound. Immersing γ-$Na_{0.7}CoO_2$ in $NaMnO_4$ (aq.) only leads to the c ≈ 19.6 Å phase of $Na_x(H_2O)_yCoO_2$. The formation mechanism and phase stability of $KMnO_4$-treated $(Na,K)_x(H_2O)_yCoO_2$ would be discussed. The hydration is a very slow process when using $KMnO_{4(aq.)}$ with a low molar ratio of $KMnO_4$/Na. Three related phases of c ≈ 19.6 Å, c ≈ 13.9 Å, and c ≈ 11.2 Å exist in the cobalt oxyhydrates, being associated with the water content in the lattice. The thermal stability of the c ≈ 19.6 Å phase of $(Na,K)_x(H_2O)_yCoO_2$ behaves differently from that of the c ≈ 13.9 Å phase, which is ascribed to the difference between the ion-dipole interaction of K^+-H_2O within the alkaline layers and that of Na^+-H_2O located between the CoO_2 layers and the alkaline layers. The thermopower behavior of three related phases of cobalt oxyhydrates will be presented. The generalized Heikes formula would be used to explain the variation of the thermopower between γ-$Na_{0.7}CoO_2$ and $Na_{0.33}K_{0.02}(H_2O)_{1.33}CoO_2$ arising from the variation of the concentration of Co^{4+}.

The large possibilities and advantages of pulsed injection metal organic vapour deposition technique (PI-MOCVD) for the growth of high-T_c superconductors and related multilayered heterostructures are demonstrated in Chapter 6. Accurate precursors micro-dosing by computer controlled electromagnetic injectors and easy variation of injections parameters allow a flexible control of the MOCVD process concerning the film growth rate, thickness and composition. Film growth can be controlled reproducibly at a level of angstrom per pulse and the film thickness is directly related with pulses number; consequently, the growth can be named as "digital." Such a feature is very favourable for the growth of sophisticated multilayered heterostructures like superlattices, where precise thickness and composition control is a key factor. In this case multiple injectors are used.

High quality superconducting films suitable for various applications can be grown by PI-MOCVD. A number of various heteroepitaxial structures in which the YBCO layer is combined with dielectric, conducting or ferromagnetic layers were deposited, including superconductor/dielectric superlattices and polarized-spin injection devices. Their properties and possible applications are discussed. The recent achievements of the PI-MOCVD technique in the field of YBCO coated conductors are also presented. The results on the ex-situ or in-situ growth and properties of various buffer layers-YBCO architectures on metallic substrates and tapes are discussed in the context of state-of-art in this field. Finally, the industrial developments of PI-MOCVD technology are reviewed.

Chapter 7 investigates the physics of the microwave response in $YBa_2Cu_3O_{7-\sigma}$, $SmBa_2Cu_3O_{7-\sigma}$ and MgB_2 in the vortex state. The authors first recall the theoretical basics of vortex-state microwave response in the London limit. They then present a wide set of measurements of the field, temperature, and frequency dependences of the vortex state microwave complex resistivity in superconducting thin films, measured by a resonant cavity and by swept-frequency Corbino disk. The combination of these techniques allows for a comprehensive description of the microwave response in the vortex state in these innovative superconductors. In all materials investigated they show that flux motion alone cannot take into account all the observed experimental features, neither in the frequency nor in the field dependence. The discrepancy can be resolved by considering the (usually neglected) contribution of quasiparticles to the response in the vortex state. The peculiar, albeit different, physics of the superconducting materials here considered, namely two-band superconductivity in MgB_2 and superconducting gap with lines of nodes in cuprates, give rise to a substantially increased contribution of quasiparticles to the field-dependent microwave response. With careful combined analysis of the data it is possible to extract or infer many interesting quantities related to the vortex state, such as the temperature-dependent characteristic vortex frequency and vortex viscosity, the field dependence of the quasiparticle density, the temperature dependence of the σ-band superfluid density in MgB_2

Chapter 8 presents the updates of the superconductivity in simple elements, as reported at the turn of the twenty first century. After almost a hundred years of superconductivity research, the non-superconducting gaps in the periodic table of elements are shrinking while the maximum superconducting temperatures achieved by simple elements is raising to values unforeseen several decades ago. The research of superconductivity in simple elements was revived by the recent developments of high pressure diamond anvil cells together with the discovery of superconductivity in magnesium diboride at a remarkable high transition temperature (40 K). If an element is not superconducting down to very low temperatures, there area several methods to transform it into a superconductor. Among the most used method to probe superconductivity of elements is by subjecting them to high pressure, irradiation, charge doping. On the other side, quenched condensed, templated, and very thin films allow amorphous, structural phase, or proximity induced superconductivity in non-superconducting elements. This review offers the readers a comprehensive picture of superconductivity, its correlations and trends for simple elements.

Unusual properties of strongly correlated liquid observed in the high-Tc superconductors and heavy-fermion (HF) metals are determined by quantum phase transitions taking place at their critical points. Therefore, direct experimental studies of these transitions and critical points are of crucial importance for understanding the physics of high-Tc superconductors and HF metals. In case of high-Tc superconductors such direct experimental studies are absent since at low temperatures corresponding critical points are occupied by the superconductivity. Recent experimental data on the behavior of HF metals illuminate both the nature of these critical points and the nature of the phase transitions. The authors show that it is of crucial importance to simultaneously carry out studies of both the high-Tc superconductivity and the anomalous behavior of HF metals. The understanding of this fact has been problematic largely because of the absence of theoretical guidance. The main features of the fermion condensation quantum phase transition (FCQPT), which are distinctive in several aspects from that of conventional quantum phase transition (CQPT), are considered. Chapter 9 deals with these fundamental problems through studies of the behavior of quasiparticles, leading to good quantitative agreement with experimental facts. They show that in contrast to CQPT, whose physics in the critical region is dominated by thermal and quantum fluctuations and characterized by the absence of quasiparticles, the physics of a Fermi system near FCQPT or undergone FCQPT is controlled by the system of Landau-type quasiparticles. However, contrary to the conventional Landau quasiparticles, the effective mass of these strongly depends on the temperature T, magnetic fields B, the number density x, etc. Our general consideration suggests that FCQPT and the emergence of novel quasiparticles near and behind FCQPT are distinctive features of strongly correlated substances such as the high-Tc superconductors and HF metals. The authors show that the main properties and universal behavior of the high-Tc superconductors and HF metals can be understood within the framework of presented here theory based on FCQPT. A large number of the experimental evidences in favor of the existence of FCQPT in high-Tc superconductors and HF metals is presented. The authors demonstrate that the essence of strongly correlated electron liquids can be controlled by both magnetic field B and temperature T. Thus, the main properties of heavy-fermion metal such as magnetoresistance, resistivity, specific heat, magnetization, volume thermal expansion, etc, are determined by its position on the $B - T$ phase diagram. The obtained results are in good agreement with recent facts and observations.

In: Recent Developments in Superconductivity Research   ISBN 978-1-60021-462-2
Editor: Barry P. Martins pp. 1-31

Chapter 1

ROUTE TO ROOM-TEMPERATURE SUPERCONDUCTIVITY FROM A PRACTICAL POINT OF VIEW

A. Mourachkine*

Cavendish Laboratory, University of Cambridge,
Madingley Road, Cambridge CB3 0HE, UK

Abstract

To synthesize a new superconductor which has a critical temperature, T_c, exceeding the room temperature, one needs to know what chemical components to start with. This chapter presents analysis of experimental data which allow one to draw a conclusion about components and the structure of a potential room-temperature superconductor. The two essential components of a room-temperature superconductor are large organic molecules (polymers, tissues) and atoms/molecules which are magnetic in the intercalated state. This conclusion is fully based on experimental facts known today, and does not require any assumptions about the mechanism of room-temperature superconductivity. This, however, does not mean that to synthesize a room-temperature superconductor is an easy task.

Never let them persuade you that things are too difficult or impossible.

—Sir Douglas Bader

1   Introduction

The superconducting state is a state of matter: it is a *quantum* state occurring on a macroscopic scale. As any state of matter, superconductivity is not a property of isolated atoms, but is a collective effect determined by the structure of a whole sample. From a classical point of view, the superconducting state is characterized by two distinctive properties:

---

*Present address: Institut de Physique des Nanostructures, Ecole Polytechnique Federal de Lausanne Batiment PH-B, Station 3, 1015 Lausanne, Switzerland *perfect electrical conductivity* ($\rho = 0$) and *perfect diamagnetism* ($\mathbf{B} = 0$ inside the super-conductor, where $\mathbf{B}$ is the magnetic field). After the discovery of the phenomenon of superconductivity in 1911 [1], humans try to derive a good deal of benefit from its peculiar properties. In spite of the fact that superconductivity is a low-temperature phenomenon, the possibility to use a superconductor at room temperature is an old dream.

It is necessary first to define the expression "a room-temperature superconductor" because some perceive it as a superconductor having a critical temperature $T_c \sim 300$ K, others as a superconductor functioning at 300 K. There is a huge difference between these two cases. From a technical point of view, superconductors only become useful when they are operated well below their critical temperature—one-half to two-third of that temperature provides a rule of thumb. Therefore, for an engineer, a room-temperature superconductor would be a compound whose resistance disappears somewhere above 450 K. Such a material could actually be used at room temperature for large-scale applications. At the same time, $T_c \sim 350$ K can already be useful for small-scale (low-power) applications. Consequently, unless specified, the expression "a room-temperature superconductor" will further be used to imply a superconductor having a critical temperature $T_c \geq 350$ K.

The invention of the transistor is directly responsible for the way in which silicon technology has so profoundly changed the world in which we live. The availability of a room-temperature superconductor may change our lives even to a greater degree. What technical marvels could we expect to see?

The benefits would range from minor improvements in existing technology to revolutionary upheavals. All devices made from the room-temperature superconductor will be reasonably cheap since its use would not involve cooling cost. Energy savings from many sources would add up to a reduced dependence on conventional power plants. Compact superconducting cables would replace unsightly power lines and revolutionize the electrical power industry. A world with room-temperature superconductivity would unquestionably be a cleaner world and a quieter world. Compact superconducting motors would replace many noisy, polluting engines. Advance transportation systems would lessen our demands on the automobile. Superconducting magnetic energy storage would become commonplace. Computers would be based on compact Josephson junctions. Thanks to the high-frequency, high-sensitivity operation of superconductive electronics, mobile phones would be so compact that could be made in the form of an earring. SQUID (Superconducting QUantum Interference Device) sensors would become ubiquitous in many areas of technology and medicine. Room-temperature superconductivity would undoubtedly trigger a revolution of scientific imagination. The effects of room-temperature superconductivity would be felt throughout society, including children who might well grow up playing with superconducting toys.

In the literature, one can find more than 20 papers reporting evidence of superconductivity near or above room temperature [2]. Most researchers in superconductivity do not accept the validity of these results because they cannot be reproduced by others. The main problem with most of these results is that superconductivity is observed in samples containing many different phases, and the superconducting fraction (if such exists at all) of these samples is usually very small. Thus, superconductivity may exist in these complex materials, but nobody knows what phase is responsible for its occurrence. In a few cases, however, the phase is known but superconductivity was observed exclusively on the surface. For any substance, the surface conditions differ from those inside the bulk, and the degree of this difference depends on many parameters, and some of them are extrinsic.

Room-temperature superconductivity was already discussed in a book [2]. The main purpose of the book was to show that it is possible to synthesize a room-temperature superconductor. It was concluded that a room-temperature superconductor should consist of large organic molecules (polymers, living tissues) and magnetic atoms/molecules (in the doped state). This outcome is based on **knowledge** of the mechanism of high-temperature superconductivity described in the other book [3]. The mechanism, in its turn, is based on experimental data, mainly, on tunneling measurements obtained in cuprates. However, the **same** conclusion about components and the structure of a room-temperature superconductor can be derived *independently*. This is exactly what we are going to do in this chapter. Thus, to draw a conclusion about components and the structure of a room-temperature superconductor, it is not necessary to make any assumptions about the mechanism of room-temperature superconductivity.

The chapter consists of ten sections and is organized as follows. In the following section, we shall briefly discuss guidelines for materials that superconduct at high temperatures, presented by Geballe in 1993. The third section describes the physical properties of known superconductors. It turns out that all superconductors can be classified into three groups according to their structural and magnetic properties. From the analysis of superconducting properties, one can easily infer that a potential room-temperature superconductor can only belong to one of these three groups, and not to the other two. Knowledge of the common physical properties of this group, which are analyzed in the fourth section, gives an opportunity to know what physical properties should we expect from a room-temperature superconductor. In the fifth section, we shall discuss the most important requirements for materials that superconduct at high temperatures. The principles of superconductivity are briefly discussed in the sixth section. Bearing in mind the information presented in the first six sections, an approach to room-temperature superconductivity is proposed in the seventh section. The presence of bipolarons (bisolitons) in some non-superconducting polymers and large organic molecules at room temperature is discussed in the following section. Components and the structure of a promising room-temperature superconductor are considered in the ninth section. The chapter ends with conclusions.

2   Geballe's Guidelines

In 1992, a diverse group of researchers gathered at a two-day workshop in Bodega Bay (California). They considered the issue of making much higher temperature superconductors. T. H. Geballe, who attended this workshop, summarized some guidelines in a two-page paper published in *Science* [4], that emerged from the discussions:

- Materials should be multicomponent structures with more than two sites per unit cell, where one or more sites not involved in the conduction band can be used to introduce itinerant charge carriers.

- Compositions should be near the metal-insulator Mott transition.

- On the insulating side of the Mott transition, the localized states should have spin-1/2 ground states and antiferromagnetic ordering of the parent compound.

- The conduction band should be formed from antibonding tight-binding states that have a high degree of cation-anion hybridization near the Fermi level. There should be no extended metal-metal bonds.

- Structural features that are desirable include two-dimensional extended sheets or clusters with controllable linkage, or both.

These hints are mainly based on knowledge of the physical properties of cuprates. In a sense, one of our tasks in this chapter is to extend these guidelines.

3   Three Groups of Superconductors

The task to synthesize a room-temperature superconductor is a materials-physics problem. Therefore, it is worthwhile to review superconducting compounds. Classify all superconducting materials into three groups. The groups consist of superconductors which are:

    1) **three-dimensional and non-magnetic,**

    2) **low-dimensional and non-magnetic,** and

    3) **low-dimensional and magnetic.**

Recently, it was shown that the mechanisms of superconductivity in compounds of these three groups are different [2, 3]. However, this issue is not important in the context of this chapter. As was mentioned in the Introduction, in order to draw a conclusion about components and the structure of a room-temperature superconductor, we shall not try to understand possible mechanisms of room-temperature superconductivity. Let us briefly review a few superconductors from these three groups (for more information, see [2]).

3.1   First Group of Superconducting Materials

The first group of superconductors incorporates non-magnetic elemental superconductors and some of their alloys. The superconducting state in these materials is well described by the BCS theory of superconductivity [5]. Thus, this group of superconductors includes all classical, conventional superconductors. The critical temperature of these superconductors **does not** exceed 10 K. Most of them are type-I superconductors. As a consequence, superconductors from this group are not suitable for applications because of their low transitional temperature and low critical field.

Ironically, many superconductors, discovered mainly before 1986, were assigned to this group by mistake. In fact, they belong to either the second or third group of superconductors. For example, the so-called A-15 superconductors, during a long period of time, were considered as conventional; in reality, they belong to the second group. The so-called Chevrel phases were first assigned also to the first group; however, superconductivity in Chevrel phases is of unconventional type, and they are representatives of the third group of superconductors.

Some elements which superconduct under high pressure belongs either to the third or the second group of superconductors. Under high pressure, their crystal structure becomes low-dimensional, often containing simultaneously two- and one-dimensional substructures. In addition, some of them are magnetic. For example, under extremely high pressure, iron exhibits superconductivity which is of unconventional type. Thus, the number of superconductors in the first group is in fact very small, and they are not suitable for applications.

3.2   Second Group of Superconducting Materials

The second group of superconductors incorporates low-dimensional and non-magnetic compounds, such as A-15 superconductors, the metal oxide $Ba_{1-x}K_xBiO_3$, the magnesium diboride MgB_2 and a large number of other binary compounds. The superconducting state in these materials is characterized by the presence of two interacting superconducting subsystems. One of them is low-dimensional and exhibits genuine superconductivity of unconventional type, while superconductivity in the second subsystem which is three-dimensional is often induced by the first one and of the BCS type. So, superconductivity in this group of materials can be called half-conventional (or alternatively, half-unconventional). The critical temperature of these superconductors is **limited** by ~ 40 K and, in some of them, T_c can be tuned. All of them are type-II superconductors with an upper critical magnetic field usually exceeding 10 T. Therefore, many superconductors from this group are suitable for different types of practical applications.

• Intermetallic compounds of transition metals of niobium (Nb) and vanadium (V) such as Nb_3B and V_3B, where B is one of the nontransitional metals, have the structure of beta-tungsten (β-W) designated in crystallography by the symbol A-15. As a consequence, superconductors having the structure A_3B (A = Nb, V, Ta, Zr and B = Sn, Ge, Al, Ga, Si) are called the A-15 superconductors. Nb_3Ge has the highest critical temperature, $T_c = 23.2$ K. The critical temperature of A-15 superconductors is very sensitive to changes in the 3:1 stoichiometry. In the crystal structure of the binary A_3B compounds, the atoms B form a body-centered cubic sublattice, while the atoms A are situated on the faces of the cube forming three sets of non-interacting orthogonal one-dimensional chains.

• Superconductivity in the metal oxide $BaPb_{1-x}Bi_xO_3$ was discovered in 1975, which has a maximum $T_c \simeq 13.7$ K at $x = 0.25$. Other members of this family are $BaPb_{0.75}Sb_{0.25}O_3$ ($T_c = 0.3$ K) and $Ba_{1-x}K_xBiO_3$ (BKBO). The metal oxide BKBO is an exceptionally interesting material and the first oxide superconductor without copper with a critical temperature above that of all the A-15 compounds. Its critical temperature is $T_c \simeq 32$ K at $x = 0.4$. At the moment of writing, BKBO still exhibits the highest T_c known for an oxide other than the cuprates. Superconducting BKBO with low potassium content exhibits a charge-density-wave ordering. The density of charge carriers in BKBO is very low. Various evidence suggests that the electron-phonon coupling is responsible for superconductivity in BKBO. A two-band model applied to BKBO accounts very well for all the available data on BKBO. Acoustic measurements performed in BKBO show that many physical properties of BKBO are quite similar to those of the A-15 superconductors.

• In January 2001, magnesium diboride MgB_2 was found to superconduct at $T_c = 39$ K. At the moment of writing, the intermetallic MgB_2 has the highest critical temperature at ambient pressure among all superconductors with the exception of superconducting cuprates. The crystal structure of MgB_2 is composed of layers of boron and magnesium, alternating along the c axis. Each boron layer has a hexagonal lattice similar to that of graphite. The magnesium atoms are arranged between the boron layers in the centers of the hexagons. Superconductivity in MgB_2 occurs in the boron layers. The electron-phonon interaction seems to be responsible for the occurrence of superconductivity in MgB_2. The density of states in MgB_2 is small. MgB_2 has a very low normal-state resistance: at 42 K the resistivity of MgB_2 is more than 20 times smaller than that of Nb_3Ge in its normal state. MgB_2 has two energy gaps, $\Delta_L/\Delta_s \simeq 2.7$. Seemingly, both the energy gaps have s-wave symmetries: the larger gap is highly anisotropic, while the smaller one is either isotropic or slightly anisotropic. The larger energy gap Δ_L occurs in the σ-orbital band, while Δ_s in the π-orbital band.

• There are a large number of binary superconductors. Non-magnetic binary compounds exhibiting high values of T_c and H_{c2} belong to the second group of superconductors, such as *nitrides*, *carbides* and *laves phases*. Nitrides and carbides are also known as B1 superconductors. Metallic AB_2 compounds that superconduct are called the laves phases. Semiconductors, e.g. GeTe and SnTe, also belongs to the second group of superconductors.

3.3 Third Group of Superconducting Materials

The third group of superconductors is the largest and incorporates superconductors which are low-dimensional and magnetic, or at least, these compounds have strong magnetic correlations. This is basically the group of unconventional superconductors. Superconductors with the highest critical temperature belong to this group—cuprates show $T_{c,max} \simeq 135$ K.

In the majority of unconventional superconductors, the magnetic correlations favor an antiferromagnetic ordering. In contrast to antiferromagnetic superconductors, ferromagnetic ones usually have a low critical temperature. The density of charge carriers in these superconductors is very low. All unconventional superconductors are of type-II. They have a very large upper critical magnetic field. As a consequence, many superconductors from this group are used for practical applications. In this subsection we shall briefly discuss the following compounds from the third group of superconductors: Chevrel phases, cuprates, charge transfer organics, fullerides, graphite intercalation compounds, non-organic polymers, carbon nanotubes, heavy fermions, nickel borocarbides, the strontium ruthenate, ruthenocuprates, $MgCNi_3$, $Cd_2Re_2O_7$, $Na_xCoO_2 \cdot yH_2O$, hydrides and deuterides. We start with the so-called Chevrel phases.

3.3.1 Chevrel Phases

In 1971, Chevrel and co-workers discovered a new class of ternary molybdenum sulfides, having the general chemical formula $M_xMo_6S_8$, where M stands for a large number of metals and rare earths (nearly 40), and $x = 1$ or 2. The Chevrel phases with S substituted by Se or Te also display superconductivity. Before the discovery of high-T_c superconductivity in cuprates, the A-15 superconductors had the highest values of T_c, but the Chevrel phases were the record holders in exhibiting the highest values of upper critical magnetic field H_{c2}. $PbMo_6S_8$ has the highest critical temperature, $T_c \simeq 15$ K, and upper critical magnetic field, $H_{c2} \simeq 60$ T. Superconductivity in the Chevrel phases coexists with antiferromagnetism of the rare earth elements. For example, a long-range antiferromagnetic order of the rare earth elements RE = Gd, Tb, Dy and Er in $(RE)Mo_6X_8$, setting in respectively at $T_N = 0.84$,

0.9, 0.4 and 0.15 K, coexists with superconductivity occurring at T_c = 1.4, 1.65, 2.1 and 1.85 K, respectively, where T_N is the Néel temperature. Superconductivity in the Chevrel phases is primarily associated with the mobile $4d$-shell electrons of Mo, while the magnetic order involves the localized $4f$-shell electrons of the rare earth atoms which occupy regular positions throughout the lattice.

The crystal structure of Chevrel phases is quite interesting. These compounds crystal-lize in a hexagonal-rhombohedral structure. The building blocks of the Chevrel-phase crys-tal structure are the M elements and Mo_6X_8 molecular clusters. Each Mo_6X_8 is a slightly deformed cube with X atoms at the corners, and Mo atoms at the face centers. The elec-tronic and superconducting properties of these compounds depend mainly on the Mo_6X_8 group, with the M ion having very little effect.

3.3.2 Cuprates

A compound is said to belong to the family of copper oxides (cuprates) if it has the CuO_2 planes. Cuprates that superconduct are also called high-T_c superconductors. The first high-T_c superconductor was discovered in 1986 by Bednorz and Müller at IBM Zurich Research Laboratory. The parent compounds of superconducting cuprates are antiferromagnetic Mott insulators. The cuprates and $Na_xCoO_2 \cdot yH_2O$ (see below) are the only Mott insulators known to superconduct. $HgBa_2Ca_2Cu_3O_{10}$ has the highest critical temperature at ambient pressure, T_c = 135 K. The crystal structure of cuprates is of a perovskite type, and it is highly anisotropic. Superconductivity in cuprates occur in CuO_2 planes. The CuO_2 layers in the cuprates are always separated by layers of other atoms such as Bi, O, Y, Ba, La etc., which provide the charge carriers into the CuO_2 planes. The ground state of CuO_2 planes is antiferromagnetic. The $T_c(p)$ dependence has a nearly bell-like shape, where p is the hole (electron) concentration in the copper oxides planes. In spite of the fact that the structure of cuprates is two-dimensional, the in-plane transport properties are quasi-one-dimensional. One compound $YBa_2Cu_3O_{6+x}$ has one-dimensional CuO chains. The majority of super-conducting cuprates are hole-doped. The number of electron-doped cuprates is very limited, e.g. $Nd_{2-x}Ce_xCuO_4$ and $Pr_{2-x}Ce_xCuO_4$.

3.3.3 Charge Transfer Organics

Organic compounds are usually insulators. It turns out that some of them superconduct at low temperatures. The first organic superconductor was discovered in 1979 by Jerome and Bechgaard: the compound $(TMTSF)_2PF_6$ was found to superconduct below T_c = 0.9 K under a pressure of 12 kbar. TMTSF denotes tetramethyltetraselenafulvalene, and PF_6 is the hexafluorophosphate. However, the ten years following this discovery saw a remarkable increase in T_c. In 1990, an organic superconductor with $T_c \approx 12$ K was synthesized. In only 10 years, T_c increased over a factor 10! All organic superconductors are layered; therefore, basically they are two-dimensional. However, the electron transport in some of them is quasi-one-dimensional.

Figure 1 shows several organic molecules that form superconductors. In general, they are flat, planar molecules. Among other elements, these molecules contain sulfur or se-lenium atoms. In a crystal, these organic molecules are arranged in stacks. The chains of other atoms or molecules (PF_6, ClO_4 etc.) are aligned in these crystals parallel to the stacks.

As an example, the crystal structure of the first organic superconductor (TMTSF)$_2$PF$_6$, a representative of the *Bechgaard salts*, is schematically shown in Fig. 2. The planar TMTSF molecules form stacks along which the electrons are most conducting (the a axis). The chains of PF$_6$ lie between the stacks, aligned parallel to them. Two molecules TMTSF donate one electron to an anion PF$_6$. The separation of charge creates electrons and holes that can become delocalized to render the compound conducting and, at low temperatures, superconducting (under pressure).

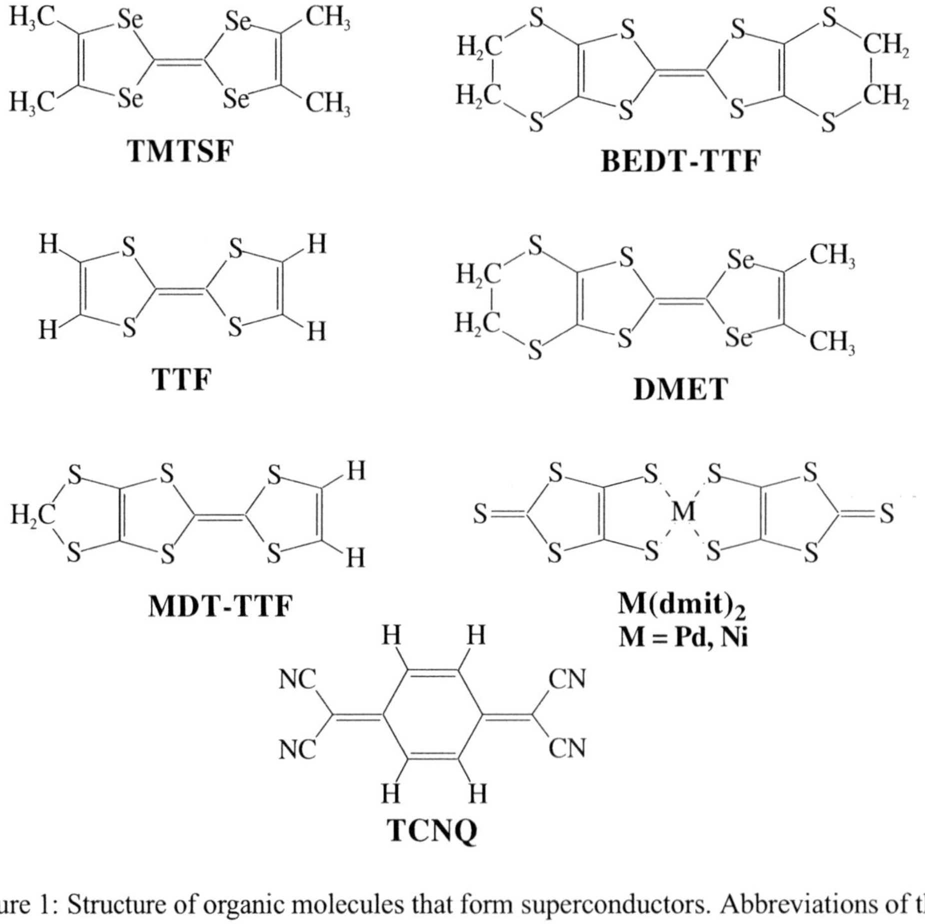

TMTSF

BEDT-TTF

TTF

DMET

MDT-TTF

M(dmit)$_2$
M = Pd, Ni

TCNQ

Figure 1: Structure of organic molecules that form superconductors. Abbreviations of their names are shown below each molecule.

After 1979, several more organic superconductors of similar structure were discovered. In all cases, some anion X$^-$ is needed to affect charge balance in order to obtain metallic properties and, at low temperature, superconductivity. So, the anions are mainly charge-compensating spacers; the conductivity is in the organic molecules. There are six different classes of organic superconductors. Two of them are the most studied—the Bechgaard salts (TMTSF)$_2$X and the organic salts (BEDT-TTF)$_2$X based on the compound BEDT-TTF shown in Fig. 1. BEDT-TTF denotes bis-ethylenedithio-tetrathiafulvalene. The members of the (BEDT-TTF)$_2$X family exhibit the highest values of T_c, and have a rich variety of crystalline structures. In contrast to the Bechgaard salts which exhibit quasi-one-dimensional electron transport, the electronic structure of the BEDT-TTF family is of two-dimensional

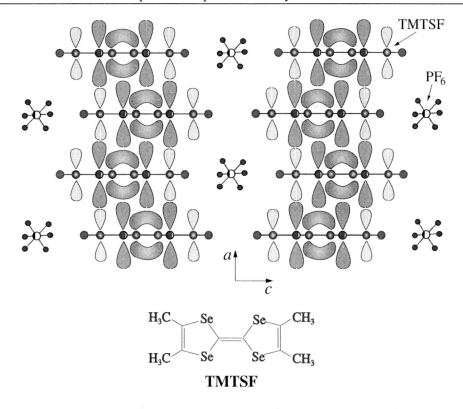

TMTSF

Figure 2: A side view of the crystal structure of the Bechgaard salt $(TMTSF)_2PF_6$. Each TMTSF molecule is shown with the electron orbitals (the hydrogen atoms are not shown). The chemical structure of the TMTSF molecule is depicted at the bottom. The organic salt $(TMTSF)_2PF_6$ is the most conductive along the TMTSF stacks (along the a axis).

nature. The highest values of T_c are observed in the $(BEDT-TTF)_2X$ salts with the anions X = $Cu(NCS)_2$; $Cu[N(CN)_2]Br$ and $Cu[N(CN)_2]Cl$. Their critical temperatures are respectively T_c = 10.4, 11.6 and 12.8 K. The first two compounds superconduct at ambient pressure, while the last one with $Cu[N(CN)_2]Cl$ becomes superconducting under a pressure of 0.3 kbar.

Depending on pressure, organic superconductors exhibit a long-range antiferromagnetic ordering. If, in the phase diagram of the Bechgaard salts, the superconducting phase evolves out of the antiferromagnetic phase, in the organic salt k-$(BEDT-TTF)_2Cu[N(CN)_2]Br$, these two phases overlap. This suggests that antiferromagnetic fluctuations—short-lived excitations of the hole-spin arrangements—are important in the mechanism of unconventional superconductivity in organic salts.

The quasi-two-dimensional organic conductor λ-$(BETS)_2FeCl_4$, superconductivity is induced by a very strong magnetic field, $18 \leq H \leq 41$ T [2]. The dependence $T_c(H)$ has a bell-like shape with a maximum $T_c \simeq 4.2$ K near 33 T. At zero field, this organic compound is an antiferromagnetic insulator below 8.5 K. The other two-dimensional compound, α-$(BEDT-TTF)_2KHg(NCS)_4$, at low magnetic fields is a charge-density-wave insulator. Thus, in these organic salts, the magnetic and electronic degrees of freedom are coupled. Furthermore, the fact that the electronic and magnetic properties of organic superconductors strongly depend on pressure indicates that their electronic, magnetic and crystal structures are strongly coupled.

3.3.4 Fullerides

Historically, any allotrope based on the element carbon has been classed as organic, but a new carbon allotrope stretches that definition. The pure element carbon forms not only graphite and diamond but a soccer-ball shaped molecule containing 60 atoms called buckminster-fullerene or buckyball. There are also lower and higher molecular weight variations such as C_{20}, C_{28}, C_{70}, C_{72}, C_{100} and so forth, which share many of the same properties. The word "fullerenes" is now used to denote all these molecules and other closed-cage molecules consisting of only carbon atoms.

C_{60} was discovered in 1985. In 1991 it was found that intercalation of alkali-metal atoms in solid C_{60} leads to metallic behavior. The alkali-doped fullerenes are called fullerides. Shortly afterwards, also in 1991, it was discovered that some of these alkali-doped C_{60} compounds are superconducting. In fullerides, the maximum critical temperature of 33 K is observed at ambient pressure in $RbCs_2C_{60}$, and $T_c = 40$ K in Cs_3C_{60} under a pressure of 12 kbar. Unfortunately, the fullerides are extremely unstable in air, burning spontaneously, so they must be prepared and kept in an inert atmosphere. The fullerides are magnetic due to spins of alkali atoms, which are ordered antiferromagnetically at low temperatures. The fullerides are electron-doped superconductors, not hole-doped as cuprates and organic salts. The values of H_{c1} in the fullerides are very small, $\sim$ 100–200 Oe, whilst those of H_{c2} are sufficiently large for electron-doped superconductors, $\sim$ 30–50 T.

3.3.5 Graphite Intercalation Compounds

The first observation of superconductivity in doped graphite goes back to 1965, when superconductivity was observed in the potassium graphite intercalation compound C_8K having a critical temperature of 0.55 K. Later, superconductivity was observed in other graphite intercalation compounds (GICs). A single layer of three-dimensional graphite is defined as a *graphene* layer. In GICs, the graphene layers are separated by the layers of intercalant atoms. According to the preparation method, the superconducting GICs can be divided into two subgroups: the stage 1 and stage 2 GICs. The stage 2 GICs are synthesized in two stages. The structures of the stage 1 and 2 GICs are different along the c axis. In the stage 1 GICs, the adjacent intercalant layers are separated from one another by *one* graphene layer, while in the stage 2 GICs, the neighboring intercalant layers are separated by *two* graphene layers. The stage 1 GICs consist of the binary C_8M, ternary C_4MHg and $C_4MTl_{1.5}$ compounds, and the stage 2 GICs are represented by the ternary C_8MHg and $C_8MTl_{1.5}$, where M = K, Rb and Cs. In the superconducting GICs, as well as in the fullerides, the charge carriers are electrons, not holes.

For binary C_8M compounds, the highest critical temperatures reported for M = K, Rb and Cs are 0.55, 0.15 and 0.135 K, respectively. In the alkali metal amalgam GICs C_8KHg and C_8RbHg, the critical temperatures are 1.93 and 1.44 K, respectively. In the potassium thallium GICs $C_4KTl_{1.5}$ and $C_8KTl_{1.5}$, respectively $T_c = 2.7$ and 1.3 K. With the potassium thallium GICs excluded, the critical temperature of the stage 2 GICs is in general higher than that of the stage 1 GICs. Under pressure, the sodium graphite intercalation compound

C_2Na superconducts below $T_c \sim 5$ K. All the GICs are two-dimensional. In $C_4KTl_{1.5}$ which has the highest T_c at ambient pressure (= 2.7 K), $H_{c2,\perp} \simeq 3$ T.

At the moment of writing, superconductivity in the isostructural graphite intercalation compounds C_6Yb and C_6Ca, with transition temperatures of 6.5 K and 11.5 K, respectively, was discovered [6].

3.3.6 $(SN)_x$ Polymer

$(SN)_x$ is a chain-like inorganic polymer in which sulphur and nitrogen atoms alternate along the chain. When doped with bromine, it becomes superconducting below $T_c = 0.3$ K. Its unit cell contains two parallel spirals of $(SN)_x$ twisted in the opposite directions. The Br_3^- and Br_5^- clusters are situated between the $(SN)_x$ spirals. Superconductivity in $(SN)_x$ was discovered in 1975. It is the first superconductor found among quasi-one-dimensional conductors and, moreover, the first that does not contain metallic elements. The single crystals have a dc electrical conductivity of about $1.7 \times 10^5 \ \Omega^{-1} \mathrm{m}^{-1}$ along the chains, and the anisotropy is of the order of 10^3. A remarkable property of $(SN)_x$ is that it does not undergo a metal-insulator (Peierls) transition at low temperatures but turns instead into a superconductor below 0.3 K.

3.3.7   Carbon Nanotubes

In addition to ball-like fullerenes, it is possible to synthesize tubular fullerenes. By rolling a graphene sheet into a cylinder and capping each end of the cylinder with a half of a fullerene molecule, a fullerene-derived tubule, one atomic layer, is formed. Depending on the wrapping angle, one can have three types of the nanotubes: zigzag, armchair and chiral. The armchair nanotubes are usually metallic, while the zigzag ones are semiconducting. The carbon nanotubes and fullerenes have a number of common features and also many differences. Carbon nanotubes can be viewed as giant conjugated molecules with a conjugated length corresponding to the whole length of the tube. The nanotubes have an impressive list of attributes. They can behave like metals or semiconductors, can conduct electricity better than copper, can transmit heat better than diamond. They rank among the strongest materials known, and they can superconduct at low temperatures.

Carbon nanotubes were found by Iijima in 1991 in Japan. In fact, they were multi-walled carbon nanotubes consisting of several concentric single-walled nanotubes nested inside each other, like a Russian doll. Two years later, single-walled nanotubes were seen for the first time. In 1999, proximity-induce superconductivity below 1 K was observed in single-walled carbon nanotubes, followed by the observation of genuine superconductivity with $T_c = 0.55$ K. In the latter case, the diameter of single-walled nanotubes was of the order of 14 Å. Soon afterwards, superconductivity below $T_c \simeq 15$ K was seen in single-walled carbon nanotubes with a diameter of 4.2 Å.

3.3.8   Heavy-Fermion Systems

This family of superconductors includes superconducting compounds which consist of one magnetic ion with $4f$ or $5f$ electrons (usually Ce or U) and other constituent or constituents being s, p, or d electron metals. The principal feature of these materials is reflected in their name: below a certain coherence temperature ($\sim$ 20–100 K), the effective mass of charge carriers in these compounds become gigantic, up to several hundred times greater than that of a free electron. A large number of heavy fermions superconduct exclusively under pressure. The T_c values of superconducting heavy fermions are in general very low; however, the family of these intermetallic compounds is one of the best examples of highly correlated condensed matter systems. The crystal structure of these compounds does not have a common pattern, but varies from case to case. For example, the crystal structure of the first discovered superconducting heavy fermions—$CeCu_2Si_2$, UBe_{13} and UPt_3—is tetragonal, cubic and hexagonal, respectively.

The first heavy fermion, $CeCu_2Si_2$, was discovered in 1979 by Steglich and co-workers, and some time passed before the heavy-fermion phenomenon was confirmed by the discovery of UBe_{13} and then UPt_3, with critical temperatures of $T_c = 0.65$, 0.9 and 0.5 K, respectively. Since then many new heavy-fermion systems that superconduct at low temperatures have been found.

Probably, the most interesting characteristic of superconducting heavy fermi-on materials is the interplay between superconductivity and magnetism. The magnetic ions are responsible for the magnetic properties of heavy fermions. For example, in the heavy fermions UPt_3, URu_2Si_2, UCu_5 and $CeRhIn_5$, magnetic correlations lead to an itinerant spin-density-wave order, while, in UPd_2Al_3 and $CeCu_2Si_2$, to a localized antiferromagnetic order. In the latter two heavy fermions, the antiferromagnetic order appears first, followed by the onset of superconductivity. In these compounds, as well as in other superconducting heavy fermions with long-range antiferromagnetic order, the Néel temperature is about $T_N \sim 10T_c$. For instance, in $CeRh_{0.5}Ir_{0.5}In_5$ and $CeRhIn_5$, the bulk superconductivity coexists *microscopically* with small-moment magnetism ($\leq 0.1\mu_B$). In the heavy fermion $CeIrIn_5$, the onset of a small magnetic field ($\sim$ 0.4 Gauss) sets in exactly at T_c.

Recently, superconductivity was discovered in $PuCoGa_5$, the first superconducting heavy fermion based on plutonium. What is even more interesting is that the superconductivity survives up to an astonishingly high temperature of 18 K.

It was a surprise when in 2000 the coexistence of superconductivity and ferromagnetism was discovered in an alloy of uranium and germanium, UGe_2. At ambient pressure, UGe_2 is known as a metallic ferromagnet with a Curie temperature of T_C = 53 K. However, as increasing pressure is applied to the ferromagnet, T_C falls monotonically, and appears to vanish at a critical pressure of $P_c \simeq$ 16–17 kbars. In a narrow range of pressure below P_c and thus *within* the ferromagnetic state, the superconducting phase appears in the millikelvin temperature range below the critical temperature. Soon after the discovery of superconductivity in itinerant ferromagnet UGe_2, two new itinerant ferromagnetic superconductors were discovered—zirconium zinc $ZrZn_2$ and uranium rhodium germanium $URhGe$.

3.3.9 Nickel Borocarbides

The nickel borocarbide class of superconductors has the general formula RNi_2B_2C, where R is a rare earth being either magnetic (Tm, Er, Ho, or Dy) or nonmagnetic (Lu and Y). In the case when R = Pr, Nd, Sm, Gd or Tb in RNi_2B_2C, the Ni borocarbides are not superconducting at low temperatures but antiferromagnetic. In the Ni borocarbides with a magnetic rare earth, superconductivity coexists at low temperatures with a long-range antiferromagnetic order. Interestingly, while in the superconducting heavy fermions with a long-range antiferromagnetic order $T_N \sim 10T_c$, in some Ni borocarbides it is just the opposite, $T_c \sim 10T_N$. Thus, antiferromagnetism appears deeply in the superconducting state.

Superconductivity in the Ni borocarbides was discovered in 1994 by Eisaki and co-workers. Transition temperatures in these quaternary intermetallic compounds can be as high as 17 K. The Ni borocarbides have a layered-tetragonal structure alternating RC sheets and Ni_2B_2 layers. As a consequence, the superconducting properties of the Ni borocarbides are also anisotropic. The layered borocarbides DyB_2C and HoB_2C without Ni also superconduct, with $T_c = 8.5$ and 7.1 K, respectively. Other related compounds, such as the Ni boronitride $La_3Ni_2B_2N_3$, are also found to superconduct.

3.3.10   Strontium Ruthenate

Nearly 40 years ago it was found that $SrRuO_3$ is a ferromagnetic metal with a Curie temperature of 160 K. In its cousin, Sr_2RuO_4, the superconducting state with $T_c \approx 1.5$ K was discovered in 1994 by Maeno and his collaborators. The crystal structure of Sr_2RuO_4 is layered perovskite, and almost isostructural to the high-T_c parent compound La_2CuO_4, in which the CuO_2 layers are substituted by the RuO_2 ones. Recently, bilayer and trilayer strontium ruthenates have been synthesized: $Sr_3Ru_2O_7$ is an enhanced paramagnetic metal, and $Sr_4Ru_3O_{10}$ is ferromagnetic with a Curie temperature of 105 K.

3.3.11   Ruthenocuprates

Ruthenocuprates are in a sense a hybrid of superconducting cuprates and the strontium ruthenate. As a consequence, they have a number of common features with the cuprates, but also many differences. Basically, there are two ruthenocuprates that superconduct at low temperatures. The general formulas of these ruthenocuprates are $RuSr_2RCu_2O_8$ and $RuSr_2R_2Cu_2O_{10}$ with R = Gd, Eu and Y. The second ruthenocuprate was discovered first in 1997. The crystal structure of $RuSr_2RCu_2O_8$ is similar to that of $YBa_2Cu_3O_7$ except for the replacement of one-dimensional CuO chains by two-dimensional RuO_2 layers. It is assumed that the RuO_2 layers act as charge reservoirs for the CuO_2 layers. The principal feature of the ruthenocuprates is that they are magnetically ordered below $T_m \sim 130$ K, and become superconducting at $T_c \sim 40$ K. For $RuSr_2RCu_2O_8$, $T_m = 130$–150 K and $T_c = 30$–45 K, while for $RuSr_2R_2Cu_2O_{10}$, $T_m = 90$–180 K and $T_c = 30$–40 K. It is believed that the magnetic order arises from ordering of Ru ions in the RuO_2 layers, while the transport occurs in the CuO_2 layers. Superconductivity and the magnetic order are found to be homogeneous. There is a consensus that in the ruthenocuprate, there is a small ferromagnetic component; however, there is no agreement on its origin. It may originate not only from the Ru moments but also, for example, from the Gd spins.

3.3.12   MgCNi$_3$

Superconductivity in $MgCNi_3$ was discovered in 2001 by Cava and co-workers, a few months later than that in MgB_2. The crystal structure of $MgCNi_3$ is cubic-perovskite, and similar to that of BKBO. The perovskite $MgCNi_3$ is special in that it is neither an oxide nor does it contain any copper. Since Ni is ferromagnetic, the discovery of superconductivity in $MgCNi_3$ was surprising. The critical temperature is near 8 K. $MgCNi_3$ is metallic, and the charge carriers are electrons which are derived predominantly from Ni. Structural studies of $MgCNi_3$ reveal structural inhomogeneity. Apparently, the perovskite cubic structure of $MgCNi_3$ is modulated locally by the variable stoichiometry on the C sites.

3.3.13 $Cd_2Re_2O_7$

Although $Cd_2Re_2O_7$ was synthesized in 1965, its physical properties remained almost unstudied. Unexpectedly, superconductivity in $Cd_2Re_2O_7$ was discovered in the second half of 2001 by Sakai and co-workers. The critical temperature of $Cd_2Re_2O_7$ is low, $T_c = 1$–1.5 K. This compound is the first superconductor found among the large family of pyrochlore oxides with the formula $A_2B_2O_7$, where A is either a rare earth or a late transition metal, and B is a transition metal. In this structure, the A and B cations are 4- and 6-coordinated by oxygen anions. The A-O_4 tetrahedra are connected as a pyrochlore lattice with straight A-O-A bonds, while B-O_6 octahedra form a pyrochlore lattice with the bent B-O-B bonds with an angle of 110–140°. Assuming that electronic structure in $Cd_2Re_2O_7$ as formally Cd^{2+} $4d^{10}$ and Re^{5+} $4f^{14}5d^2$, the electronic and magnetic properties are primarily dominated by the Re $5d$ electrons.

Oxide superconductors with non-perovskite structure are rare. Previous studies indicate that the pyrochlores, like the spinels, are geometrically frustrated. The effect of geometric frustration on the physical properties of spinel materials is drastic, resulting in, for example, heavy-fermion behavior in LiV_2O_4. Another spinel compound $LiTi_2O_4$ is a superconductor below $T_c = 13.7$ K. Indeed, x-ray diffraction studies performed under high pressure showed that superconductivity in $Cd_2Re_2O_7$ is detected only for the phases with a structural distortion. It was suggested that the charge fluctuations of Re ions play a crucial role in determining the electronic properties of $Cd_2Re_2O_7$.

3.3.14 $Na_xCoO_2 \cdot yH_2O$

One of the newest superconductors is the layered cobalt oxyhydrate $Na_xCoO_2 \cdot yH_2O$ ($\frac{1}{4} < x < \frac{1}{3}$ and $y = 1.3$–1.4). The structure of the parent compound Na_xCoO_2 consists of alternating layers of CoO_2 and Na. In the hydrated Na_xCoO_2, the water molecules form additional layers, intercalating all CoO_2 and Na layers. After the hydration of Na_xCoO_2, the c-axis lattice parameter increases from 11.16 Å to 19.5 Å. Thus, the elementary cell of $Na_xCoO_2 \cdot yH_2O$ consists of three layers of CoO_2, two layer of Na^+ ions and four layers of H_2O. Within each CoO_2 layer, the Co ions occupy the sites of a triangular lattice. The $1 - x$ fraction of Co ions is in the low spin $S = \frac{1}{2}$ Co^{4+} state, while the x fraction is in the $S = 0$ Co^{3+} state. In the triangular lattice, the spins of Co^{4+} ions are ordered antiferromagnetically.

Superconductivity in $Na_xCoO_2 \cdot yH_2O$ occurs in the CoO_2 layers. The superconducting phase as a function of x has a bell-like shape, situated between 0.25 and 0.33 with a maximum $T_c \simeq 4.5$ K near $x = 0.3$. $Na_xCoO_2 \cdot yH_2O$ is the first superconductor containing water (ice). All experimental facts indicate that the presence of water is crucial to superconductivity.

3.3.15   Hydrides and Deuterides

In addition to the nitrides and carbides from the second group of superconductors, another class of superconducting compounds that also have the NaCl structure includes hydrides and deuterides (i.e. compounds containing hydrogen or deuterium). However, in contrast to the nitrides and carbides, superconducting hydrides and deuterides are magnetic. In the seventies it was discovered that some metals and alloys, not being superconducting in pure form, become relatively good superconductors when they form alloys or compounds with hydrogen or deuterium. These metals include the transition elements palladium (Pd) and thorium (Th) that have unoccupied $4d$- and $5f$-electron shells, respectively.

In 1972, Skoskewitz discovered that the transition element Pd which has a small magnetic moment normally preventing the pairing of electrons, joins hydrogen and forms the PdH compound that superconducts at $T_c = 9$ K. Later on, it was found that by doping such a system with noble metals the critical temperature increases up to 17 K. Interestingly, the palladium-deuterium compound also superconducts, and its critical temperature equal to 11 K is higher than that of PdH. So the hydrogen isotope effect in PdH is reverse (negative). In contrast, the critical temperatures of the ThH and ThD compounds do not differ drastically from each other like those of PdH and PdD.

4   Physical Properties of the Third Group of Superconductors

In the previous section, superconducting materials were classified into three groups. The first group consists of conventional superconductors; the second comprises half-conventional ones, and unconventional superconductors form the third group. Assume that one day a room-temperature superconductor will become available. What group will it belong to? The answer is more or less obvious: to the third group of unconventional superconductors. Indeed, the highest critical temperature in third group of superconductors, $T_{c,max} \simeq 135$ K, is more than three times higher than that of the second group and more than one order of magnitude higher than that of group of conventional superconductors. Moreover, it seems that the critical temperatures of superconductors of the first and the second groups cannot exceed 10 K and 40 K, respectively. The third group of superconductors is the largest and has the highest rate of growth in the last twenty five years. If the rate of the growth of the third group will remain in the future at the same level, then, one will soon need to make an internal classification of this group.

From the discussion in the previous paragraph, it is evident that, in order to have guide to the properties of room-temperature superconductors, it is necessary to analyze the physical properties of superconductors of the third group, and not those of the first and second groups. One may argue, however, that room-temperature superconductors can form a separate group of superconductors, i.e. the fourth one. Generally speaking such a situation may occur, but there are only two options for a low-dimensional compound to be either magnetic or non-magnetic. As a consequence, every low-dimensional superconductor belongs either to the second or third group of superconductors.

Let us enumerate the common physical properties of superconductors of the third group. All superconductors of the third group:

- are magnetic or, at least, have strong magnetic correlations,

- are low-dimensional,

- have strongly correlated electrons/holes,

- are near a metal-insulator transition,

- are apparently near a quantum critical point,

- are type-II superconductors,

- have small-size Cooper pairs,

- have a low density of charge carriers, n_s,

- have a universal $n_s \propto \sigma(T_c)\dot{T}_c$ dependence [7], where $\sigma(T_c)$ is the *dc* conductivity just above T_c,

- have large values of H_{c2} and λ (magnetic penetration depth) and a large gap ratio $2\Delta_p/(k_B T_c)$, where Δ_p is the *pairing* energy gap (see Section 6),

- have anisotropic transport and magnetic properties,

- have a complex phase diagram (if there is a parameter to vary),

- have the moderately strong electron-phonon interaction,

- have an unstable lattice,

- have charge-donor and charge-acceptor sites (i.e. there is a charge transfer), and

- have a complex structure (with the exception of hydrides, deuterides and a few heavy fermions).

Among superconductors of the third group:

- the T_c value correlates with the behavior of spin fluctuations—the more dynamic the fluctuations are, the higher the T_c value is,

- the localized states in *undoped* material have spin-1/2 ground states,

- the average T_c value of layered compounds is higher than that of one-dimensional ones (even so, the transport properties of the layered compounds are quasi-one-dimensional),

- superconductors with $T_c > 20$ K have no metal-metal bonds (only heavy fermions have metal-metal bonds), and

- oxides and organic superconductors represent the absolute majority of the group.

At the same time, there are differences among superconductors of the third group. The two main differences are:

- the T_c value of hole-doped superconductors is on average a few times higher than that of electron-doped superconductors,

- the average T_c value of antiferromagnetic compounds is at least one order of magnitude higher than that of ferromagnetic superconductors.

Considering the common features of superconductors of the third group, one should realize that some of these features are direct consequences of the others. For example, the anisotropic character of transport and magnetic properties of superconductors of the third group is a direct consequence of a low-dimensional structure of these superconductors. The expression "systems with strongly correlated electrons" partially assumes that the electron-phonon interaction in these systems is sufficiently strong, and they have a complex phase diagram. The moderately strong electron-phonon interaction results in a large value of the pairing energy gap and, therefore, in a large value of the gap ratio $2\Delta_p/(k_B T_c)$. Since in superconductors of the third group, the Cooper pairs have a small size and a low density, this leads to the penetration depth and, consequently, the ratio λ/ξ to be large, where ξ is the coherence length. Therefore, all superconductors of the third group are type-II. The so-called Homes law, $n_s \propto \sigma(T_c)\dot{T}_c$, [7] literally means that an high-temperature superconductor should be a bad conductor just above T_c. The same conclusion follows from the other experimental fact that superconductivity with an high T_c occurs near a metal-insulator transition. Hence, some of these common features of superconductors of the third group are more important than others. Let us select the most important ones.

5   Requirements for High-T_c Materials

We are now in a position to discuss the most important requirements for materials that superconduct at high temperatures. From the previous section, one can conclude that a room-temperature superconductor should:

- be hole-doped,

- be low-dimensional (preferentially, layered but with quasi-one-dimensional transport properties),

- be antiferromagnetic or, at least, have strong magnetic correlations,

- have strongly correlated holes,

- be near a metal-insulator Mott transition,

- have an *undoped* parent compound with the spin-1/2 localized states,

- have **dynamic** spin fluctuations,

- be in a state above a quantum critical point,

- have an unstable lattice,

- have a complex structure (i.e. with more than two sites per unit cell),

- have electron-acceptor sites, and

- have no metal-metal bonds.

It is necessary to comment on the issue of quantum critical point. In a quantum critical point where a magnetic order is about to form or to disappear, the spin fluctuations are the strongest. At the moment of writing, we do not know yet how to determine from a single measurement the presence/absence of a quantum critical point in a certain compound. So, this can be the first intermediate goal: how to determine quickly the presence/absence of a quantum critical point in a given compound.

In addition to these common features of the third group of superconductors, consider one more observation concerning the structure of good superconductors. In cuprates, the unit cell has three *interacting* subsystems: a quasi-metallic, a magnetic ones and charge reservoirs. The charge reservoirs in cuprates are the layers that intercalate the CuO_2 layers. After accepting/donating electrons, the charge reservoirs become semiconducting or insulating. The quasi-metallic and magnetic subsystems in cuprates are located into the CuO_2 planes, resulting in the phase separation. In organic superconductors, however, the second and the third subsystems coincide: the charge reservoirs, after donating/accepting electrons to/from organic molecules, become magnetic. For instance, the structure of a Bechgaard salt shown in Fig. 2 is a good example. Thus, in organic superconductors, one subsystem performs two functions. To summarize, a potential room-temperature superconductor should have:

    1) **a quasi-metallic subsystem**,
    2) **charge reservoirs**, and
    3) **magnetic atoms/molecules**.

Each subsystem should be coupled to the two others. Experimentally, the second and the third subsystems can be represented by the same atoms/molecules.

The requirements summarized in these section basically include Geballe's guidelines, described in Section 2. We shall use these hints in Sections 7–9.

6 Principles of Superconductivity

Considering magnetic and structural requirements for materials, we should remember that the materials must superconduct after all. What does superconductivity as a phenomenon require? Let us discuss in this section the main principles of superconductivity as a phenomenon, valid for every superconductor independently of its characteristic properties and material. The underlying mechanisms of superconductivity can be different in various materials, but certain principles must be satisfied. One should however realize that the principles of superconductivity are not limited to those discussed in this section: it is possible that there are others which we do not know yet about. More detailed description of the principles of superconductivity can be found elsewhere [2, 3].

The first principle of superconductivity is:

Principle 1:   **Superconductivity requires quasiparticle pairing**

The electron (hole) pairs are known as Cooper pairs. In solids, superconductivity as a quantum state cannot occur without the presence of bosons. Fermions are not suitable for forming a quantum state since they have spin and, therefore, they obey the Pauli exclusion principle according to which two identical fermions cannot occupy the same quantum state. Electrons are fermions with a spin of 1/2, while the Cooper pairs are already composite bosons since the value of their total spin is either 0 or 1. Therefore, the electron (hole) pairing is an inseparable part of the phenomenon of superconductivity and, in any material, superconductivity cannot occur without quasiparticle pairing.

In the framework of the BCS theory [5], the electron pairing occurs in momentum space. However, for the occurrence of superconductivity in the general case, the electron pairing may take place not only in momentum space but also in real space. The electron pairing in momentum space can be considered as a *collective* phenomenon, while that in real space as *individual*. Independently of the space where they are paired—momentum or real—two electrons can form a bound state **only if** *the net force acting between them is attractive*.

Superconductivity requires the electron pairing and the Cooper-pair condensation. The second principle of superconductivity deals with the Cooper-pair condensation taking place at T_c. This process is also known as the onset of long-range phase coherence.

Principle 2: **The transition into the superconducting state is the Bose-Einstein-like condensation and occurs in momentum space**

The two processes—the electron pairing and the onset of phase coherence—are independent of one another. Superconductivity requires both. In conventional superconductors, the pairing and the onset of phase coherence take place simultaneously at T_c. In many unconventional superconductors, quasiparticles become paired above T_c and start forming the superconducting condensate only at T_c.

In the 1920s, Einstein predicted that if an ideal gas of identical atoms, i.e. bosons, at thermal equilibrium is trapped in a box, at sufficiently low temperatures the particles can in principle accumulate in the lowest energy level. This may take place only if the quantum wave packets of the particles overlap. In other words, the wavelengths of the matter waves associated with the particles—the *Broglie waves*—become similar in size to the mean particle distances in the box. If this happens, the particles condense, almost motionless, into the lowest quantum state, forming a Bose-Einstein condensate. The two condensates—superconducting and Bose-Einstein—have common quantum properties, but also, they have a few differences, which have been discussed elsewhere [2].

The third principle of superconductivity is:

Principle 3: **The mechanism of electron pairing and the mechanism of Cooper-pair condensation must be different**

The validity of the third principle of superconductivity will be evident after the presentation of the fourth principle. Historically, this principle was introduced first. To recall, in conventional superconductors, phonons mediate the electron pairing, while the overlap of wavefunctions ensures the Cooper-pair condensation. Generally speaking, if in a superconductor, the same "mediator" (for example, phonons) is responsible for the electron pairing and for the onset of long-range phase coherence (Cooper-pair condensation), this will lead to the collapse of superconductivity.

If the first three principles of superconductivity do not deal with numbers, the forth principle can be used for making various estimations.

Generally speaking, a superconductor is characterized by a pairing energy gap Δ_p and a phase-coherence gap Δ_c. For genuine (not proximity-induced) superconductivity, the phase-coherence gap is proportional to T_c:

$$2\Delta_c = \Lambda\, k_B T_c, \tag{1}$$

where Λ is the coefficient proportionality, and k_B is the Boltzmann constant. At the same time, the pairing energy gap is proportional to the pairing temperature T_{pair}:

$$2\Delta_p = \Lambda'\, k_B T_{pair}. \tag{2}$$

Since the formation of Cooper pairs must precede the onset of long-range phase coherence, then in the general case, $T_{pair} \geq T_c$.

In conventional superconductors, however, there is only one energy gap Δ which is in fact a pairing gap but proportional to T_c:

$$2\Delta = \Lambda\, k_B T_c, \tag{3}$$

This is because, in conventional superconductors, the electron pairing and the onset of long-range phase coherence take place at the same temperature—at T_c. In all known cases, the coefficients Λ and Λ' lie in the interval between 3.2 and 6 (in one heavy fermion, ~ 9) [2, 3]. We are now in position to discuss the fourth principle of superconductivity:

Principle 4: **For genuine, homogeneous superconductivity, $\Delta_p > \Delta_c > \frac{3}{4}k_B T_c$ always (in conventional superconductors, $\Delta > \frac{3}{4}k_B T_c$)**

The reason why superconductivity occurs exclusively at low temperatures is the presence of substantial thermal fluctuations at high temperatures. In conventional superconductors, the energy of electron binding, 2Δ, must be larger than the thermal energy; otherwise, the pairs will be broken up by thermal fluctuations. In the case of unconventional superconductors, the same reasoning is also applicable for the phase-coherence energy gap, Δ_c. The last inequality, $\Delta_p > \Delta_c$ for unconventional superconductors, means that the pairing energy $2\Delta_p$ of the Cooper pairs must be larger than the strength of the coupling of bosonic excitations responsible for mediating the long-range phase coherence with the Cooper pairs, which is measured by the energy $2\Delta_c$. If the strength of this coupling will exceed the pairing energy $2\Delta_p$, the Cooper pairs will immediately be broken up.

In the case $\Delta_p = \Delta_c$ occurring at some temperature $T < T_c$ remaining constant, locally there will be superconducting fluctuations due to thermal fluctuations, thus, a kind of inhomogeneous superconductivity. It is necessary to mention that the case $\Delta_p = \Delta_c$ must not be confused with the case $T_{pair} = T_c$, because usually $\Lambda' > \Lambda$.

Finally, let us go back to the third principle of superconductivity to show its validity. The case in which the same bosonic excitations mediate the electron pairing **and** the phase coherence is equivalent to the case $\Delta_p = \Delta_c$ discussed above. Since, in this particular case, the equality $\Delta_p = \Delta_c$ is independent of temperature, the occurrence of homogeneous superconductivity is impossible.

7   An Approach to Room-Temperature Superconductivity

Approaching the problem of room-temperature superconductivity, different researchers can use the information presented above in different ways: human imagination does not have limits. One approach, however, is more or less obvious. The structure of cuprates basically satisfies all the requirements described above. Then, one can replace the CuO_2 planes by planes of different type, which can accommodate, depending on the doping level, the Cooper pairs and antiferromagnetic ordering above room temperature. The only problem is that one should know what type of planes to use.

Another possibility is to use the structure of organic salts shown in Fig. 2 as the basis for the structure of new superconductors. Luckily, the issue of the presence of Cooper pairs in some organic compounds above room temperature is known already for some time. In addition to substitution of the organic molecules, the magnetic molecules of PF_6 in Fig. 2 should be replaced by other type of molecules which are ordered antiferromagnetically above room temperature. To the end of this chapter, we shall explore this approach to the problem.

8   Organic Molecules, Polymers and Tissues with Electron Pairs above Room Temperature

According to the first principle of superconductivity, superconductivity requires electron pairing. Indeed, the electron pairing is the keystone of superconductivity. Therefore, in quest of compounds that superconduct above room temperature, one should first look at materials which tolerate the presence of Cooper pairs at high temperatures. Fortunately, it is known already for some time that the Cooper pairs exist in some organic compounds at and above room temperature. In this section we are going to discuss these organic materials. It is worth to mention, however, that superconductivity does not occur in these compounds because superconductivity requires not only electron pairing but also the establishment of long-range phase coherence.

The idea to use organic compounds as superconducting materials is not new. In 1964, Little proposed a model in which a high T_c is obtained due to a non-phonon mediated mechanism of electron attraction, namely, an *exciton* model for Cooper-pair formation in long chainlike organic molecules [8]. In the framework of his model, the maximum critical temperature was estimated to be around 2200 K! Little's paper has encouraged the search for room-temperature superconductivity, especially in organic compounds.

8.1   Organic Molecules

In 1975, Kresin and co-workers showed that the superconducting-like state exists *locally* in complex organic molecules with conjugate bonds [9]. Figure 3 shows a few examples of such molecules. Their main building blocks are carbon and hydrogen atoms. The characteristic feature of these conjugated hydrocarbons is the presence of a large number of π electrons. These collectivized electrons are in the field of the so-called σ electrons which are located close to the atomic nuclei and not much different from the ordinary atomic electrons. At the same time, the π electrons are not localized near any particular atom, and they

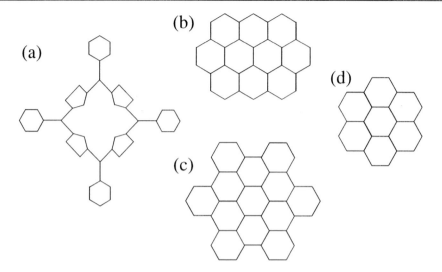

Figure 3: Organic molecules with delocalized π electrons: (a) tetraphenylporphin; (b) ovalene; (c) hexabenzocoronene, and (d) coronene [9].

can travel throughout the entire molecular frame. This makes the molecule very similar to a metal. The framework of atoms plays the role of a crystal lattice, while the π electrons that of the conduction electrons. It turns out, in fact, that the conjugated hydrocarbons with even number of carbon atoms are more than just similar to a metal, but are actually small superconductors [9]. Experimentally, conjugated hydrocarbons with even number of carbon atoms (thus, with even number of π electrons) exhibit properties similar to those of a superconductor: the Meissner-like effect, zero resistivity and the presence of an energy gap. The π electrons form bound pairs analogous to the Cooper pairs in an ordinary superconductor. The pair correlation mechanism is principally due to two effects: (i) the polarization of the σ electrons, and (ii) $\sigma - \pi$ virtual electron transitions. However, if the number of π electrons is odd, the properties of such conjugated hydrocarbons are different from those of a superconductor.

• The alkali-doped fullerenes discussed in Section 3 are able to superconduct but they are electron-doped. To exhibit room-temperature superconductivity, the single crystals of C_{60} must be doped by holes. Thus, one should find suitable dopant species for this purpose. Theoretical calculations show that fullerenes having a diameter smaller than that of C_{60}, such as C_{28} and C_{20}, are able to exhibit a higher value of T_c relative to that of C_{60} [2]. Unlike graphene and other long organic polymers, the fullerenes have an advantage to be packed into any form.

One can use fullerenes not only in pure but also in polymerized form. As an example, Figure 4 shows various one- and two-dimensional polymeric solids formed from C_{60}.

• As discussed in Section 3, carbon nanotubes can be viewed as giant conjugated molecules with a conjugated length corresponding to the whole length of the tube. The nanotubes are also a promising candidate with which to form a room-temperature superconductor.

The single-walled carbon nanotubes with a diameter of 4.2 ± 0.2 Å exhibit *bulk* superconductivity below $T_c \simeq 15$ K. The nanotubes with a smaller diameter may display a

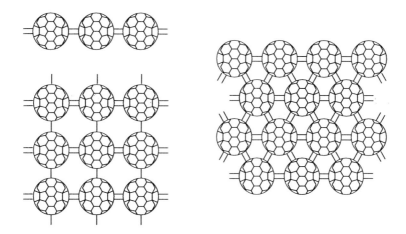

Figure 4: Various one- and two-dimensional polymeric solids formed from C_{60} [10]. The C_{60} balls are shown schematically.

higher T_c. The onset of *local* superconductivity at 645 K was observed in single-walled carbon nanotubes containing a small amount of the magnetic impurities Ni : Co ($\leq$ 1.3 %) [11]. This unconfirmed evidence is based on transport, magnetoresistance, tunneling and Raman measurements. In single-walled carbon nanotubes, the energy gap obtained in tunneling measurements is about $\Delta_p \simeq 100$ meV [11]. By embedding these nanotubes into a *dynamic* magnetic medium, one can witness bulk superconductivity above 450 K. Thus, in the general case, in order to accommodate the pairs at high temperatures, nanotubes should be *doped* or contain *defects*.

8.2   Organic Conjugated Polymers

Polythiophene is a one-dimensional conjugated polymer. Figure 5(a) shows its structure. It has been known already for some time that, in polythiophene, the dominant nonlinear excitations are positively-charged polarons and bipolarons [12]. This means that the Cooper pairs with a charge of $+2|e|$ exist at room temperature in polythiophene. Figure 5(b) depicts a schematic structural diagram of a bipolaron on a polythiophene chain. In a thiophene ring, the four carbon p electrons and the two sulfur p electrons provide the six p electrons that satisfy the $(4n + 2)$ condition necessary for aromatic stabilization. Polythiophene has a few derivatives and one of them shown in Fig. 5(c) is called poly(3-alkylthienylenes) or P3AT for short. In contrast to polythiophene, P3AT is soluble.

Polythiophene chains have the infinite length. Polythiophene chains having a finite length can be used too. However, the ends of the pieces of the polymer must be "closed." Alternatively, the two ends of a polymer piece can be attached one to another leading to the formation of a ring. Taking into account that the width of a bipolaron is a few lattice constants; then, the length of pieces of polythiophene chains, ℓ, should be 2–3 times larger; thus $\ell \sim 15a$, where a is the lattice constant.

Other conjugated polymers also contain bipolarons. It is known that positively-charged bipolarons exist, for example, in polyparaphenylene, polypyrrole and poly(2,5-diheptyl-1,4-phenylene-alt-2,5-thienylene) (PDHPT) [12]. The structure of polyparaphenylene is

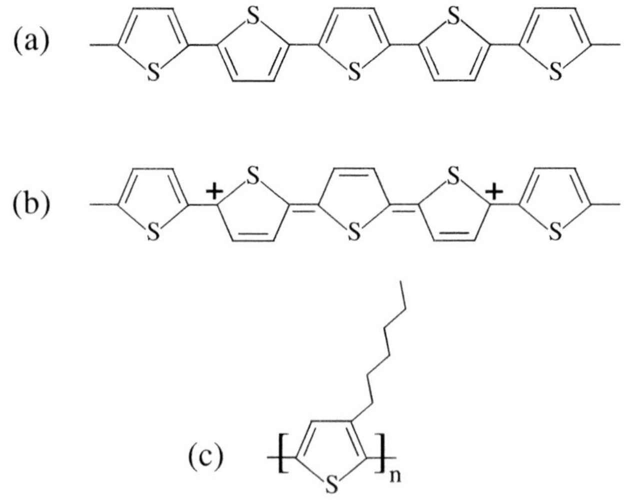

Figure 5: (a) Chemical structure of polythiophene. (b) Schematic structural diagram of a positively-charged polaron on a polythiophene chain. (c) Chemical structure of soluble poly(3-alkylthienylenes) [12].

depicted in Fig. 6(a). A bipolaron on a polyparaphenylene chain is schematically shown in Fig. 6(b). One of the derivatives of polyparaphenylene, p-sexiphenyl depicted in Fig. 6(c). The structure of PDHPT is illustrated in Fig. 6(d). Bipolarons have also been observed in other one-dimensional conjugated organic polymers such as polyparaphenylene and polypyrrole [12]. A large number of conjugated polymers used in electroluminescent diodes at room temperature contain bipolarons, such as poly(p-phenylene vinylene) etc., (see Fig. 2 in [13]). All these polymers are commercially available.

8.3   Living Tissues

It is known that (i) in redox reactions occurring in living organisms, electrons are transferred from one molecule to another in pairs with opposite spins; and (ii) electron transport in the synthesis process of ATP (adenosine triphosphate) molecules in conjugate membranes of mitochondria and chloroplasts is realized by pairs, not individually [14, 15, 16].

In living tissues, the electron pairing simplifies their propagation because the calculations show that, for electrons, it is more profitable *energetically* to propagate together than separately, one by one [15]. So, the electron pairing occurs in living tissues first of all because of an energy gain; the electron spin is a secondary reason for the pairing. However, we are more interested in electron pairing because of spin. The body temperature of living creatures is usually near the room temperature. Therefore, one can use these materials to form a room-temperature superconductor. As discussed earlier, superconductivity does not occur in living tissues because, according to the principles of superconductivity, it requires not only the electron pairing but also the onset of long-range phase coherence.

Recently, living organisms have been found in extreme conditions: some survive without sunlight, some survive in water near the boiling point. Probably the most extreme case is the discovery of so-called black smokers or chimneys on the ocean floor. Deep-sea vents

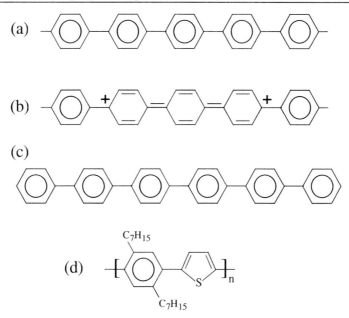

Figure 6: (a) Chemical structure of polyparaphenylene. (b) Schematic structural diagram of a positively-charged bisoliton on a polyparaphenylene chain [12]. (c) Molecular structure of p-sexiphenyl, and (d) the chemical structure of soluble poly(2,5-diheptyl-1,4-phenylene-alt-2,5-thienylene) (PDHPT).

provide an unusual habitat for some primitive forms of extremophile bacteria and deep-sea crabs that can survive extreme conditions. For example, the *Spire* vent is located at the Broken Spur Vent Field in the Mid-Atlantic Ridge, 3080 m below sea level. The measurement of water temperature at which the crabs reside there yielded $T \sim 365$ C. This means that the electron pairs exist in certain organic materials at temperatures above 600 K.

Mitochondria and chloroplasts are integral parts of almost every living cell. In principle, one can easily use their membranes to form a room-temperature superconductor. The redox reactions occur practically in every cell. One should find out what parts of the cells are responsible for the redox reactions, and then use these tissues. The living tissues are usually one- or two-dimensional.

Other living tissues may also contain bipolarons. Nowadays it is known that, in the living matter, the signal transfer occurs due to charge (electron) transfer. It is possible that, in *some* cases, the electron transfer occurs in pairs with opposite spins.

Pullman and Pullman already in 1963 emphasized that "the essential fluidity of life agrees with the fluidity of the electronic cloud in conjugated molecules. Such systems may thus be considered as both the cradle and the main backbone of life" [17]. Indeed, all natural molecules are conjugated [18], and it is possible that, some of them can be used to synthesize a room temperature superconductor.

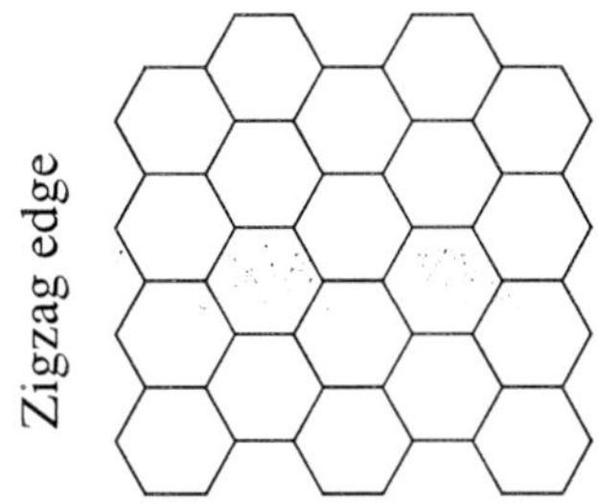

Armchair edge

Figure 7: Two basic types of graphite edges.

8.3.1 Graphite

For the last forty years, graphite is one of the most studied materials (see, e.g. [19]). It is also one of the **most** promising superconducting materials. Graphite intercalation compounds (GICs) able to superconduct were discussed in Section 3. All superconducting GICs are alkali-doped and, therefore, magnetic due to alkali spins ordered antiferromagnetically. In the superconducting GICs, the charge carriers are however electrons, not holes.

There exist both theoretical predictions and experimental evidence that electronic instabilities in pure graphite can lead to the occurrence of superconductivity and ferromagnetism, even at room temperature [20]. In graphite, an intrinsic origin of high-temperature superconductivity relates to a **topological disorder** in graphene layers [20]. This disorder enhances the density of states at the Fermi level. For example, four hexagons in graphene can in principle be replaced by two pentagons and two heptagons. Such a defect in graphene modifies its band structure. In real space, the disorder in graphene transforms an ideal two-dimensional layer into a network of quasi-one-dimensional channels.

In practice, the graphene sheets are always finite. Their electronic properties are drastically different from those of bulk graphite. It is experimentally established that the electronic properties of nanometer-scale graphite are strongly affected by the structure of its edges [20]. The graphene edges induce electronic states near the Fermi level. Any graphene edge can be presented by a linear combination of the two basic edges: zigzag and armchair, shown in Fig. 7. The free energy of an armchair edge is lower than that of a zigzag edge.

The appearance of bipolarons at high temperatures in graphene depends on the graphene structure: graphene sheets must be topologically disordered. As mentioned above, one can replace some hexagons in graphene by pentagons and heptagons. Alternatively, some carbon atoms in graphene can be substituted for B, N or Al (B and Al are probably better than N because each N adds an additional electron to graphene, while room-temperature superconductivity requires holes). Instead of graphene sheets, one can for instance use nanostripes of graphene. Since the bonds between adjacent layers in graphite are weak, individual atomic graphene planes can be pulled out of bulk crystals [21] and be used further.

In principle, it should be not a problem for bipolarons to occur in graphite and graphene-based compounds above room temperature. The main problem for the occurrence of bulk superconductivity in graphite and other organic compounds above room temperature is the onset of long-range phase coherence. One can dope graphite by atoms/molecules, spins of which are ordered antiferromagnetically after the diffusion. This issue will be discussed in the following section.

At the end, it is worth noting that materials which contain bipolarons above room temperature are not limited to those discussed in this section. It is possible that there are compounds which tolerate the presence of bipolarons above room temperature but are unknown to us at the moment of the writing.

9   Structure and Constituents of a Potential Room-Temperature Superconductor

As discussed in Section 7, the structure of a Bechgaard salt shown in Fig. 2 basically satisfies all the requirements for materials that superconduct at high temperatures, which are described in Section 5. In superconducting organic salts, the Cooper pairs reside on organic molecules organized in stacks, which are shown in Fig. 1. The stacks form one-dimensional structures similar to those in Fig. 2. In the crystal, the chains of other atoms or molecules (e.g. PF_6, ClO_4 etc.) are situated between the stacks and aligned parallel to them, as shown in Fig. 2. The organic molecules donate electrons to the anions, which prefer to order antiferromagnetically. In some organic salts, e.g. in the BEDT-TTF family, the arrays of organic molecules form conducting layers separated by insulating anion sheets. So, in contrast to the Bechgaard salts which exhibit quasi-one-dimensional electron transport, the electronic structure of the BEDT-TTF family is two-dimensional.

As suggested in Section 7, the idea to create a compound able to superconduct above room temperature is straightforward. Taking the crystal structure of the Bechgaard salt shown in Fig. 2 as a basis, one should replace the stacks of the organic TMTSF molecules by molecules/polymers/layers of a certain material which contains the Cooper pairs above room temperature. At the same time, the molecules of PF_6 in Fig. 2 should be substituted for other atoms or molecules which are able to accept electrons from the conducting counterparts and to become antiferromagnetically ordered above room temperature. Also, one can add a small amount of atoms/molecules which stand duty *exclusively* as charge reservoirs. So, the basic idea is more or less obvious; the main question is what materials to use and how to achieve the right intercalation.

Materials which contain bipolarons above room temperature have been discussed in the previous section. As a matter of fact, they are all organic. What is about atoms/molecules which are magnetic in the intercalated state?

A superconductor of the third group must be magnetic or, at least, have strong magnetic correlations. While oxides can be magnetic naturally, like cuprates for example, organic and living-tissue-based compounds must be doped by magnetic species. Unfortunately, during evolution, Nature did not need to develop such magnetic materials. Hence, one should only rely on accumulated scientific experience and work by trial and error.

By doping organic materials or living tissues, one should take into account that, after the diffusion, the dopant species must not be situated too close to the organic molecules/tissues. Otherwise, they will have a strong influence on bipolaron wavefunctions and may even break up the bipolarons. On the other hand, the dopant species cannot be situated too far from the organic molecules/tissues because, as the common logic suggests, bipolarons should be coupled to spin fluctuations. Therefore, synthesizing a room-temperature superconductor, one must pay attention to its structure: the "distance" between failure and success can be as small as 0.01 Å in the lattice constant.

What materials can be used as the acceptors of electrons? From experience [2], materials able to accept electrons from organic molecules are the following atoms and molecules: Cs, I, Br, PF_6, ClO_4, $FeCl_4$, $Cu(NCS)_2$, $Cu[N(CN)_2]Br$ and $Cu[N(CN)_2]Cl$. However, in existing superconducting organic salts, they are ordered antiferromagnetically at low temperatures, and it is not obvious at all, if they are able to behave in the same way above room temperature. As discussed in Section 5, in a superconductor of the third group, spin fluctuations should be dynamic. As a matter of fact, the dynamic character of spin fluctuations can be achieved artificially, as suggested elsewhere [2].

At the end of this section, a few remarks about the structure of a potential room-temperature superconductor should be made. As an example, consider polythiophene. Polythiophene, its derivatives and other organic conjugated polymers are usually doped by using the so-called electrochemical method [12]. The reaction is carried out at room temperature in an electrochemical cell with the polymer as one electrode. To remove electrons from organic polymers, oxidation is usually used. Through doping, one can control the chemical potential. In practice, however, it is impossible to foresee the structure of a doped organic compound, even knowing materials before the beginning of the doping procedure. Depending on their origin, concentration and size, the dopant species after the diffusion can take different positions relative to the polythiophene chains. Figure 8 shows several examples of possible positions of dopant species relative to polythiophene chains. Upon doping, the polythiophene chains and dopant species can, for example, be in-plane, as sketched in Fig. 8(a), or alternate, as illustrated in Fig. 8(b). They can, for example, form a checker-board pattern shown schematically in Fig. 8(c). For instance, in Na-doped polyacetylene, the Na^+ ions and polyacetylene chains form a modulated lattice with a "triangular" pattern depicted in Fig. 8(d). In Na-doped polyacetylene, such a lattice appears exclusively at moderate doping levels.

Using various magnetic atoms/molecules, one should not forget to control the doping level of the organic polymers. It can be done, for example, by adding a small amount of atoms/molecules of another type, which may or may not be magnetic after the diffusion. Undoubtedly, some of these doped polythiophene-chain materials will superconduct. The main question is what maximum value of T_c can be attained in these organic compounds.

From the discussions in this section, one can make an important conclusion, namely that, *for the occurrence of superconductivity at room temperature, the onset of long-range phase coherence will be the bottleneck, not the quasiparticle pairing*.

In the case of graphene or other two-dimensional materials, an alternative way to create a room-temperature superconductor can be suggested. One can dope graphene directly by species which become antiferromagnetically ordered in the doped state above room temperature, and induce the states at the Fermi level.

10 Conclusions

This chapter presents analysis of experimental data. On the basis of this analysis, it is possible to draw conclusions about components and the structure of a superconductor with a crit-

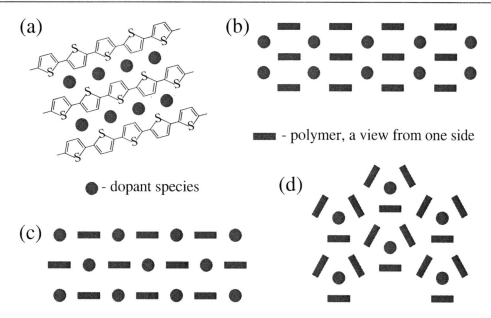

Figure 8: Possible positions of dopant species relative to infinite polythiophene chains: (a) in the plane of polythiophene chains; (b) between the planes; (c) a checker-border pattern, and (d) a "triangular" pattern realized in Na-doped polyacetylene [12].

ical temperature which may exceed the room temperature. The two essential components of a promising room-temperature superconductor are large organic molecules (polymers, tissues) and atoms/molecules which are magnetic in the intercalated state. To reach this conclusion, one does not require knowledge of the mechanism of room-temperature super-conductivity. However, to synthesize a room-temperature superconductor is a very difficult task and, in a first approximation, equivalent to the task of finding a needle in a haystack. Nevertheless, the importance of this chapter is in that it indicates in which "haystack" to search.

11   A Note about the Mechanism of Superconductivity in Cuprates

It is possible that in cuprates there are two more or less independent processes, leading to the occurrence of high-T_c superconductivity. One of them is the formation of incoherent electron (hole) pairs, and the second process is the Bose-Einstein condensation of magnetic excitations, for example, magnons. Recently, the Bose-Einstein condensation of magnons was observed in a number of antiferromagnetic compounds [22]. If the Cooper pairs are coupled to magnetic excitations, then the pairs can adjust their phases, i.e. establish the long-range phase coherence, through the long-range phase coherence of the magnon Bose-Einstein condensate. The Cooper pairs can be coupled to magnetic excitations through the amplitude of their wavefunctions, or only through the phase, or through both.

References

[1] Kamerlingh Onnes, H. *Commun. Phys. Lab. Univ. Leiden* 1911, 124c.

[2] Mourachkine, A. *Room-Temperature Superconductivity*; Cambridge International Science Publishing: Cambridge, 2004, pp 12, 71, 81, 129, 136, 285, 107, 293 (also http://arxiv.org/ftp/cond-mat/papers/0606/0606187.pdf).

[3] Mourachkine, A. *High-Temperature Superconductivity in Cuprates: The Nonlinear Mechanism and Tunneling Measurements*; Kluwer Academic Publishers: Dordrecht, 2002, pp 247, 246,

[4] Geballe, T. H. *Science* 1993, 259, 1550.

[5] Bardeen, J.; Cooper, L. N.; Schrieffer, J. R. *Phys. Rev.* 1957, 108, 1175.

[6] Weller, T. E.; Ellerby, M.; Saxena, S. S.; Smith, R. P.; Skipper N. T. *Nature Physics* 2005, 1, 39.

[7] Homes, C. C.; Dordevic, S. V.; Strongin, M.; Bonn, D. A.; Liang Ruixing; Hardy W. N.; Komiya Seiki; Ando Y.; Yu, G.; Kaneko, N.; Zhao, X.; Creven, M.; Basov, D. N.; Timusk, T. *Nature* 2004, 430, 539.

[8] Little, W. A. *Phys. Rev.* 1964, 134, A1416.

[9] Kresin, V. Z.; Litovchenko, V. A.; Panasenko, A. G. *J. Chem. Phys.* 1975, 63, 3613.

[10] Andriotis, A. N.; Menon, M.; Sheetz, R. M.; Chernozatonskii, L. *Phys. Rev. Lett.* 2003, 90, 026801.

[11] Zhao, G.-m. preprint 2000, cond-mat/0208200.

[12] Heeger, A. J.; Kivelson, S.; Schrieffer, J. R.; Su, W.-P. *Rev. Mod. Phys.* 1988, 60, 781.

[13] Friend R. H.; Gymer R. W.; Holmes A. B.; Burroughes J. H.; Marks R. N.; Taliani C.; Bradley D. D. C.; Dos Santos D. A.; Brédas J. L.; Lögdlund M.; Salaneck W. R. *Nature* 1999, 397, 121.

[14] Davydov, A. S. *Solitons in Molecular Systems*; *Naukova Dumka*. Kiev, 1988 (in Russian), p. 103.

[15] Davydov, A. S. *Phys. Rep.* 1990, 190, 191.

[16] Davydov, A. S. *Solitons in Molecular Systems*; Kluwer Academic Publishers: Dordrecht, 1991, p. 124.

[17] Pullman, B.; Pullman, A. *Quantum Biochemistry*; Interscience: New York, 1963.

[18] Clardy, J.; Walsh, C. *Nature* 2004, 432, 829.

[19] *Graphite Intercalation Compounds*, Tanuma, S. and Kamimura, H.; Ed.; World Scientific: Singapore, 1985.

[20] Kopelevih, Y.; Esquinazi, P.; Torres, J. H.; da Silva, R. R.; Kempa, H.; Mrowka, F.; Ocana, R. preprint 2002 (a chapter in a book), cond-mat/0209442, and references therein.

[21] Novoselov, K. S.; Jiang, D.; Booth, T.; Khotkevich, V. V.; Morozov, S. M.; Geim, A. K. *PNAS* 2005, 102, 10451.

[22] See, for example, Ruegg, Ch.; Cavadini, N.; Furrer, A.; Gudel, H.-U.; Kramer, K.; Mutka, H.; Wildes, A.; Habicht, K.; Vorderwisch, V. *Nature* 2003, 423, 62.

In: Recent Developments in Superconductivity Research   ISBN 978-1-60021-462-2
Editor: Barry P. Martins, pp. 33-64   © 2007 Nova Science Publishers, Inc.

Chapter 2

SYNTHESIS AND THERMOPHYSICAL PERFORMANCE OF BARIUM DOPED BI-BASED SUPERCONDUCTORS

Muhammad Anis-ur-Rehman *and Asghari Maqsood*

Thermal Physics Laboratory, Department of Physics,
Quaid-i-Azam University, Islamabad (45320), Pakistan.[*]

Abstract

The discovery of ceramic-based high temperature superconductors (HTS) during 1980's opened the possibility of applying the technology to electric power devices such as power transmission cable, high magnetic field magnets, SQUID, motor and magnetic levitation trains etc. Dissipation phenomena in high temperature superconductors are governed by the microstructure that develops during the preparation process. Therefore, superconductors prepared either in the form of single crystals, thin films or polycrystalline plays an important role for understanding superconductivity as well as for practical applications. Also various doping elements play an impressive role in any superconducting systems.

In superconductors where the dc electrical conductivity diverges to infinity below T_c, the thermal conduction is almost a unique measurement to study the transport properties below T_c. The figure of merit is the deciding factor for the quality of thermoelectric materials. In order to increase the whole figure of merit, it is of interest to replace the p-type leg of the Peltier junction by a thermoelectrically passive material with a figure of merit close to zero. This is why it is interesting to study the figure of merit of the ceramic superconductors. One of the important thermomagnetic transport quantities is the electrothermal conductivity and is shown to be one of the powerful probes of high-temperature superconductors. Potent applications for electrothermal conductivity of superconductors are actuators in MEMS technologies, superconducting bolometric detectors, electrothermal rockets etc.

The present study investigates the relationships among macroscopic physical properties and features at atomic level for the high-T_c superconducting (Bi(Pb)Sr(Ba)2223) material, which was prepared by a solid state reaction method. The samples were analyzed by dc electrical resistivity, ac susceptibility, thermal transport, electrothermal conductivity and thermoelectric properties all as a function of temperature (from room down to LN_2 temperature). Room temperature X-ray diffraction studies were also done. Samples are

---

[*] E-mail address: tpl.qau@usa.net & tpl@qau.edu.pk

investigated for thermal transport properties, i.e. thermal conductivity, thermal diffusivity and heat capacity per unit volume, by an Advantageous Transient Plane Source method. Simultaneous measurement of thermal conductivity and thermal diffusivity makes it possible to estimate specific heat and the Debye temperature Θ_D, as well as to separate the electron and phonon contributions to thermal conductivity and diffusivity. Thermoelectric power (Seebeck Coefficient) and electrothermal conductivity are also measured. Using electrical resistivity, thermal conductivity and thermoelectric power, the figure of merit factor is estimated.

Introduction

Due to the ever-increasing number of materials being used in a wide temperature range applications, knowledge of their thermophysical properties, especially thermal conductivity and thermal diffusivity, are of paramount importance.

The modified TPS technique referred to as Advantageous Transient Plane Source (ATPS) technique [1, 2], offers the possibility to measure thermal properties which are directly related to heat conduction such as thermal conductivity, thermal diffusivity and heat capacity per unit volume, with more ease and improved precision.

The selection of suitable materials for thermophysical properties measurement depends on the possibilities of exploring the genuine understanding of the physical phenomena, the processes involved during the transport of thermal energy, and the effect of different mechanisms on these processes. It also relies on the practical applications of the selected materials, which involves many factors such as economics, safety etc. depending on the conditions required by the nature of that particular application.

The discovery of ceramic-based high temperature superconductors (HTS) during 1980's opened the possibility of applying the technology to various devices. The HTS has the ability to achieve the superconducting state above temperatures of liquid nitrogen (around 77K), rather than the liquid helium (around 4K) required by the low temperature superconductors.

In the years since the discovery of high-temperature (high-T_c) cuprate superconductors by Bednorz and Muller [3], more than 40 new high-T_c substances have been found. Among them, superconductors of Y-based (Y-Ba-Cu-O) [4], Bi-based (Bi-Sr-Ca-Cu-O) [5], and Tl-based (Tl-Ba-Ca-Cu-O) cuprates, which can be used at liquid nitrogen temperature, are attracting much interest for not only their basic physical properties but also for their applications.

Bi-based cuprate superconductors encompass the series $Bi_2Sr_2Ca_{n-1}Cu_nO_{2n+4+\delta}$ with n = 1 - 3. The critical temperature, T_c, of the Bi-cuprates rises with increasing the number of CuO_2 layers, n [6]; 20K at n = 1, 80-90K at n = 2 and 110K at n = 3. Among the Bi-cuprates much attention has been paid to the oxides with n = 2 and 3, namely, $Bi_2Sr_2CaCu_2O_{8+\delta}$ (Bi-2212 or low T_c phase) [7, 8] and $Bi_2Sr_2Ca_2Cu_3O_{10+\delta}$ (Bi-2223 or high-T_c phase) [9], which have T_c exceeding the temperature of liquid nitrogen. After the discovery of high-T_c superconductivity above 100K in the Bi-Sr-Ca-Cu-O system one still could not isolate the n = 3 oxide (high-T_c phase) from the n = 2 oxide (low-T_c phase).

In the Pb-free BSCCO system, it is well known that it is relatively easy to prepare the 80K, Bi-2212 compound in nominal single-phase form. In contrast, the processing window for the Pb-free Bi-2223 phase is narrow, and obtaining a commercially significant amount of single phase material is difficult. Partial doping of Bi with Pb, on the other hand, can stabilize the phase formation of the Pb-doped 110K, Bi-2223 or ([Bi, Pb]:Sr:Ca:Cu) superconductors [10-16].

Substitutions may play an important role on the properties of high temperature superconducting materials. The existence of doping elements in substitution lattice sites lead to the formation of a superconductor with altered properties. Lead (Pb) is the most important doping element that influences the microstructure, phase composition and related superconducting properties of the Bi-Sr-Ca-Cu-O system [17-27]. The presence of high valence cations V^{5+}, Nb^{5+} and Ta^{5+} in the initial stoichiometry can significantly enhance the formation of high-T_c Bi-2223 phase [28]. The B_2O_3 doping is effective to result in the faster growth and better alignment of the Bi-2223 grains in the superconducting core and improve the magnetic field dependence of the critical current density, J_c, value [29]. Many other doping elements and substitutions like Sm, V, Ti, Zr, Hf, Y, La, Pr, Nd, Eu, Gd, Tb, Dy, Ho, Er and Tm are also undertaken by different workers [16, 30-33]. It was also interesting to study the effect of barium addition on the system Bi-Pb-Sr-Ca-Cu-O, because barium belongs to the same alkaline earth metal group as strontium and calcium, having different ionic radii. The appropriate amount of added Ba in Bi-2223 had the affect of raising T_c to a higher temperature region, a single transition phase and an improved critical current density, J_c [34].

The Bi-system was found to be sensitive to the Cu composition [35, 36]. Maximum T_c is observed for n=3 and for higher n values results are almost same [37]. T_c varies depending on the number of holes provided on the CuO_2 layers in addition to the number of CuO_2 layers, n [38-41]. The number of holes can be artificially adjusted by replacing the cation element, which varies in valence (such as Ca^{2+} and to Y^{3+}) but usually the adjustment is done by changing the amount of oxygen, δ in the Bi double layers by heat treatment.

Bi-based cuprate superconductors have some interesting characteristics from the viewpoint of both basic science and applications: (1) Bi-cuprates have a highly waved modulated crystal structure, which have not been experienced so far to be incommensurate with the lattice, and (2) highly anisotropic or highly two-dimensional superconductivity, which strongly demonstrates a new type of vortex motion appearing in high-T_c superconductors and induces a new mechanism for superconducting phase change. (3) One can easily obtain high grain alignment even in polycrystalline samples by using mechanical and/or heating treatments, solving one of the serious problems of weak-link behavior between grains, which is promising for future applications of high field superconducting magnets. (4) Bi-cuprates are very stable in the atmosphere and much less poisonous. These characteristics are useful in shielding applications such as in microwave devices, SQUID sensors, and magnetic shielding. (Bi, $Pb)_2Sr_2Ca_2Cu_3O_x$ (Bi-2223), the leading high temperature superconductor with T_c of 110K, has been demonstrated to be the most technically viable present materials for superconducting application to electric power transmission lines, fault current limiters, transformers, electromagnets and motors [42]. Moreover, silver sheathed Bi-2223 is the only high-Tc superconducting material that can be fabricated in long lengths suitable for large-scale engineering applications.

It is now recognized that for many high-temperature superconductor applications the materials must be highly textured to nearly single-crystalline / single phase form, posing problems for wire fabrication; on the other hand, interfaces need to be controlled at the near-atomic level. The linear temperature dependence of the electrical resistivity is one of the most important properties of the normal phase kinetics of high-T_c layered cuprates [43]. Apart from the usual Pb substitution on the Bi sites, only a few substitutions, such as Ba on Sr sites or Ni on Cu sites, have been realized in Bi-2223 [34, 44]. The appropriate amount of added Ba in Bi-2223 had the affect of raising T_c to a higher temperature region and a single transition phase.

In superconductors where the dc electrical conductivity diverges to infinity below T_c, the thermal conduction is almost a unique measurement to study the transport properties below T_c. There is not only an obvious technological interest, in how efficiently and by what means the heat flows in these solids but also a deep theoretical desire to understand the electronic and vibrational properties of these materials. The magnitude and temperature dependence of the thermal conductivity are parameters which have an impact on a broad spectrum of devices. From a theoretical point of view, the thermal conductivity of superconductors offers important clues about the nature of their charge carriers and phonons and scattering processes between them. In high-T_C superconductors, such information is even more valuable to know how the free carriers and lattice vibrations contribute to the transport of heat. A phase transition, represented by the onset of charge carrier condensation as the temperature is swept through the superconducting transition point, is responsible for a sharp change in the electromagnetic and kinetic properties of superconductors and, among other things, leads to a drastic modification of the heat flow pattern in these substances.

An applied thermal gradient to a solid is accompanied by an electric field in the opposite direction; i.e. the thermoelectric effect. Thermoelectric material applications include refrigeration or electrical power generation. Peltier refrigerators use the thermoelectric materials for refrigeration. Peltier thermoelectrics are more reliable than compressor based refrigerators, and are used in situations where reliability is critical like deep space probes. Thermoelectric materials used in the present refrigeration or power generation devices are heavily doped semiconductors. The metals are poor thermoelectric materials with low Seebeck coefficient and large electronic contribution to the thermal conductivity. Insulators have a large Seebeck coefficient and a small contribution to the thermal conductivity, but have too few carriers, which result in a large electrical resistivity. The figure of merit is the deciding factor for the quality of thermoelectric materials. In order to increase the whole figure of merit, it is of interest to replace the p-type leg of the Peltier junction by a thermoelectrically passive material with a figure of merit close to zero [45]. This is why it is interesting to study the figure of merit of the ceramic superconductors.

One of the important thermomagnetic transport quantities is the electrothermal conductivity and is shown to be one of the powerful probes of high-temperature superconductors. Cryogenic bolometers are sensitive detectors of infrared and millimeter wave radiation and are widely used in laboratory experiments as well as ground-based, airborne, and space-based astronomical observations [46]. In many applications, bolometer performance is limited by a trade off between speed and sensitivity. Superconducting transition-edge bolometer can give a large increase in speed and a significant increase in sensitivity over technologies now in use. This combination of speed with sensitivity should open new applications for superconducting bolometric detectors [47]. Other potent applications for electrothermal conductivity of superconductors is actuators in MEMS technologies, electrothermal rockets etc.

The present study investigates the relationships among macroscopic physical properties and features at atomic level for the high-T_c superconducting (Bi(Pb)Sr(Ba)2223) material, which was prepared by a solid state reaction method. The samples were analyzed by dc electrical resistivity, ac susceptibility, thermal transport, electrothermal conductivity and thermoelectric properties all as a function of temperature (from room down to LN_2 temperature). Room temperature X-ray diffraction studies were also done. Samples are investigated for thermal transport properties, i.e. thermal conductivity, thermal diffusivity and heat capacity per unit volume, by an Advantageous Transient Plane Source method. Simultaneous measurement of thermal conductivity and thermal diffusivity makes it possible to estimate specific heat and the Debye temperature Θ_D, as well as to separate the electron and phonon contributions to thermal conductivity and diffusivity. Thermoelectric power (Seebeck Coefficient) and electrothermal conductivity are also measured. Using the data of electrical resistivity, thermal conductivity and thermoelectric power, the figure of merit factor was calculated.

High-temperature superconductors are recent innovations from scientific research laboratories and was supposed to break new ground in power industry. New commercial innovations begin with the existing technological knowledge generated by the research scientists. The work of commercialization centers on the development of new products and the engineering needed to implement the new technology. Superconductivity has had a long history as a specialized field of physics. Through the collaborative efforts of government funded research, independent research groups and commercial industries, applications of new high–temperature superconductors was thought to be in the not so distant future. Time lags however, between new discoveries and practical applications are often great. The rapid progress in the field of superconductivity had lead one to believe that applications of superconductors is limited only by one's imagination and time. But no major break through, i. e. superconductor at room temperature, was observed after late 90's, due to this the interest of commercial industries is reduced little bit. Still the work is going on in research laboratories to understand the theory of high -T_c superconductors and their future applications.

Preparation of Superconducting Samples

In the Bi-based high-T_c superconductors the Bi-2212 phase is known to be thermodynamically stable over a wide temperature range and in the presence of most of the compounds existing in this system. In contrast, the Bi-2223 phase is stable within a narrow temperature range and exhibits phase equilibria with only a few of the compounds existing in the system [16].

Two major issues seem to complicate the development of the Bi-2223 phase; it is stable in only a very narrow temperature range and the kinetics of its formation are so slow that it is almost impossible to obtain the phase-pure material [48]. Because of this, precise control over the processing parameters is required so the desired properties can be realized.

All the specimens were prepared from 99.9% pure powders of Bi_2O_3, PbO, $SrCO_3$, $BaCO_3$, $CaCO_3$ and CuO. The powders were mixed to give nominal composition of $Bi_{1.6}Pb_{0.4}Sr_{1.6}Ba_{0.4}Ca_2Cu_3O_y$ and were thoroughly ground in an agate mortar to give very fine powder. The grind powder was calcined for 21 hours in air at $800^{\circ}C$.

The starting composition of $Bi_{1.6}Pb_{0.4}Sr_{1.6}Ba_{0.4}Ca_2Cu_3O_Y$ compound is

$$0.8Bi_2O + 0.4PbO + 1.6SrCO_3 + 2CaCO_3 + 0.4BaCO_3 + 3CuO \rightarrow$$
$$Bi_{1.6}Pb_{0.4}Sr_{1.6}Ba_{0.4}Ca_2Cu_3Oy + 4CO_2 \qquad (1)$$

A series of pellets were produced in two sizes, from this well mixed material and controlled heating and cooling carried out, in air, using a horizontal tube furnace. Poly Vinyl Alcohol (PVA) was used as binder in the samples. PVA is one of the few high molecular weight commercial polymers, which is water soluble and is dry solid, commercially available in granular or powder form. The properties of Poly Vinyl Alcohol vary according to the molecular weight of the parent poly vinyl acetate and the degree of hydrolysis. Fully hydrolyzed form with medium viscosity grade PVA was used in our case. Samples were in the shape of cylindrical disks having diameters 13mm and 28mm, and lengths 3mm and 11mm respectively. These samples were sintered at 830^{0}C for the intervals of 24 hours in each sintering step as sintering procedures do affect the properties [37].

Advantageous Transient Plane Source (ATPS) Technique

Over the past decade there has been a rapid development and application of new and improved materials for a broad range of physical, chemical, biological and medical applications and this has led to a huge increase in requirements for thermal performance data. However, established thermal properties measurement methods may require large samples, or they are too complex and time-consuming, and so alternative methods have been developed. These have the virtues of being fast, apparently precise, measure more than one thermal property and are flexible, covering a wide range of materials, temperatures and thermal property values. Advantageous Transient Plane Source Technique (ATPS) is an improved form of the Transient Plane Source (TPS) method to achieve the above mentioned aims.

Transient Plane Source (TPS) method is one of the contact transient methods and was developed by a number of workers [49-57].

The TPS method utilises a thin disk shaped temperature dependent resistor simultaneously as the temperature sensor and as the heat source for the thermal conductivity and thermal diffusivity measurements. The sensor is sandwiched between two specimen halves, as indicated in Figure 1. A direct current is passed through the sensor, this current is sufficiently large to increase the sensor temperature by 1-2K [50, 51]. Due to the temperature increment, the resistance of the sensor will change and there will be a corresponding detention in voltage drop over the sensor. By recording the voltage variation over a certain time period from the onset of the heating current, it is possible to obtain precise information on the heat flow between the sensor and the test specimen.

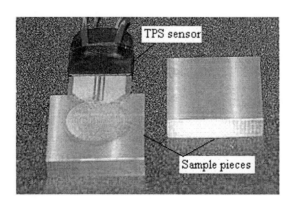

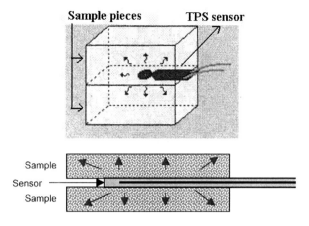

Figure 1: TPS sensor sandwiched between sample pieces.

There will, however, be a small temperature drop, ΔT_i, over the electrically insulating mica layer [53, 54, 57]. After a short initialization period this temperature drop will stay constant due to the liberation of constant power. The resistance of the sensor can then be expressed by:

$$R(t) = R_o \left(1 + \alpha \Delta T_i + \alpha \overline{\Delta T(\tau)}\right)$$

(2)

where R_o is the resistance of the TPS element before the transient recording has been initiated, α is the temperature coefficient of the resistance and $\overline{\Delta T(\tau)}$ is the average temperature increase of the TPS element assuming perfect thermal contact with the test specimen, $\overline{\Delta T(\tau)}$ may be expressed as:

$$\overline{\Delta T(\tau)} = \frac{P_o}{\pi^{3/2} r \lambda} D(\tau)$$

(3)

In equation 3, P_o is the heat liberation from the sensor, λ is the thermal conductivity of the test specimen, r is the radius of the TPS sensor and τ is given by:

$$\tau = \frac{\sqrt{\kappa t}}{r} = \sqrt{\frac{t}{\theta}}$$

(4)

Here t is the measurement time and θ is the characteristic time of the measurement

$$\left(\theta = \frac{r^2}{\kappa}\right).$$

The D(τ) function is the design parameter. Inserting equation 3 into equation 2 and setting and

$$C = \frac{\alpha R_O P_O}{\pi^{3/2} r \lambda} \; ;$$

Equation 2 may be rewritten as:

$$R(t) = R^* + CD(\tau) \tag{5}$$

The thermal diffusivity can be calculated once θ is determined ($\kappa = r^2 / \theta$ and the thermal conductivity is obtained from the slope 'C' of the straight line [58].

Using a measurement time much shorter than θ ($\tau < 0.5$), only the parameter κ can be calculated since the mathematical solution approaches the situation of a semi-infinite slab exposed to a constant heat flux at the surface. Using a measurement time much longer than θ ($\tau > 2$) only the thermal conductivity can be calculated since the mathematical solution approaches the situation of an infinite solid internally heated by a constant point source. Hence in order to obtain both the thermal diffusivity and thermal conductivity from one transient run, the time period of heating should not deviate too much from the characteristic time of the measurement [58].

To improve the above mentioned technique, some improvements in the circuit were made. For convenience this improved technique is named Advantageous Transient Plane Source Technique (ATPS).

The ATPS method considers three dimensional heat flow inside the specimen and is regarded as an infinite medium. The technique uses a 'resistive element' similar to TPS method (the TPS sensor) both as heat source and temperature sensor as shown in Figure 1.

Starting from the above mentioned theory, the thermal transport properties can be found using an appropriate curve-fitting technique for the experimentally measured temperature versus time points. The ideal model presupposes that the double spiral sensor, assumed to consist of a set of equally spaced, concentric, and circular line heat sources, is sandwiched in specimens of infinite dimensions. In practice all real specimens do have finite dimensions. However, by restricting the time of the transient, which relates to the thermal penetration depth of the transient heating, a measurement can still be analyzed as if it was performed in an infinite medium. This means that the ideal theoretical model is still valid within a properly selected time window for the evaluation.

The sensitivity coefficients (β_p) [59] are the theoretical foundation for determining a suitable time window to be used in the curve-fitting procedure. The mathematical formula for the sensitivity coefficients is given by

$$\beta_p = p \frac{\partial T(y,t)}{\partial p} \tag{6}$$

where p is any of the thermal transport parameters and T(y, t), is the temperature function given by equation 3. The optimal time window for determining both the thermal conductivity and the thermal diffusivity from a single transient recording has been identified as the interval $[t_{min}, t_{max}]$ where $t_{min} \geq (\delta_{ins})^2 / \kappa_{ins}$ (here δ_{ins} is the thickness and κ_{ins} is the thermal diffusivity of the sensor insulation material) and $0.3 \leq t_{max} / \theta \leq 1.1$ [60].

To record the potential difference variations, that correspond to temperature change, during the transient recording, a modified bridge arrangement was used as shown in Figure 2. The circuit components are reduced with this new arrangement as compared to the bridge used earlier [57].

The bridge is in the balanced mode shortly before the transient recording, during transient run the bridge is working a bit off-balanced. If we assume that resistance increase will cause a potential difference $\Delta U(t)$ measured by the voltmeter in the bridge, the analysis of the bridge indicates that temperature will vary as described below;

$$\frac{\Delta U(t)}{\Delta T(\tau)} = \frac{\Delta U(t)}{1.0 - \Delta U(t) / r_s i_0} * \frac{(r_t + r_s)}{(\alpha r_0 r_s i_0)} , \tag{7}$$

where r_s is the high power standard resistor and r_t, between points b and e (Figure 2), is the total resistance of the TPS sensor (r_o) and connecting leads (r_L) before transient run and is given by

$$r_t = \frac{r_s U_{rt}}{U_{rs}} \tag{8}$$

Here, U_{rt} is the voltage across resistance r_t and between points b & e; $(r_t = r_o + r_L)$; U_{rs} is the voltage across standard resistance r_s ; and i_o is the heating current through the sensor during transient run.

It is to be noted that the resistance of the TPS sensor is calculated from the same experiment and no separate experiment is required, (for TPS method previously used, resistance of the TPS-sensor was measured separately) can be obtained as;

$$r_0 = \frac{r_s U_{r0}}{U_{rs}} \tag{9}$$

U_{r0} being the voltage across the TPS sensor.

Accuracy of the resistance measurement of the TPS sensor plays an important role in the calculation of thermal parameters, so an exact value is achieved by the above procedure.

A TPS sensor is in principle a very sensitive (resolution better than millikelvin) resistance thermometer, which is recording temperature during a transient run. The collected data are vulnerable to distortion if there is any temperature drift in the specimen or in the whole specimen assembly, resulting in erroneous calculations of thermal parameters. This could be compensated either by measuring temperature drift or giving time of relaxation to the system. The detail of factors influencing the experimental measurements is as follows.

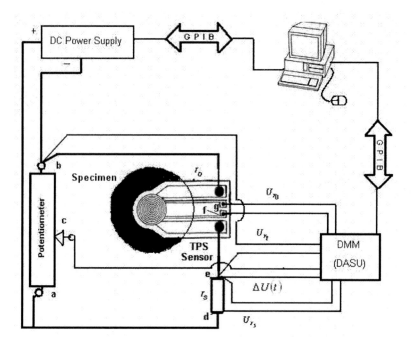

Figure 2: Principal electrical circuit for the modified bridge arrangement, showing TPS sensor (r_o) with specimen, standard resistance (r_s), potentiometer, dc power supply, DASU and points of measuring voltages.

The first important factor is the temperature drift. In certain cases it is difficult to achieve a homogeneous temperature distribution around the specimen assemblage (the sensor and specimen pieces) particularly when cooling down to liquid Nitrogen temperature. This could be compensated by the following procedure; the temperature variation, $\Delta T_d(\tau)$, of the element, owing to the temperature drift, should be monitored. Immediately after temperature drift measurement $\Delta T_d(\tau)$, the transient experiment ($\Delta T(\tau)$) is performed. The value of actual change in temperature $\Delta T_a(\tau)$ is then obtained by subtraction of $\Delta T_d(\tau)$ from measured $\Delta T(\tau)$, i.e.

$$\Delta T_a(\tau) = \Delta T(\tau) - \Delta T_d(\tau). \tag{10}$$

Second factor to be considered is the time of relaxation. Initially the specimen is considered isothermal and placed in a perfect thermostat so that any temperature variation from the surroundings would not distort data. The time gap between two consecutive transient experiments, for reliable results, is termed as time of relaxation and is at least a^2/κ, where a = radius of specimen or a = height of specimen, depending which ever is greater. κ is the thermal diffusivity of the material under study and it is assumed that probing depth is equal to the radius of the sensor. Variation in sensor radius does not make any difference. Relaxation time is reduced by six times with the modified circuit (ATPS) arrangement [1, 2].

For the low temperature measurements inside a cryostat [55, 61] or for the high temperature measurements inside a furnace [51, 57, 62] special attention is needed to eliminate any heating effect of the long leads used i. e. lead resistance (r_L). Instead of measuring resistance r_o directly by using four probe method [61], voltages at points b-e and f-g as shown in Figure 2 are measured to determine,

$$Ur_t = Ur_o + Ur_L$$ (11)

Using i_o and Ohms law, r_L is determined as $r_L = r_t - r_o$.

Calibration of the ATPS Apparatus

With the purpose of demonstration that ATPS technique, the modified setup, could be used down to temperature of liquid nitrogen, measurements on fused quartz, carbon steel and silver chloride crystals were performed for calibration purpose. Main aim of the whole practice was to develop an easy, with an improved accuracy and reduced limitations, system that could be used for the measurement of thermal transport properties of insulators, conductors and high-T_c superconductors as well. The choice of fused quartz was due to the fact that it is an insulator and that its thermal conductivity data are well established at low temperatures [63]. Carbon steel fell in conductors and silver chloride is a crystalline material so their information will help to determine the reliability for conductors and crystalline materials. Measurements are done as a function of temperature for all the three specimens. Due to availability of data for thermal conductivities of the studied specimens from other sources (with different techniques), only comparison of thermal conductivities is made. The results of these reference samples are given in the following sections.

Data for the fused quartz specimens were taken by using two sensors namely, Sensor-1 (radius = 9.734 ± 0.001mm) and Sensor-2 (radius = 3.300 ± 0.001mm). Variation of thermal conductivity with temperature is shown in Figure 3. The results obtained are compared with the recommended experimental points [63] and also with the previously calibrated TPS bridge circuit [55].

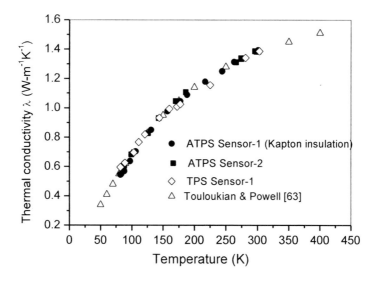

Figure 3: Variation of thermal conductivity of fused quartz samples with temperature.

It has been observed that error is not more than 1% at any temperature. Previously reported errors were 1% at room temperature and 4% at low temperatures [55]. These results prove that the performance of ATPS is better than TPS. Measurements were made with two different sensor sizes and values for thermal conductivity are in very good agreement. However the results for the thermal diffusivity show some differences. This difference is due to the following factors. The Sensor-2 requires that the total experimental time should be short enough. With the total time selected within the optimum time window, the thermal diffusivity is to some extent on the lower side, and similar observations are reported by other workers [65]. Also a higher-resistance sensor is preferable to minimize the error in measuring $\Delta r_o/r_o$, which is nearly as good as that claimed by [66], here r_o is the resistance of the sensor. Resistance of the Sensor-1 is higher than that of the Sensor-2.

Carbon steel specimens SS215/3 obtained from Analytical Standards AB, were analyzed from room temperature down to boiling point of liquid nitrogen. Plot of thermal conductivity (measured by ATPS technique) variation with temperature is shown in Figure 4 along with composition of the specimen. A similar trend has been observed by an earlier worker [67] for an almost similar composition. They made measurements from room temperature to 573K.The values for thermal conductivity are in good agreement with this report for the identical temperature range. Detailed comparison is made with the data taken by Maqsood et al [55] by TPS method. The thermal conductivity values decreased at low temperatures due to the reduction of electronic contribution towards the conduction.

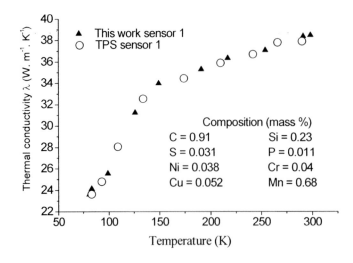

Figure 4: Variation of thermal conductivity with temperature for the carbon steel SS 215/3

Silver chloride crystals are useful material for deep IR applications where sensitivity to moisture is a problem. A major use is in the manufacture of small disposable cell windows for spectroscopy, known as 'mini-cell'. Structure of these crystals is cubic FCC. These specimens were obtained from Crystran, UK [64].

The thermal conductivity and thermal diffusivity values determined experimentally, by using Sensor-2, for AgCl are plotted in Figure 5. These data will be helpful for application to optical windows used at low temperatures. The available data from other sources are also indicated in the figure. Our results on thermal conductivity are in excellent agreement with these reports. For thermal diffusivity no directly measured data could be found. The calculated values of κ based on the thermal conductivity, specific heat [64] and density of AgCl at 273K agreed with our measurements. Good agreement of our data with already published experimental data and data calculated from different theoretical models, show the strength of ATPS technique [1, 2, 66-79].

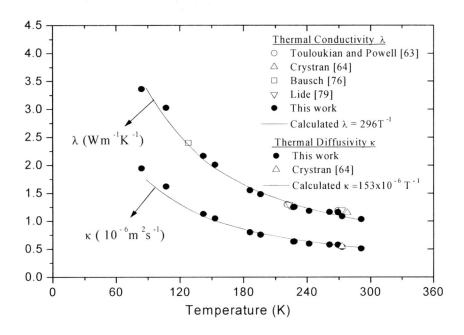

Figure 5: Thermal conductivity and thermal diffusivity of AgCl as a function of temperature along with the reported data.

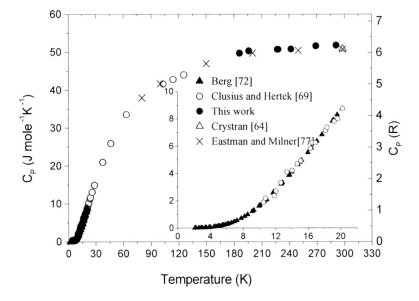

Figure 6: Heat capacity at constant pressure (Cp) of AgCl as a function of temperature (R = 8.31 J mol^{-1} K^{-1}). Expanded view of Cp in the temperature range 0-22 K is also shown.

Thermoelectric Power (TEP) Measurements Apparatus

The thermoelectric power or Seebeck coefficient S is the central material property for thermoelectric applications. It is decisive, e. g., for the responsivity of thermal detectors. It is closely related to the concentration of charge carriers which has to be optimized for thermoelectric conversion purposes. Therefore, it contains sufficient information for material choice, whenever the interdependence between S and the thermal and electrical conductivity is known.

An easy to use and simple apparatus was designed and developed for thermoelectric power (S) measurements. Circuit diagram along with the sample holder assembly is shown in Figure 7.

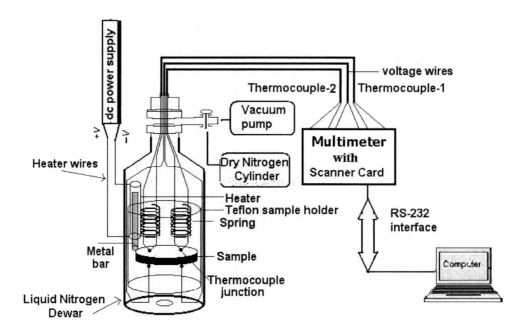

Figure 7: Block diagram of the apparatus developed for thermoelectric power measurements. Scanner card is used with the multimeter for simultaneous measurements at different points as shown in this figure.

The sample is subjected to a temperature difference ΔT by using a heating resistor and corresponding voltage difference ΔV across the sample is measured. Thermoelectric power is obtained by taking ratio of the voltage difference to the temperature difference. Chromel-alumel thermocouples are used for measuring the temperature difference, ΔT. The thermocouples are electrically isolated from the sample and thermally connected to the sample. Heat losses through the electrical connections are minimized by using long leads wrapped around a Teflon tube. The voltage leads are then silver pasted to the sample in the vicinity of thermocouples to assure that the voltage and temperature gradients are measured at the same locations on the sample for accurate thermoelectric power measurements. The next step includes loading the sample assembly into the sample chamber and evacuation of the chamber. The chamber is evacuated to eliminate any water vapour condensation on the sample, which can result in erroneous measurements. Dry nitrogen gas is then filled in the chamber as a conducting media between chamber walls and the sample. This sample chamber is then inserted in liquid nitrogen container for cooling. Data are collected under the computer control. By incorporating multiple measurements in a single run, considerable time is saved by avoiding remounting, and recooling of the samples. In this technique the surface mount resistor (50Ω) was used to heat one end of the sample to establish a measured temperature gradient of approximately 1K.

To check the calibration of this new apparatus (Figure 7), thermoelectric power of copper was measured in the temperature range 85-310K. Results of our measurements are shown in Figure 8 indicating an agreement with the already published data [80]. The standard deviation in the data was between 0.01-0.22 μVK^{-1} and the difference between measurements done in this work and the already published [80] data are within 5%. Polycrystalline samples of superconductors are studied in the present case as polycrystalline High Temperature Superconductors (HTSC) are being used in technological applications [16, 48].

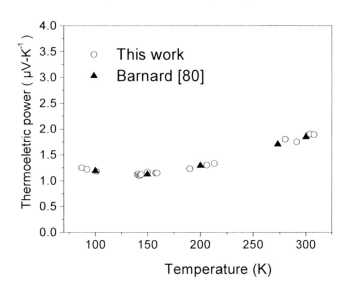

Figure 8: Thermoelectric power of the copper sample with temperature.

Thermophysical Properties of Bɪ(Pʙ)Sʀ(Bᴀ)2223 Superconductors

The high-T_c superconducting material with nominal composition $Bi_{1.6}Pb_{0.4}Sr_{1.6}Ba_{0.4}Ca_2Cu_3O_y$ was prepared by a solid state reaction method as described in chapter 2. The samples were analyzed by dc electrical resistivity, ac susceptibility, thermal transport and thermoelectric properties all as a function of temperature (from room down to LN_2 temperature). Room temperature X-ray diffraction studies were also done. Samples are investigated for thermal transport properties, i.e. thermal conductivity, thermal diffusivity and heat capacity per unit volume, by the Advantageous Transient Plane Source method [1, 2]. Thermoelectric power (Seebeck Coefficient) is also measured with a newly developed and calibrated apparatus as described in previous sections. Electrothermal conductivity was estimated and also by using electrical resistivity, thermal conductivity and thermoelectric power, the figure of merit factor was calculated.

The experimental results and discussion on them are given below.

- ### Dc Electrical Resistivity

Variation of resistivity with change in temperature is recorded for all the samples after each sintering step and the plots for one of the samples are given in Figure 9. One of the most striking features about the cuprate superconductors is the behavior of the resistivity of the normal state that is found above the transition temperature of the optimally doped materials. The resistivity vs temperature plots above the critical temperature exhibit a linear behavior, it may suggest that the scattering of the charge carriers is mainly by phonons in this temperature range. After the final sintering the measured density of the sample was 3.48 gcm^{-3} and $T_{c,0}$ was $110 \pm 1K$. Residual resistivity was 0.19 mΩ-cm and the intrinsic resistivity was 5.9 $\mu\Omega$-cmK^{-1}. The ratio $\rho(273K)/\rho(4.2K)$ is the residual resistivity ratio (RRR), an important parameter in the deign of superconductive applications. In the case of a superconductor, the denominator has to be taken at a temperature slightly above the critical temperature [81]. RRR in our case was in the range 23-33.

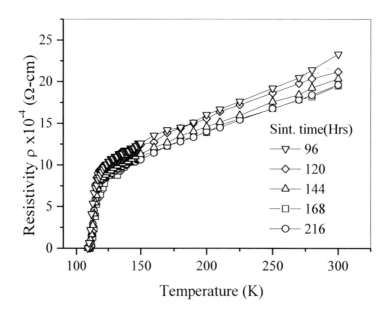

Figure 9: Dc electrical resistivity as a function of temperature for the sample after each sintering step.

- ### Ac Susceptibility

Ac susceptibility measurements were done after each sintering step and are plotted in Figure 10. Initially two transition phases were present. One of the identified phases is the Bi-2212(low-T_c) phase and the other Bi-2223(high-T_c) phase. With sintering, the low T_c phase was smoothed out and only phase left is the Bi-2223(high-T_c) phase. Although the resistivity variation with temperature was smoothed out after the third sintering but slight kinks were observed in the susceptibility against temperature plot showing the more sensitivity of the measuring method.

- **X-ray Diffraction Studies**

Almost all the peaks are indexed. The only phase is the orthorhombic high-T_c Bi-2223 phase. Lattice parameters were calculated from the (h k l) values of the indexed peaks. The lattice parameters are a = 5.42 (1) Å, b = 5.37 (1) Å and c = 37.12 (8) Å. No peaks were found matching the Bi-2212 low T_c phase. Indexed X-ray diffractograph is shown in Figure 11. The lattice constants agreed with the previous reports [6, 55].

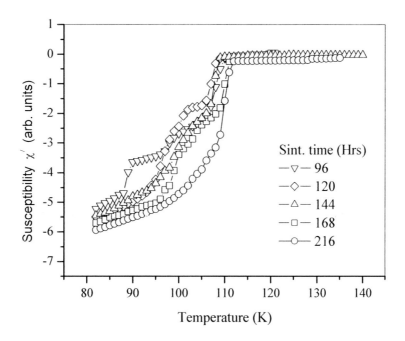

Figure 10: Variation of ac susceptibility (real part), with temperature after each sintering step.

The size of the grains in polycrystalline materials has pronounced effects on many of its properties. Using Scherrer's equation [82];

$$B = \frac{0.9\lambda}{t\cos\theta},$$ (12)

where; B = Broadening of diffraction line measured at half its maximum intensity (radians), λ = Radiation source wavelength, t = Diameter of crystal particle, particle sizes are determined and are given in table 6.1.The diameter of the crystal particles lies between 172 – 512 Å.

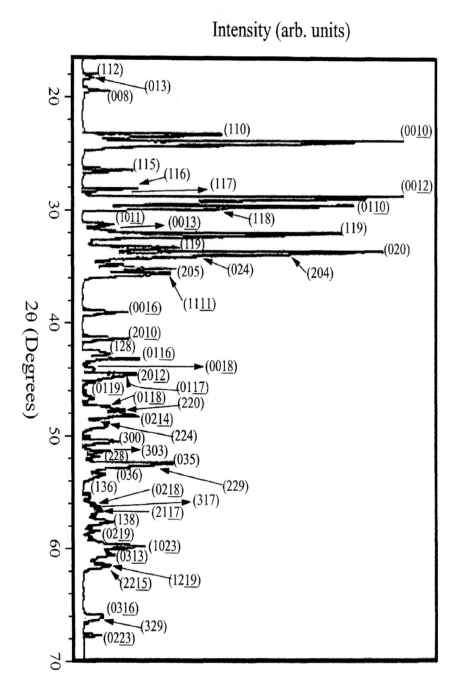

Figure 11: Indexed X-ray diffraction pattern of the sample after the final sintering at room temperature.

Table 1. Indexed X-ray diffraction peaks of superconductor along with the diameter (t) of crystallites.

Line No.	2θ (Deg.)	d (Å)	h	k	l	t (Å)	Line No.	2θ (Deg.)	d (Å)	h	k	l	t (Å)
1	17.68	5.012	1	1	2	335.08	27	44.86	2.018	0	1	17	409.37
2	18.00	4.924	0	1	3	335.23	28	46.48	1.952	0	0	19	360.34
3	19.14	4.633	0	0	8	206.63	29	47.24	1.922	0	1	18	361.37
4	23.18	3.834	1	1	0	270.39	30	47.62	1.908	2	2	0	222.71
5	24.00	3.704	0	0	10	225.66	31	48.16	1.887	0	2	14	322.36
6	26.26	3.39	1	1	5	209.22	32	48.72	1.867	2	2	4	264.33
7	27.56	3.233	1	1	6	272.73	33	50.48	1.806	3	0	0	172.26
8	27.92	3.193	1	1	7	227.45	34	51.26	1.78	3	0	3	326.43
9	28.90	3.086	0	0	12	341.91	35	51.76	1.764	2	2	8	226.46
10	29.24	3.051	0	1	10	248.86	36	52.52	1.741	0	3	5	210.97
11	30.08	2.968	1	1	8	210.98	37	52.74	1.734	2	2	9	369.55
12	31.08	2.875	1	0	11	196.38	38	53.42	1.713	0	3	6	269.56
13	31.64	2.825	0	0	13	305.90	39	55.96	1.641	1	3	6	299.94
14	32.14	2.782	1	1	9	229.71	40	56.28	1.633	0	2	18	375.48
15	33.18	2.697	1	1	9	230.32	41	56.46	1.628	3	1	7	501.06
16	33.84	2.646	0	2	0	197.76	42	56.62	1.624	2	2	12	376.08
17	34.30	2.612	2	0	4	462.01	43	56.84	1.618	2	1	17	430.25
18	34.66	2.585	0	2	4	277.47	44	57.62	1.598	1	3	8	335.88
19	35.04	2.558	2	0	5	252.51	45	58.32	1.58	0	2	19	303.32
20	35.52	2.525	1	1	11	173.83	46	59.78	1.545	1	0	23	339.46
21	38.90	2.313	0	0	16	199.45	47	60.94	1.519	0	3	13	512.20
22	41.22	2.188	2	0	10	188.66	48	61.14	1.514	1	2	19	439.48
23	42.28	2.135	1	2	8	354.99	49	61.46	1.507	2	2	15	385.18
24	42.62	2.119	0	1	16	284.32	50	66.02	1.413	0	3	16	225.62
25	44.06	2.053	0	0	18	317.49	51	66.32	1.408	3	2	9	351.56
26	44.42	2.037	2	0	12	190.74	52	67.72	1.382	0	2	23	354.42

• **Thermal Properties Measurements**

After the preliminary characterization of the samples and existence of almost a single phase, large disc-shaped samples (28 mm diameter and 11 mm thickness) were prepared by the standard solid state reaction method as already described. The size of the samples used for the thermal properties measurements was indeed the largest reported for this kind of measurement. Figure 12 shows the temperature dependence of the thermal conductivity λ. As the temperature decreases, the conductivity gradually decreases down to near T_c then remarkably increases below T_c. Further decrease in temperature was not possible, due to limitation of the cryostat used, to take the maximum in λ. This temperature dependence agrees with the widely observed behavior of λ for the oxide superconductors [83 - 87].

Comparing the results between different laboratories, one notes that the thermal conductivity depends on a particular sample preparation process. The temperature dependence of the conductivity is really similar for all the samples [88, 89]. So the order of magnitude of thermal conductivity (measured by non-steady state method in our case) is comparable to the results obtained by different authors (measured by steady state methods) [90 - 94].

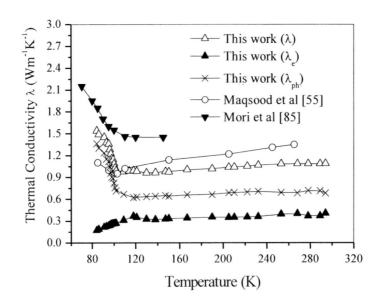

Figure 12: Thermal conductivity (λ) as a function of temperature. Electronic and phonon part of λ are also presented.

As is well known, the thermal conductivity of metals is given by two contributions,

$$\lambda = \lambda_{ph} + \lambda_e \tag{13}$$

Here λ_{ph} is the conductivity due to phonons and λ_e is the conductivity due to electrons. In simple metals, the separation of the two components of the thermal conductivity is made by use of the Wiedemann-Franz (WF) law [95, 96],

$$\frac{\lambda_{en}\rho}{T} = \frac{1}{3}\pi^2 \left(\frac{k_B^2}{e^2}\right), \tag{14}$$

where λ_{en} is the electronic thermal conductivity in the normal state, ρ is the electrical resistivity, k_B is the Boltzman's constant. In ordinary metals, the WF law fails at intermediate temperatures where the electrical resistivity ρ deviates from the T linear dependence. The resistivity of oxide superconductors shows the characteristic T linear dependence over quite a wide temperature range as can be seen in Figure 9. Accordingly, the WF law is expected to hold over the entire temperature range above T_c for the oxide superconductors and to result in constant and small λ_{en}, making a marked contrast to ordinary metals.

Below T_c the charge carriers which have condensed in the ground state do not contribute to the heat conduction and the electronic component λ_{es} (the electronic thermal conductivity in the superconducting state) is expected to decrease with lowering temperature. Among the several theories [97 - 99], which treat λ_{es}, we refer to revised Kadanoff's formulation adapted by Ikebe et al [89] to observe the linear dependence of ρ,

$$\frac{\lambda_{es}}{\lambda_{en}} = \frac{3}{2\pi^2} \int_0^\infty d\varepsilon\, \varepsilon^2 \sec h^2 \left\{ \frac{1}{2}[\varepsilon^2 + (\beta\Delta)^2]^{1/2} \right\} \left[\frac{1 + a\dfrac{T}{T_c}}{\dfrac{\varepsilon}{[\varepsilon^2 + (\beta\Delta)^2]^{1/2}} + a\dfrac{T}{T_c}} \right], \quad (15)$$

Here, a represents the ratio of the T-linear electrical resistance at T_c to the residual resistance, $\beta = \dfrac{1}{k_B T}$ and Δ is the BCS energy gap in the Buckingham [100, 101] form,

$$\Delta = 3.2 k_B T_c \left[1 - \frac{T}{T_c} \right]^{1/2}. \quad (16)$$

The electronic and phonon contribution to thermal conductivity estimated by equations 14 and 15 in normal and superconducting state respectively is also shown in Figure 12.

Figure 13 shows the variation of measured thermal diffusivity κ with temperature for the sample. Between 294K and T_c, κ increases very gradually with decreasing temperature. The increase of κ becomes very large around T_c and become very steep. This increase in κ may be due to its proportionality to λ and the present observation is in rough agreement with those reported by Onuki et al [102, 103]. Correspondingly, two contributions to thermal diffusivity κ are defined by the following relation,

$$\kappa = \frac{\lambda}{\rho C_p} = \frac{\lambda_{ph}}{\rho C_p} + \frac{\lambda_e}{\rho C_p} = \kappa_{ph} + \kappa_e, \quad (17)$$

where ρC_p is the heat capacity per unit volume. The separation of the phonon and the carrier contribution to the thermal diffusivity κ can easily be achieved on the basis of the corresponding separation for λ by the following equation,

$$\frac{\kappa_e}{\kappa_{ph}} = \frac{\lambda_e}{\lambda_{ph}}, \quad (18)$$

The temperature dependence of κ_{ph} and κ_e estimated in this way is shown in Figure 13.

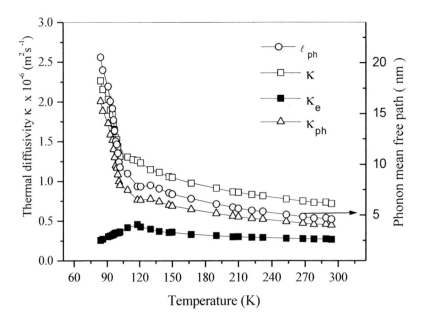

Figure 13: Variation of thermal diffusivity and mean free path of phonons with temperature for the samples.

Figure 14 shows the heat capacity per unit volume, ρC_p, calculated from the thermal conductivity measurements and thermal diffusivity measurements by using equation

$$\kappa = \frac{\lambda}{\rho C_p}$$

(19)

ρC_p decreases with decrease in temperature and near T_c a sizeable kink is observed. This jump is mostly due to the improved sharpness of the transition related to the reduction of intergrowth structure by adding Pb [104] and is improved by adding Ba in our case. Since the calculated lattice constants of our sample are similar to Bi-2223 composition so it is assumed that oxygen is 10, and then the composition becomes $Bi_{1.6}Pb_{0.4}Sr_{1.6}Ba_{0.4}Ca_2Cu_3O_{10}$. Also there is no change in the density of the superconducting sample in the studied temperature range so the value of specific heat C_p is calculated and is shown in Figure 15. The specific heat jump at T_c is apparently more clearly seen by plotting C_p / T against T, as shown in inset of Figure 15. This abrupt change of specific heat may be explained by the presence of the energy gap [105].

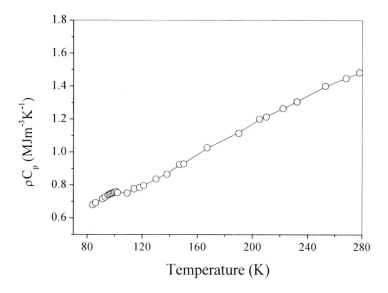

Figure 14: Variation of heat capacity per unit volume (ρC_p) with temperature for $Bi_{1.6}Pb_{0.4}Sr_{1.6}Ba_{0.4}Ca_2Cu_3O_y$ samples.

During the normal to superconducting transition, the electronic structure is reconstructed to allow the developed electron pairs to construct the superconducting current. This new electronic structure results in an anomaly in the continuity of the permitted energy levels, creating therefore, a forbidden energy gap. When the energy gap is exceeded, the breaking of electron pairs takes place, leading to the destruction of superconductivity [106]. The absolute value of C_p is 320 $Jmol^{-1}K^{-1}$ at 180K and that is similar to already reported value of a similar composition [104, 107].

Because the phonon contribution is by far dominant than the electronic contribution in the temperature range studied, the specific heat data were fitted to the following Debye formula,

$$C_{p-ph} = 9nR \frac{T^3}{\Theta_D^3} \int_0^{\Theta_D/T} \frac{x^4 e^x}{(e^x - 1)^2} dx \qquad (20)$$

where C_{p-ph} is molar specific heat, x is the reduced phonon frequency, n (= 19) the number of atoms composing Bi(Pb)Sr(Ba)2223 molecules, R the gas constant and Θ_D is the Debye temperature. Although a single Θ_D fitting fails to give a unified strict fitting over the entire temperature range, but Θ_D = 510K gives a satisfactory fitting between T = 130 to 190K as is shown in Figure 16.

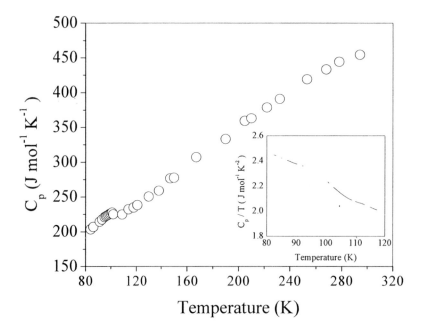

Figure 15: Temperature dependence of specific heat (C_p) for $Bi_{1.6}Pb_{0.4}Sr_{1.6}Ba_{0.4}Ca_2Cu_3O_{10}$ sample (Under the assumption that density remains constant within this temperature region). Inset is showing the C_p/T for the enhanced view of jump in specific heat.

If the electronic contribution to the specific heat ($C_{p, e}$), which is quite small compared to specific heat due to phonons ($C_{p, ph}$), except at extremely low temperatures, is neglected (i.e., $C_p = C_{p, ph}$) then κ_{ph} defined by equation 17 is given by

$$\kappa_{ph} = \frac{1}{3}\upsilon\ell_{ph},$$

(21)

where ℓ_{ph} is the mean free path of the phonon and υ is the sound velocity [106]. The mean free path ℓ_{ph} can be directly estimated from the values of κ_{ph} if we know the sound velocity υ. The value of υ is taken to be 2.93×10^2 m/sec [108]. Estimated mean free path as a function of temperature is shown in Figure 13.

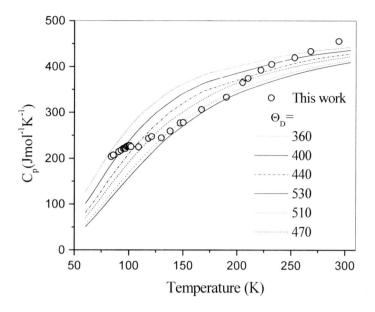

Figure 16: The specific heat estimated from phonon contribution of thermal conductivity (λ) and thermal diffusivity (κ).Calculated values for different values of Θ_D are also shown.

- **Thermoelectric Power**

The thermoelectric power of the high-T_C superconducting sample was measured in temperature range 85-300K. The thermoelectric power (S) reached zero within experimental uncertainty in superconducting state. The thermoelectric power increased with decrease in temperature and after reaching T_c value, thermoelectric power decreased strongly to zero value and is shown in Figure 17. At high temperatures the thermoelectric power is almost linear. Thus we can use the Mott expression to determine the Fermi level [80, 109]:

$$S = S_0 - \frac{\pi^2 k_B{}^2}{3|e|E_F} T$$

(22)

where S_0 is a constant. From the slope (-0.03145 μVK^{-2}) estimated by a linear extrapolation we have found the Fermi level to be 0.78 eV.

Similar profile for the same kind of superconductors is reported [110 - 114].

- **Electrothermal Conductivity**

The electrothermal conductivity (P) is the thermoelectric power divided by the dc electrical resistivity [115] and is given as,

$$P = \frac{S}{\rho},$$

(23)

where S is the thermoelectric power and ρ is the dc electrical resistivity.

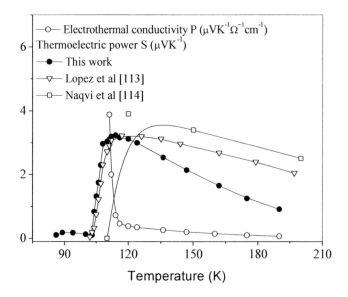

Figure 17: Variation of thermoelectric power (S) and electrothermal conductivity (P), with temperature.

In the mixed state of a superconductor, the electrothermal conductivity is also defined as the measure of the electrical current density produced by a thermal gradient and is supposed to be independent of the magnetic field. We have utilized the former definition to calculate electrothermal conductivity and is shown in Figure 17.

- **Figure of Merit**

The figure of merit is the deciding factor for the quality of thermoelectric materials. In order to increase the whole figure of merit, it is of interest to replace the p-type leg of the Peltier junction by a thermoelectrically passive material with a figure of merit close to zero [116]. This is why it is interesting to study the figure of merit of the ceramic superconductors.

Using the data of electrical resistivity, thermal conductivity and thermoelectric power, figure of merit factor is calculated and is plotted in Figure 18.

The figure of merit is calculated from the expression [109],

$$Z(T) = \frac{S^2(T)}{\lambda(T)\rho(T)}$$

(24)

Where; Z(T) is the figure of merit factor, S(T) is the thermoelectric power, λ(T) is the thermal conductivity and ρ(T) is the electrical resistivity.

Near critical temperature the figure of merit presents a remarkable peak for the samples. This peak is due to a quick drop in the electrical resistivity which occurs about 3K before the drop in thermoelectric power. Outside the critical temperature region one can see that the curves of the figure of merit and thermoelectric power are characterized by a similar behaviour near T_c. Similar trend of the figure of merit is observed in the Bi-based high-T_c superconductors [109]

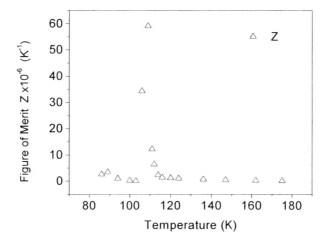

Figure 18: Variation of Figure of Merit Factor (Z), with temperature.

Conclusion

Bridge circuit used for thermal analysis in the TPS technique is modified and improved results are achieved. New bridge, called as Advantageous Transient Plane Source (ATPS), is calibrated with the fused quartz, carbon steel and silver chloride specimens. Error in the evaluated results of thermal conductivity is less than 2% with already reported values in any case and at any temperature. With the modified arrangement in the present case the bridge circuit components are reduced, temperature drift of the specimens is compensated very well and also time of relaxation (time between two reliable readings) is minimized. No separate experiment is required to determine the resistance of the TPS sensor at different temperatures and is calculated very accurately during the same experiment. Thermal conductivity values for the TPS sensor with smaller radius are in an excellent agreement with that of TPS sensor with larger radius The new setup can be used for multiproperty measurements with more accuracy. Use of small sized sensors is helpful for measurements on high-T_c superconductors as preparation of large sized specimens is a cumbersome job.

The thermal conductivity, thermal diffusivity and volumetric heat capacity of monocrystalline AgCl as a function of temperature (77–300K) are measured simultaneously using the ATPS technique. The thermal conductivity and thermal diffusivity data follow the λ $\alpha \kappa \alpha \, T^{-1}$ law. The observed volumetric heat capacity follows a usual trend with temperature. The experimental C_p obtained from the volumetric heat capacity is in agreement with the calculation at room temperature, and remains almost constant above 200K. The available data on C_p of AgCl in the temperature 2–30K are reported. C_p follows the Debye T^3 law below 5K. The data on thermal diffusivity are new. It is also possible to get the information about C_p by very simple experiment.

The samples with nominal composition $Bi_{1.6}Pb_{0.4}Sr_{1.6}Ba_{0.4}Ca_2Cu_3O_Y$ were prepared by a solid state reaction method with controlled synthesis process to get the desired single phase. This composition was selected on the basis of experiments conducted. The samples were almost a single phase with Bi-2223 high-T_c phase identified, other phases, if any, were so less in proportion that these were not detected. All the three types of tests i.e. dc electrical resistivity, ac magnetic susceptibility and X-ray diffraction are in agreement with each other, all confirming almost a single Bi-2223 high-T_c phase. Linear dependence of the electrical resistivity above transition temperature is showing a good quality of sample composition and application for further study. Single transition phase in the material and production of homogenous samples in large sizes shored up the Advantageous Transient Plane Source (ATPS) technique for thermal transport measurements. Simultaneous measurement of thermal conductivity and thermal diffusivity makes it possible to estimate specific heat and the Debye temperature Θ_D, as well as to separate the electron and phonon contributions to thermal conductivity and diffusivity in normal (Wiedemann-Franz law) and superconducting (Kadanoff's formulation) state. The simultaneous measurement also provides a useful check on the reliability and the consistency of the analysis. Thermal conductivity variation with temperature shows slight decrease initially and then a pronounced increase around T_c. Although the expected theoretical trend is similar but the peak near $T_c/2$ was not observed due to temperature limitations of the temperature range used. A similar behaviour is observed in all hole-type CuO_2-plane superconductors and in all their structural forms. This effect is due to phonon or quasiparticle scattering. Thermal diffusivity shows a similar trend as that of the thermal conductivity. Heat capacity per unit volume decreases with decrease in temperature. Assuming density of the sample to be constant in the studied temperature range molar specific heat is also calculated. Specific heat jump around T_c is also very prominent. These results indicate a good crystalline structure and an optimal doping. Thermoelectric power measurements are done with a newly designed and calibrated arrangement. Thermoelectric power was positive in this bismuth-based superconductor. The behavior of thermoelectric power of the sample was approximately linear with temperature as observed in other bismuth-based high-T_C superconductors. The superconducting transition started at 114±1K and after that, thermoelectric power decreased almost to zero value at 103±1K. The known value of the transition temperature of this sample measured from electrical resistivity was 110±1K. Therefore, the difference between thermoelectric transition temperature and resistivity transition temperature were almost in agreement within experimental errors. The figure of merit of this ceramic superconductor showed maximum around the superconducting transition temperature and is then reduced to zero below critical temperature. Electrothermal conductivity of the samples is also calculated. This system can be useful for application in low-temperature Peltier devices in order to reach temperatures lower than the temperature of liquid nitrogen.

References

[1]   Rehman, M. A.; Maqsood, A. *J. Phys D: Appl. Phys*. 2002, 35, 2040.
[2]   Rehman, M. A.; Maqsood, A. *Int. J. Thermophys*. 2003, 24, 867.
[3]   Bednorz, J. G.; Muller. K. A. *Z. Phys*. 1986, B64, 189.

[4]   Wu, M. K.; Ashburn, J. R.; Torng, C. J.; Hor, P. R.; Meng, R. L.; Gao, L.; Haang, Z. J.; Wang, Y. Q.; Chu, C. W. *Phys. Rev. Lett*. 1987, 58, 908.

[5]   Maeda, H.; Tanaka, Y.; Fukutomi, M.; Asano, T. Jpn. *J. Appl. Phys*. 1988, 27, L209.

[6]   Jasiolek, G.; Gorecka, J.; Majewski, J.; Yuan, S.; Jin, S.; Liang, R. *Supercond. Sci. Technol*. 1990, 3, 194.

[7]   Matsui, Y.; Maeda, H.; Tanaka, Y.; Horiuchi, S. *Jpn. J. Appl. Phys*. 1988, 27, L36l, L372.

[8]   Maqsood, A.; Bhatti, N. M.; Ali, S.; Haq, I. *Mater. Res. Bull*. 1990, 25, 779.

[9]   Ikeda, S.; Ichinose, H.; Kimura, T.; Matsumoto, T.; Maeda, H.; Ishida, Y.; Ogawa, K. *Jpn. J. Appl. Phys*. 1988, 27, L999.

[10]   Ikeda, Y.; Hiroi, Z.; Ito, H.; Shimomura, S.; Takano, M.; Bando, Y. *Physica C* 1989, 165, 189.

[11]   Iwai, Y.; Hoshi, Y.; Saito, H.; Takata, M. *Physica C* 1990, 170, 319.

[12]   Boekhlt, M.; Gotz, D.; Idink, H.; Fleuster, M.; Hah, T.; Woermann, E.; Guntherodt, G. *Physica C* 1991, 176, 4207.

[13]   Majewski, P.; Kaesche, S.; Su, H. L.; Aldinger, F. *Physica C* 1994, 221, 295.

[14]   Wakata, Y.; Namba, T.; Takada, J.; Egi, T. *Physica C* 1994, 219, 366.

[15]   Dorris, S. E.; Pitz, M. A.; Dawley, J. T.; Trapp, D. J. *J. Elect. Mater*. 1995, 24, 832.

[16]   Majewski, P. *J. Mater. Res*. 2000, 15, 854.

[17]   Kim, S. H.; Kim, H. S.; Lee, S. H.; Kim, K. H. *Solid State Commun*. 1992, 83, 127.

[18]   Hudakova, N.; Plechacek, V.; Dordor, P.; Flachbart, K.; Knizek, K.; Kovac, J.; Reiffers, M. *Supercond. Sci. Technol*. 1995, 8, 324.

[19]   Kim, S. H.; Kim, Y. Y.; Lee, S. H.; Kim, K. H. *Physica C* 1992, 196, 27.

[20]   Mori, K.; Cao, S.; Nishimura, K. *J. Phys. Soc. Jpn*. 1996, 65, 193.

[21]   Takano, M.; Takada, J.; Oda, K.; Kitaguchi, H.; Miura, Y.; Ikeda, Y.; Tomii, Y.; Mazaki, H. Jpn. *J. Appl. Phys*. 1988, 27, L1041.

[22]   Pissas, M.; Niarehos, D.; Christides, C.; Anagnostou, M. *Supercond. Sci. Technol*. 1990, 3, 126.

[23]   Pissas, M.; Niarchos, G. *Physica C* 1989, 159, 643.

[24]   Maqsood, A.; Ali, S.; Maqsood, M.; Haq, I.; Khaliq, M. *J. Mater. Sci*. 1992, 27, 2363.

[25]   Maeda, T.; Sakuyama, K.; Yamauchi, H.; Tanaka, S. ibid. 1989, 159, 784.

[26]   Green, S. M.; Jiang, C.; Mei, Y.; Luo, H. L.; Politis, C. *Phys. Rev. B* 1988, 38, 5016.

[27]   Tacano, K.; Kumakura, H.; Maeda, H.; Yanagisawa, E.; Takahashi, K. *Appl. Phys. Lett*. 1988, 53, 1329.

[28]   Li, Y.; Yang, B. *J. Mater. Sci. Lett*. 1994, 13, 594.

[29]   Jiang, L.; Sun, Y.; Wan, X.; Wang, K.; Xu, G.; Chen, X.; Ruan, K.; Du, J. *Physica C* 1998, 300, 61.

[30]   Kim, H. S.; Lee, G. J.; Lee, J. Y.; Lee, D. H.; Kim, K. H. *Mat. Chem. Phys*. 1997, 49, 12.

[31]   Maeda, H.; Kakimoto, K.; Kikuchi, M.; Willis, J. O.; Watanabe, K.; Tanaka, Y.; Kumakura, H. *Appl. Supercon*. 1997, 5, 151.

[32]   Tanaka, Y.; Ishizuka, M.; He, L. L.; Horiuchi, S.; Maeda, H. Physica C 1996, 268, 133.

[33]   Kaesche, S.; Majewski, P.; Aldinger, F. *J. Am. Ceram. Soc*. 1999, 82, 197.

[34]   Maqsood, A.; Khaliq, M.; Maqsood, M. *J. Mat. Sci*. 1992, 27, 5330.

[35]   Ali, Z.; Maqsood, A.; Maqsood, M.; Ramay, S. M.; Yousaf, M.; Haq, A. *Supercond. Sci. Technol*. 1996, 9, 197.

[36]   Rehman, M. A.; Maqsood, M.; Akbar, Z.; Ahmed, N.; Maqsood A. *J. Mater. Sci. Lett.* 1997, 16, 1281.

[37]   Rehman, M. A.; Maqsood, M.; Ahmed, N.; Maqsood, A.; Haq, A. *J. Mater. Sci.* 1998, 33, 1789.

[38]   Shen, Z. Z.; Hernann, A. M. *Nature* 1988, 332, 138.

[39]   Cava, R. J.; Batlogg, B.; Krajewski, J. J.; Rupp, L. W.; Schneemeyer, L. F.; Siegrist, T.; van Dover, R. R.; Marsh, P.; Peck, W. F.; Gallagher, P. K.; Glarum, S. H.; Marshall, J. H.; Farrow, R. C.; Waszczak, J. V.; Hull, R.; Trevor, P. *Nature* 1988, 336, 211.

[40]   Schilling, A.; Cantoni, M.; Gao, D.; Ott, H. R. *Nature* 1993, 365, 56.

[41]   Kawashima, T.; Matsui, Y.; Takayama-Muroinachi, E. *Physica C* 1994, 233, 143.

[42]   Larbalestier, D. C.; Gurevich, A.; Feldmann, D. M.; Polyanskii, A. A. *Nature* 2001, 414, 368.

[43]   Batlogg, B. High temperature superconductivity; Addison Wesley: Redwood city CA, 1990.

[44]   Kim, S. H.; Kim, H. S.; Lee, S. H.; Kim, K. H. *Solid State Commun.* 1992, 83, 127.

[45]   Fee, M. *Appl. Phys. Lett.* 1993, 62, 1161.

[46]   Richards, P. L. *J. Appl. Phys.* 1994, 76, 1.

[47]   Leea, A. T.; Richards, L. P.; Nam, S. W.; Cabrera, B.; Irwin, K. D. *Appl. Phys. Lett.* 1996, 69, 12.

[48]   Balachandran, U.; Iyer, A. N.; Haldar, P.; Hoehn, J. G.; Motowidlo, L. R. In Book Bi-Based High-T_c Superconductors; Editors, Maeda, H. and Togano, K.; Marcel Decker: New York, 1996.

[49]   Gustafsson, S. E.; Karwacki, E.; Khan, M. N. J. Phys. D: *Appl. Phys.* 1979, 12, 1411.

[50]   Gustafsson, S. E.; Ahmed, K.; Hamdani, A. J.; Maqsood, A. *J. Appl. Phys.* 1982, 53, 6064.

[51]   Gustafsson, S. E. *Rev. Sci. Instrum.* 1991, 62, 797.

[52]   Suleiman, B. M.; Haq, I.; Karawacki, E.; Gustafsson, S. E. *J. Phys. D: Appl. Phys.* 1992, 25, 813.

[53]   Shabbir, G.; Maqsood, A.; Maqsood, M.; Haq, I.; Amin, N.; Gustafsson, S. E. *J. Phys. D: Appl. Phys.* 1993, 26, 1576.

[54]   Maqsood, A.; Amin, N.; Maqsood, M.; Shabbir, G.; Mahmood, A.; Gustafsson, S. E. *Int. J. Energy. Res.* 1994, 18, 777.

[55]   Maqsood, M.; Arshad, M.; Zafarullah, M.; Maqsood, A. *J. Supercond. Sci. Technol.* 1996, 9, 321.

[56]   Rehman, M. A.; Rasool, A.; Maqsood, A.; *J. Phys. D: Appl. Phys.* 1999, 32, 2442.

[57]   Maqsood, A.; Rehman, M. A.; Gumen, V.; Haq, I. *J. Phy. D: Appl. Phys.* 2000, 32, 2442.

[58]   Log, T.; Gustafsson, S. E. *Fire Mater.* 1995, 19, 43.

[59]   Beck, J. V.; Arnold, K. J. *Parameters estimation in engineering and science*; Wiley: New York, 1977.

[60]   Bohac, V.; Gustavsson, M. K.; Kubicar, L.; Gustafsson, S. E. *Rev. Sci. Instrum.* 2000, 71, 2452.

[61]   Suleiman, B. M.; Karwacki, E.; Gustafsson, S. E. *High Temp.- High Pressures* 1993, 25, 205.

[62]   Gustafsson, S. E.; Suleiman, B.; Saxaena, N. S.; Haq, I. *High Temp.-High Pressures* 1991, 23, 289.

[63] Touloukian, Y. S.; Powell, R. W. Thermal Conductivity: *Nonmetallic Solids Thermophysical Properties of Matter*; IFI / Plenum: New York, 1970.

[64] *Technical Hand Book of Materials*; Crystran: UK.

[65] Gustavsson, M.; Gustavsson, I. S.; Gustafsson, S. E.; Halldahl, L. *High-Temp. High Pressures* 2000, 32, 47.

[66] Suleiman, B. M.; Gustafsson, S. E.; Borjesson, L. *Sensors and Actuators A* 1996, 57, 15.

[67] Saxena, N. S.; Haq, I. *Int. J. of Energy Research* 1992, 16, 489.

[68] Eucken, A. Ann. d. *Physik* 1911, 34, 185.

[69] Clusius, K.; Hartek, P. Z. *Physik Chem.* 1928, 134, 243.

[70] Arenberg, D. L. *J. Appl. Phys.* 1950, 21, 941.

[71] Quimby, S. L.; Sutton, P. M. *Phys. Rev.* 1953, 91, 1122.

[72] Berg, W. T. *Phys. Rev. B* 1976, 13, 2641.

[73] Klemens, P. G. In Book *Solid State Physics*; Editors, Seitz, F. and Turnbull, D.; Academic: New York, 1958, Vol. 7.

[74] Chau, C. K.; Klein, M. V. *Phys. Rev. B* 1970, 1, 2642.

[75] Donecker, J. Phys. *Status Solidi* 1968, 26, K131.

[76] Bausch, W.; Guckenbiehl, F.; Waidlich, W. *Phys. Lett. A* 1968, 28, 38.

[77] Eastman, E. D.; Milner, R. T. *J. Chem. Phys.* 1933, 1, 444.

[78] Seitz, F. *The Modern Theory of Solids*; McGraw-Hill: New York and London, 1940.

[79] Lide, D. R. CRC *Hand book of Chemistry and Physics*; CRC Press: Florida, 2003, 83, 12-224.

[80] Barnard, R. D. *Thermoelectricity in metals and alloys*; Taylor & Francis: London, 1972.

[81] Seeber, B. *Handbook of Applied Superconductivity*; Institute of Physics Publishing: Bristol and Philadelphia, 1998, Vol. 1.

[82] Cullity, B. D. *Elements of X-ray diffraction*; Addison-Wesley: London, 1967, 3rd ed.

[83] Uher, C.; Kaiser, A. B. *Phys. Rev. B* 1987, 36, 5680.

[84] Peacor, S. D.; Uher, C. *Phys. Rev. B* 1989, 39, 11559.

[85] Mori, K.; Sasakawa, M.; Igarashi, T.; Isikawa, Y.; Sato, K.; Noto, K.; Muto, Y.; *Physica C* 1989, 162, 512.

[86] Crommie, M. F.; Zettle, A. *Phys. Rev. B* 1990, 41, 10978.

[87] Cohn, J. L.; Wolf, S. A.; Vanderah, T. A. *Phys. Rev. B* 1992, 45, 511.

[88] Ginsberg, D. M. High Temperature Superconductivity; World Scientific: Singapore, 1992.

[89] Ikebe, M.; Fujishiro, H.; Naito, T.; Noto, K. *J. Phys. Soc. Japan* 1994, 63, 3107.

[90] Mori, K.; Cao, S.; Nishimura, K. *J. Phys. Soc. Jpn.* 1996, 65, 193.

[91] Ginsberg, D. M. High Temperature Superconductivity; World Scientific: Singapore, 1992.

[92] Herrmann, P. F.; Albrecht, C.; Bock, J.; Cottevieille, C.; Elschner, S.; Herkert, W.; Lafon, M-O.; Lauvray, H.; Nick, W.; Preisler, E.; Salzburger, H.; Tourre, J-M.; Verhaege, T.; *IEEE Trans. Appl. Superconductivity* **AS-3** 1993, 876.

[93] Matsukawa, M.; Noto, K.; Fujishiro, H.; Todate, T.; Mori, K.; Yamada, Y.; Ishihara, I. *Physica B* 1994, 194, 2217.

[94] Uher, C. T. In Book Physical properties of high temperature superconductors; Editor Ginsberg D. M., *World Scientific*: Singapore, 1992, Vol. 3.

[95] Kittel, C. *Introduction to Solid State Physics*; John Wiley & Sons: New York, 1996, 7th Edn.

[96] Ziman, J. M. *Electrons and phonons*, Oxford: Clarendon CA, 1963.

[97] Kadanoff, L. P.; Martin, P. C. *Phys. Rev.* 1962, 124, 670.

[98] Bardeen, J.; Rickayzen, G.; Tewordt, L. *Phys. Rev.* 1959, 113, 982.

[99] Tewordt, L. *Phys. Rev.* 1963, 129, 657.

[100] Buckingham, M. J. *Phys. Rev.* 1956, 101, 1431.

[101] Bardeen, J.; Cooper, L. N.; Schrieffer, J. R. *Phys. Rev.* 1957, 108, 1175.

[102] Onuki, M.; Higashi, T.; Fujiyoshi, T.; Miyahara, K. In: Proc. Beijing *Int. Conf. High-Temp. Supercon.* (Bhtsc'92), Beijing, 1992, p. 217.

[103] Highashi, T.; Onuki, M.; Ishii, S.; Kubota, H.; Fujiyoshi, T. *Physica C* 1991, 185, 1257.

[104] Okazaki, N.; Hasegawa, T.; Kishio, K.; Kitazawa, K.; Kishi, A.; Ikeda, Y.; Takano, M.; Oda, K.; Kitaguchi, H.; Takada, J.; Miura, Y. *Phys. Rev. B* 1990, 41, 4296.

[105] Doss, J. D. Engineer's Guide to High-Temperature Superconductivity; John Wiley & Sons: New York, 1989.

[106] Mamalis, A. G.; Manolakos, D. E.; Szalay, A.; Pantazopoulos, G. In Book Processing of High-Temperature superconductors at high strain rates; Technomic Publishing Company Inc.: Pennsylvania, 2000.

[107] Gordon, J. E.; Prigge, S.; Collocott, S. J.; Driver, R. *Physica C* 1991, 185-189, 1351.

[108] Yusheng, H.; Jiong, X.; Jincang, Z.; Aisheng, H.; Fangao, C.; Fuxue, L. *Modern Phys. Lett. B* 1990, 4, 651.

[109] Bougrine, H.; Ausloos, M.; Cloots, R.; Pekala, M. *17th International conference on thermoelectrics; IEEE*: Singapore, 1998, 273.

[110] Mitra, N.; Trefny, J.; Yarar, B.; Pine, G.; Sheng, Z. Z.; Hermann, A. M. *Phys. Rev. B* 1988, 38, 7064.

[111] Chen, G. H.; Yang, G.; Yan, Y. F.; Jia, S. L.; Ni, Y. M.; Zheng, D, N.; Yang, Q. S.; Zhou, Z. X. Mod. *Phy. Lett. B* 1989, 3, 1045.

[112] Laurent, C.; Patapi, S. K.; Green, S. M.; Luo, L.; Politis, C.; Durczewski, K.; Ausloos, M. Mod. *Phy. Lett B* 1989, 3, 241.

[113] Lopez, A. J.; Maza, J.; Yadava, Y. P.; Vidal, F.; Garcia-Alvarado, F.; Morán, E.; Senaris-Rodriguez, M. A. *Supercond. Sci. Technol.* 1991, 4, S292.

[114] Naqvi, S. M. M. R.; Rizvi, S. D. H.; Rizvi S.; Raza S. M. *Proc. 5th International Symposium on Advanced Materials*, KRL; Rawalpindi, 1997, 179.

[115] Pekala, M.; Tampieri, A.; Celotti, G.; Houssa, M.; Ausloos, M.; *Supercond. Sci. Technol.* 1996, 9, 644.

[116] Fee, M. *Appl. Phys. Lett.* 1993, 62, 1161.

In: Recent Developments in Superconductivity Research
Editor: Barry P. Martins, pp. 65-84

ISBN 978-1-60021-462-2
© 2007 Nova Science Publishers, Inc.

Chapter 3

DEVELOPMENT AND CHARACTERISTICS OF JOINT PROPERTIES IN BI-2223/AG TAPE AND COIL

Jung Ho Kim and Jinho Joo[*]

School of Advanced Materials Science Engineering, Sungkyunkwan University,
Suwon, Republic of Korea, 440-746

Abstract

$(Bi,Pb)_2Sr_2Ca_2Cu_3O_x$ (BSCCO) high-temperature superconductor (HTS) tape has been widely studied for use in the insert coil of 1 GHz NMR magnets because of its high upper critical magnetic field at 4.2 K. This coil system is required to have a high magnetic field with excellent field stability and field homogeneity, in order to improve its spectral resolution and sensitivity. To obtain this high magnetic field, multiple superconducting coils are assembled vertically and thus require many joints to interconnect them. In addition, to maintain the field stability, the coil system must be operated in the persistent current mode, thereby necessitating the connection of the coils to a persistent current switch or a flux pump. Therefore, it is considered that the joint between the superconducting tapes is an important factor in designing and developing HTS magnets for industrial applications.

In the present study, we joined Bi-2223/Ag multifilamentary superconductor tape and fabricated closed double-pancake coils using the resistive-joint and superconducting-joint methods. The critical current ratio (CCR) of the jointed tape and the decay characteristics, joint resistance, and n-value of the closed pancake coils were estimated by the standard four-point probe technique and field decay technique. In addition, the joint resistance of the closed coil was evaluated as a function of the critical current of coil, contact length, sweep time, and external magnetic flux density. It was observed that the measured value of the CCR was higher in the jointed tape made by the resistive-joint method than in that made by the superconducting-joint method. On the other hand, the joint resistance was measured to be 4 orders of magnitude smaller in the superconducting-joint coil, in which approximately 40% of the critical current was retained in the persistent current mode and the joint resistance was 0.18 nΩ. This better and longer retention of the magnetic field in the superconducting-joint coil is believed to be due to the direct connection between the superconducting cores. In addition, the measured results were compared with the numerical values calculated by the 4th order Runge-Kutta technique.

---

[*] E-mail address: Jinho@skku.ac.kr (J. Joo). Corresponding author. Tel.: +82-31-290-7358; fax: +82-31-290-7371

1 Introduction

For the application of Ag-sheathed Bi-2223 tapes in superconducting magnets, it is necessary to establish joining technology as a way to minimize the degradation of the superconducting properties in the joint region [1-2]. It is widely known that the electrical, mechanical, and thermal properties are degraded at the joint region, which consequently deteriorates the characteristics of the magnet integrated in the superconducting system [3-7]. Specifically, the non-uniformity in the joint part can result in "hot spot formation" that leads to localized thermal quenching, and the resultant increase in the joint resistance causes excessive loss of liquid nitrogen or helium during the operation of the magnet. Therefore, it is considered that the joint between superconducting tapes is an important factor in designing and developing HTS magnets for industrial applications.

In general, the joining between tapes is performed using either the superconducting-joint or resistive-joint method [8-9]. In the superconducting-joint method, part of Ag sheath on the superconductor tape is chemically etched and the exposed superconducting cores are overlapped in order to make a direct contact between them [10]. In this way, a joint resistance was obtained as low as 10^{-10} Ω - 10^{-15} Ω, which facilitates the use of this method to produce NMR and MRI magnets that employ the persistent current mode. In the resistive-joint method, on the other hand, the two tapes are joined by soldering with filler material. This method is widely used because its process is simple and easily applicable. However, the joint properties can be degraded, causing resistive dissipation in the joint part due to the solder material.

The joint resistance of the high temperature superconductor (HTS) magnet must be kept as low as 10^{-8} to 10^{-15} Ω below its critical current [11-12]. Generally, the joint resistance can be characterized by either the standard four-probe method or the field decay technique [9]. It is known that, in the case of the former method, it is difficult to measure resistances below 10^{-9} Ω precisely. On the other hand, the latter technique is more complicated but has the advantage of being able to estimate the resistance accurately over wide range of 10^{-7} to 10^{-15} Ω. In order to develop HTS magnets, therefore, establishment of the joint technology to minimize joint resistance together with its precise measurement techniques is necessary.

This chapter describes the joint processes applicable to the Bi-2223 tape, the measurements of joint resistance, and the characterizations of the joint properties. Specifically, we fabricated Bi-2223 jointed tapes and closed coils by the superconducting-joint and resistive-joint methods and evaluated their current carrying capacity and joint resistance at 77.3 K. The joint resistance was measured by the two different techniques mentioned above and the results were compared with each other. In addition, we fabricated a prototype HTS magnet with a persistent current switch (PCS) and evaluated the decay properties in persistent current mode at 77.3 K.

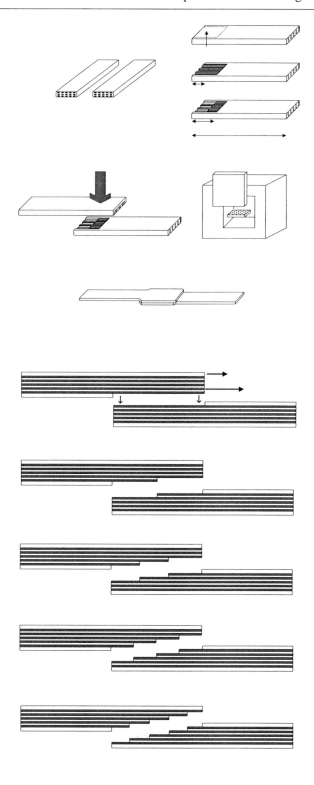

Fig. 1. (a)-(e) The superconducting-joint process for Bi-2223 multifilamentary tape and (f) schematic of the contacting surface

2   Joining Process: Superconducting-Joint and Resistive-Joint Methods

Multifilamentary Bi-2223 tapes were fabricated by the conventional powder-in-tube technique and a detailed explanation of the process is described in our previous study [8]. In the superconducting-joint method, a lap-joint was used with the contact length of 1 cm. A schematic of the joint process is illustrated in Fig. 1(a)-(e). Tapes were selectively etched out to expose the superconducting core. In order that more filaments from the tapes connected to each other, the contacting surface (window) was modified by repeating the chemical etching and the mechanical removal of both Ag sheath and the core. As a result, the shape of the exposed surface became stepped, and the number of steps in the window was zero (no stepped), one, two, three, and four, respectively, as shown in Fig. 1(f) (hereafter, called no-stepped, 1-, 2-, 3- and 4-stepped specimens). The exposed cores of the two tapes were brought into contact with each other and only the joined region was uniaxially pressed at 1,000~2,500 MPa. The jointed tapes were then sintered at 840°C in an ambient atmosphere for 50 h. In resistive-joint method, both ends of tapes were aligned and joined by soldering with 40Pb-60Sn as an insert material. The contact lengths of jointed tape were prepared for 1, 3, 7, and 10 cm.

By using both superconducting- and resistive-joint methods, we fabricated the Bi-2223 closed double pancake coils (DPC). To fabricate the DPC, Bi-2223 tape of 1.5-2 m length was wound on an alumina bobbin with a diameter of 35~45 mm yielding a coil with 8~10 turns. Alumina or Kapton^{TM} film was inserted between each turn for electrical insulation. Both ends of the tape were joined with a "praying-hands" joint configuration [13]. Critical current was measured by the standard four-probe method with 1 μV/cm criterion at 77.3 K. The current carrying capacity of jointed tape was represented by critical current ratio (CCR), which was defined as the ratio of the critical current in the whole region across the joint to that in the unjoined region. Joint resistance of the coils was evaluated by both the standard four-probe method and the field decay technique.

3   Evaluation of Joint Resistance

a)   Standard Four-Probe Method

Joint resistance of the tape can be measured by the standard four-probe method based on the nonlinear voltage-current characteristic of superconductor. Power law is commonly used for calculating and modeling in the nonlinear transition of superconductor and can be expressed by

$$V = V_c \left(\frac{I}{I_c} \right)^n ,$$

(1)

where V is the voltage, V_c is the critical voltage, I is the transport current, I_c is the critical current, and n is the index number which is calculated in the range of 0.1-1 μV/cm. In case of dissipation due to solder material in resistive-joint tape, voltage equation is expressed using ohm's law as follows

$$V = IR,$$ (2)

where R is the resistance component. Thus, we need to account for both equations (1) and (2) for voltage-current characteristic of the jointed tape. This characteristic is described in terms of relation between power and ohm's law, and the voltage equation has to be modified as follows

$$V = V_c \left(\frac{I}{I_c} \right)^n + IR_j,$$ (3)

where R_j is the joint resistance in unit of Ω and is given by

$$R_j = \frac{R_{ct}}{A_{ct}} = \frac{R_{ct}}{al_j},$$ (4)

where R_{ct} is the contact resistance in unit of Ωm^2, A_{ct} is the contact area, a is the width of tape, and l_j is the contact length. Theoretically, $R_{ct} \approx \rho_s \delta_s$, where ρ_s is the solder resistivity and δ_s is the solder layer thickness. Finally, we can express the voltage equation of jointed tape as follows

$$V = V_c \left(\frac{I}{I_c} \right)^n + I \left(\frac{\rho_s \delta_s}{al_j} \right).$$ (5)

b)   Field Decay Technique

Joint resistance of the tape can also be measured by the field decay technique [14]. Fig. 2 shows a schematic diagram of the equivalent circuit of joint resistance measurement and operating procedures of the field decay technique. The circuit consists of two kinds of coils, i.e., the excitation coil generating external magnetic fields and the closed DPC. The measurement was carried out as follows: (1) turn on heat to place the closed DPC coil in the normal state, (2) apply current to the excitation coil to generate an external magnetic field (10 ~ 150 gauss), (3) turn off heat, (4) reduce external magnetic field gradually to zero, and (5) measure the induced field trapped in the closed DPC as a function of time until the persistent current went out. Decay characteristic was observed by measuring the magnetic field with a Hall sensor located at the center of the coil. The joint resistance was calculated in two ways applying the following equations; since the characteristics of a closed loop are similar to a simple R-L circuit, the decay field is expressed by

$$B(t) = B(t_0) \exp\left(-\frac{R}{L}t\right),$$

(6)

where $B(t)$ is the induced magnetic field at t, $B(t_0)$ is the initial magnetic field, τ is the time constant, and t is the operation time. The joint resistance can be obtained as the ratio between the inductance (L) of the loop and the time constant, i.e., $R = L/\tau$. On the other hand, according to the studies of Leupold $et\ al.$, [15] and Iwasa $et\ al.$, [16], the variation of induced field can be expressed by the linear relationship;

$$B(t) = B(t_0)\left(1-\frac{R}{L}t\right).$$

(7)

In order to measure the field decay behavior, in our study, superconducting closed DPC was fabricated by using both the resistive- and the superconducting-joint method. The fabrication of the closed DPC was described in section 2 and the magnetic constant of the fabricated loop was 2.5 gauss/A. For the excitation coil, copper wire was wound on a fiber reinforced plastic (FRP) bobbin with a diameter of 60 mm. The excitation coil had 300 turns and the generated magnetic field was 28 gauss when a current of 1 A was applied.

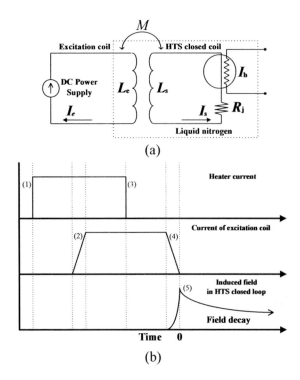

Fig. 2. (a) Schematic of equivalent circuit of joint resistance measurement and (b) operating procedures of field decay technique

4   Joint Properties of Bi-2223 Tapes and Coils

Figure 3 shows the variations of CCR for the Bi-2223 tapes joined by the two methods: in the resistive-joint technique, the CCR was evaluated as a function of the contact length, while, it was done as a function of the pressing pressure and number of step in the superconducting-joint technique. In the former technique, it was observed that the CCR was significantly dependent on the contact length as shown in Fig. 3 (a). For the tape with a contact length of 1 cm, the CCR was 40~50%, and the value increased to approximately 90% for a contact length of 3 cm, and then slowly increased as the contact length increased to 7 cm. Slightly larger CCR in the jointed tape than 1.0 is thought to have resulted from the smooth transition behavior (or low index n) under 1 μV/cm criterion.

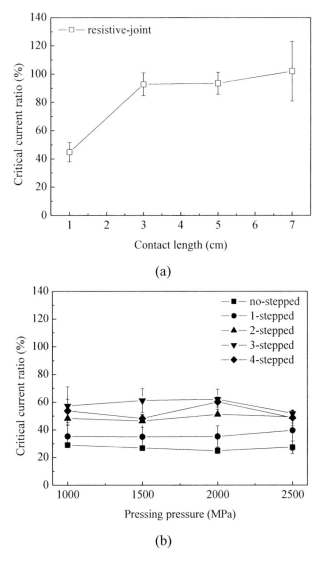

(a)

(b)

Fig. 3. Dependence of critical current ratio (a) on the contact length and (b) pressing pressure of the jointed tapes

In the superconducting-joint technique, the dependence of CCR on pressing pressure and number of step was shown in Fig. 3 (b). It was observed that the CCR was 24.8~29.0% in the range of uniaxial pressure for the no-stepped specimen, and the CCR value increased to 35.0-39.5%, 46.4~51.1%, 51.8~61.9%, and 48.2~60.0% for the 1-, 2-, 3-, and 4-stepped specimens, respectively. The CCR did not significantly depend on uniaxial pressure. The result is quite different from that of monofilamentary tape. In our previous work [10], we observed that the CCR was significantly dependent on uniaxial pressure; the value ranged from 45 to 90% and there was optimal pressure for improved CCR value. This difference is thought to have resulted from the dissimilarity in morphology, i.e., the size and shape of the superconducting core in mono- and multifilamentary tapes. Because each filament is separated from the next by Ag, application of uniaxial pressure does not seem to significantly improve the contacting area and filament interconnection in multifilamentary tape.

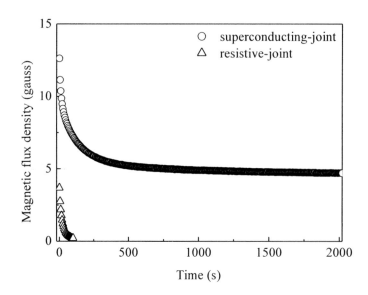

Fig. 4. The decay behaviors of induced field in the coils made by the resistive-joint and superconducting-joint technique

In addition, the CCR increased steadily as the number of steps increased from zero to 3 and reached its peak value of 61.9% in the 3-stepped specimen that had undergone a uniaxial pressure of 2,000 MPa. This improvement likely occurred because the contacting area in the window increased and more filaments from the two tapes were therefore connected to each other in the joined region. On the other hand, CCR did not increase further but decreased slightly in the 4-stepped specimen. It is suggested that the slightly lower CCR is related to the fact that each set of steps in the two tapes was not precisely matched because the length of each step became shorter as the number of steps increased.

Figure 4 shows the decay behavior of induced field in closed DPC made by the both joint methods. The contact length was 1 cm for resistive-joint and no-stepped specimen with the uniaxial pressure of 1,000 MPa was used for superconducting-joint method. In the coil with resistive-joint, it can be seen that the induced field decayed exponentially and disappeared after 120 s. By contrast, in the coil with superconducting-joint, the decay behavior is quite different; the induced field initially decayed rapidly for 100 s after which its degradation rate became slower until about 500 s of operation time. Subsequently, the field linearly decayed very slowly and this linear decay lasted until the test was terminated after 2 hours. This implies that the mechanism of the field decay in the superconducting-joint coil is different from that in the resistive-joint. Better and longer retention of the magnetic field in the superconducting-joint coil is believed to be due to the direct connection between the superconducting cores.

The retained field was approximately 5 gauss and this value corresponds to a current of 2 A. By assuming that the I_c of the coil is 25% of that in the tape itself (21.5 A) based on the CCR data, the retained (persistent) current is approximately 40% of the I_c of the coil. It is to be noted that several groups have observed similar decay behavior but mechanisms of the rapid decay in the early stage have not been clearly understood. Leupold *et al.*, [15] pointed out that the occurrence of initial decay is related to the fact that the current induced from the external field is larger than the capability of the superconducting tape. Tominaka *et al.*, [17] suggested that the change of current distributions in Ag sheath and superconductor filaments causes the initial decay.

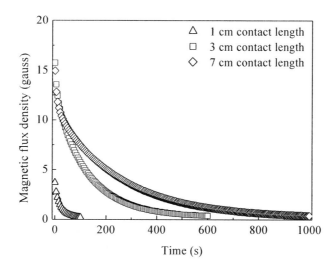

Fig. 5. Variation of decay behaviors of induced field with the contact length in the resistive-joint coil

Because the two decay behaviors were quite different from each other, joint resistance in the coils made by the two techniques should be calculated by using different equations. For the resistive-joint coil, the time constant of the induced field was estimated to be 28 s from the figure and the joint resistance was estimated to be $107 \times 10^{-9} \, \Omega$ by using equation (6). On the other hand, in the superconducting-joint coil, the joint resistance was evaluated and the value was $0.18 \times 10^{-9} \, \Omega$ from the relaxation range of $500 < \Delta t < 2000$ s by using equation (7), which is approximately 4 orders of magnitude smaller than that of resistive-joint coil. The inductance of superconducting coil and excitation coil were measured to 3.0 µH and 5.3 mH, respectively, using a *L.C.R* meter.

Figure 5 shows the dependence of decay behavior in resistive-joint closed DPC on the contact length. The decay curve of the contact length of 1 cm in figure 4 was inserted for comparison. The decay properties improved with increasing contact length; the decay time was measured to 600 and 1,000 s, respectively, for the coils of contact lengths of 3 and 7 cm. The time constant and joint resistances were estimated to 131 s and 263 s and $22.9 \times 10^{-9}\,\Omega$ and $11.4 \times 10^{-9}\,\Omega$, respectively, for the corresponding coils of 3 and 7 cm. This result of joint resistance is reasonable because joint resistance can be obtained as the ratio of contact resistance in unit of Ωm^2 (R_{ct}) to contact area between tapes (A_{ct}), i.e., $R = R_{ct}\,/\,A_{ct}$, as pointed out by Iwasa et al [2].

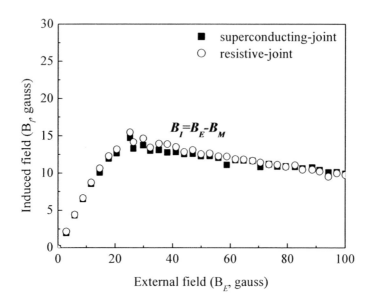

Fig. 6. Dependence of the induced field in coils on external field strength

It is also to be noted that the initial magnitude of induced field of the coil for contact length of 1 cm was relatively lower than that of others. Similar observation had been reported by Tanaka et al., [18] that the initial field was remarkably decreased when the contact length was 0.5 cm, compared to that of 2 cm. On the other hand, the initial induced field of resistive-joint coils for contact length of 3 and 7 cm was similar to that of the superconducting-joint in figure 4. This result seems to be reasonable because the initial field is mainly dependent on its geometry such as the cross sectional area of the tape, the radius of the loop, and the height of the pancake coil, etc., and independent of critical current of coil. Thus, further study is needed to understand the nature of initial field and decay and the occurrence of lower values in resistive-joint coils for contact length of shorter than 1 cm.

Figure 6 shows dependence of the induced field in the closed DPC on external field strength. The induced field (B_I) was derived by subtracting B_M from B_E, where B_E and B_M are the measured induced field at the center of excitation coil without and with superconducting coil, respectively. The induced field in the coils made by superconducting-joint and resistive-joint (contact length of 7 cm) was similar to each other; both the induced field increased with increasing external field, and reached a peak value of approximately 15 gauss at the external field of 25 gauss. After that, the value slightly decreased with further increase in external field. The maximum induced field of 15 gauss corresponds to the initial magnitude of induced field in the coil, suggesting that the initial field does not depend on the joint method, which is mostly consistent with the results in figures. 4 and 5.

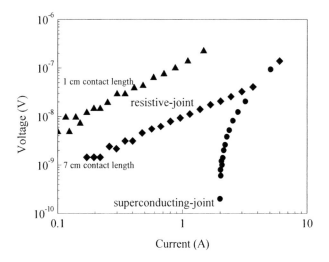

Fig. 7. The V-I characteristic of the coils derived by field decay behavior

The *V-I* characteristic was derived from the field decay behavior (figures 4 and 5) using $V = -L(di/dt)$ relations and shown in figure 7. In the superconducting-joint DPC, voltage increased more rapidly as current increased, compared to that of the resistive-joint. The voltage increased rapidly initially (in the range of 2.0~2.3 A) and the index *n* was measured to be 22.0 from the slope in the range. Subsequently, the voltage increase became smaller and the corresponding value was 3.9. By contrast, for the resistive-joint DPC, the voltage increased linearly with increasing current and the *n*-value was measured to be approximately 1.3, for both contact length of 1 and 7 cm. The higher *n*-value is related to the sharper transition from normal to superconducting state, and the coil made by superconducting-joint is expected to reduce resistance components during operation in persistent current mode.

From the results of the CCR and decay characteristics, it is to be noted that resistive-joint coil with contact length of longer than 3 cm can be effectively applied for the power systems such as cables and motors where electric current is continuously supplied from an external power source. On the other hand, in the persistent current mode, superconducting-joint should be applied but the current capacity and *n*-value of the coil should be improved further for practical applications.

The joint resistance was measured by both standard four-probe method and field decay technique. Figure 8 shows the dependence of joint resistance on contact length. The joint resistance was calculated to be 128.0×10^{-9} Ω, 54.2×10^{-9} Ω, and 38.3×10^{-9} Ω with contact lengths of 1, 3, and 7 cm, respectively, by using the standard four-probe method, while the corresponding value from the field decay technique using equation (6) was 107.0×10^{-9} Ω, 22.9×10^{-9} Ω, and 11.4×10^{-9} Ω. It is to be noted that joint resistance decreased as contact length increased; this result is reasonable as discussed before. In addition, it is observed that the joint resistance measured by the standard four-probe method is slightly higher than that by the field decay technique. This difference is thought to have resulted from equipment noise and the limit of sensitivity in the standard-four-probe method.

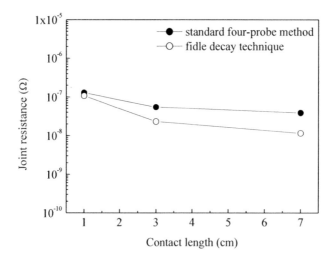

Fig. 8. Dependence of the joint resistance on contact length and measurement techniques

5   Numerical Estimation

A resistive-joint closed DPC was also fabricated by using tape of high critical current which was supplied by American Superconductor Corporation (AMSC) and its decay behavior was compared with that of closed DPC by using our tape. Coil made of AMSC tape was denoted as Coil 1, while, coil made of our tape was denoted as Coil 2. Specifications of both tapes and coils are shown in Tables 1 and 2.

Table 1. Parameters of superconductor tapes

	Tape 1	Tape 2
Thickness [mm]	0.21	0.26
Width [mm]	4.10	3.40
Critical current [A] @ 77 K	125.0	21.5
Number of filaments	55	19
Manufacturer	AMSC	SKKU univ.

Table 2. The main parameters of coils

	Coil 1	Coil 2
Superconductor tape	Tape 1	Tape 2
I.D. [mm]	45	45
O.D. [mm]	49.05	49.65
Height [mm]	8.25	6.90
Total turns	10	10
Magnetic constant [gauss/A]	2.64	2.62
Critical current [A] @ 77 K	58.0	11.6
Self inductance [μH]	7.10	7.32
Mutual inductance [μH]	49.2	50.7

Fig. 9 shows the variation of induced magnetic field with time in a closed DPC with a contact length of 7 cm. It can be seen that the magnetic field generated by induced current decayed exponentially in both coils with increased time. The initially induced field of Coil 1 was measured to be 64.73 gauss, considerably higher than that of Coil 2 (18.07 gauss). The lower value for Coil 2 is thought to occur because the induced current was higher than the critical current of the coil. From the decay curve, the time constant was estimated to be 892 s and 615 s, and joint resistance was calculated by equation (6) to be 8.0×10^{-9} Ω and 11.9×10^{-9} Ω, respectively, for Coils 1 and 2. Joint resistance is similar in each of the coils, and this seems to be a result of applying the same resistive-joint method.

The initial current can be obtained from the formula of $I_{s0} = (M / L) \cdot I_e$, which is derived from R-L circuit. When the current of the excitation coil (I_e) was 3.5 A, and M and L_s were taken from table 2, the initial currents of Coils 1 and 2 were calculated to be 24.25 and 24.24 A, respectively. The corresponding fields were 64.0 and 63.5 gauss, as determined by using the magnetic constant in the table 2. The measured field is similar to the calculated field for Coil 1. On the other hand, the measured field is remarkably lower than that of the calculated one for Coil 2, and this discrepancy is probably due to the higher induced current than I_c of the coil, as noted before.

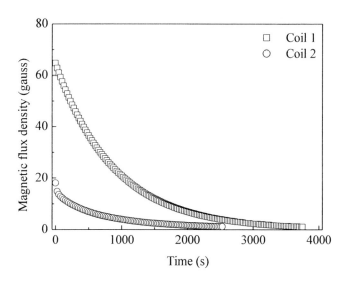

Fig. 9. The decay behavior of induced field in the coils

We estimated the field decay behavior numerically and compared with the measured behavior. The numerical method is as follows: once the current, I_e, is applied in the excitation coil, the superconducting closed DPC is exposed to the external field generated by the excitation coil, resulting in induced current I_s. The voltage equation in the circuit is given by

$$L_s \frac{dI_s}{dt} + IR + M \frac{dI_e}{dt} = 0 \tag{8}$$

where L_s, I_s, and R is the inductance, transport current, and total resistive component in HTS coil, M is the mutual inductance, I_e is the transport current in an excitation coil, and t is the operating time.

If we assume that the resistance component in the closed DPC is a constant, we can deduce the solution in the form of an exponential function and the current decay behavior is easily understood. On the other hand, the resistance in the closed DPC consists of both joint resistance and index resistance and the latter is not a constant. Thus, we need to account for both resistances, and the voltage in (8) in the closed DPC has to be modified as follows

$$L_s \frac{dI_s}{dt} + I\left[R_j + R_n(I)\right] + M \frac{dI_e}{dt} = 0 \tag{9}$$

where R_n is the index resistance component and expressed by

$$R_n(I) = \frac{V_c}{I}\left(\frac{I}{I_c}\right)^n \tag{10}$$

We used the 4th order Runge-Kutta method to solve (9).

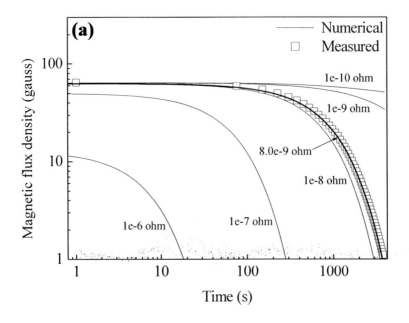

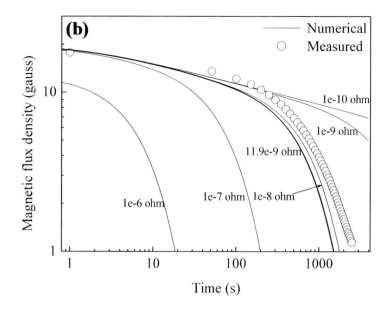

Fig. 10. Variations of the induced field with time for (a) coil 1 and (b) coil 2

From the numerical result, the variations of the field with time for various joint resistances were plotted as shown in figure 10. The induced field varied remarkably with joint resistance, as expected. The measured field decays for the coils were inserted for comparison in the figure, and agreement is good between the measured and calculated decays at joint resistances of $8.0 \times 10^{-9}\,\Omega$ and $11.9 \times 10^{-9}\,\Omega$ for Coils 1 and 2, respectively.

6   HTS Coil with Persistent Current Switch (PCS)

Bi-2223 tape has been widely studied for use in the insert of 1 GHz nuclear magnetic resonance (NMR) coil because of its high upper critical magnetic field of more than 21 T at 4.2 K [19-23]. In general, the insert coil is operated in persistent mode to maintain a field stability of 0.01 ppm/hr [24]. To improve field stability in persistent mode, the coil must minimize dissipation, which is mainly due to index n and joint resistance as noted before. Dissipation from the low index n of Bi-2223 tape is reported to limit the use of its coil in practical application for NMR: index n of BSCCO (10-20 for both 2212 and 2223) is lower than those of LTS (40-100 for NbTi and Nb$_3$Sn) [25-29], leading to that the HTS coil must be operated at a lower operating current to its critical current than that of the LTS coil. The joint resistance between coils also degrades the stability and the operating current of coil, therefore, the nature of the relative contributors of both index and joint resistance to decay behavior is critical in designing an HTS coil–PCS system [30].

Table 3. Specifications of the HTS magnet and PCS

HTS magnet (DPC)	
I.D. [mm]	75.0
O.D. [mm]	90.8
Height [mm]	8.8
Turns/DPC	54
Inductance [mH]	0.32
Magnetic constant [gauss/A]	10
Tape length [m]	14
Critical current [A] @ 77 K	48
Index n	6.3
Persistent current switch (PCS)	
Tape length [m]	1
Winding method	Inductive

We fabricated the HTS magnet with the PCS and evaluated the effect of two resistance components on the decay characteristics in persistent mode at 77.3 K. Figure 11 shows a schematic diagram of the HTS magnet circuit with PCS. To fabricate the magnet and the PCS, the Bi-2223 tape made by AMSC was wound in a G-10 bobbin, and Kapton[TM] film was inserted after each turn for electrical insulation. For a persistent magnet system, both ends of the PCS were jointed to the HTS magnet by a resistive-joint method with 40Pb/60Sn. The contact length was 7 cm and a "shaking-hands" type joint configuration was used. Specifications of the HTS magnet and PCS are given in Table 3.

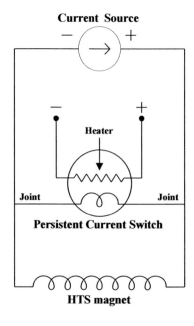

Fig. 11. Schematic diagram of circuit of HTS magnet with persistent current switch

The operating procedure for persistent mode is as follows: (1) turn off the PCS (in normal state) by applying current to the heater, (2) apply current to the magnet, (3) turn on PCS, and (4) remove current from the magnet. The persistent current will circulate in the circuit between the magnet and the PCS. The decay characteristic of the magnetic field generated by persistent current can be expressed by equation (6).

Figure 12 shows dependence of decay behavior of the trapped magnetic field in the HTS magnet with PCS on operating current at 77.3 K. Operating currents of 6 A and 24 A were applied and the ramping rate was 0.5 A/s. As shown in the figure, magnetic fields decayed in different manners at operating currents. Decay was linear at an operating current of 6 A, mainly because joint resistance became dominant rather than index resistance, as noted earlier. At an operating current of 24 A, however, the field began to decrease rapidly within approximately 3000 s, after which it then decayed linearly. The rapid decrease at the initial stage is likely due to a resistive component caused by index resistance. For an operating current of 6 A, circuit resistance was estimated by equation (6) to be 2.0×10^{-8} Ω, which is similar to that of Coil 1. This result is reasonable because the same joint method and contact length were employed in those coils. In addition, the field decay rate was 20%/hr during the first hour, considerably higher than that of the corresponding coil made of LTS.

By using equation (9), we calculated the field decay, and the result is presented in figure 12. The calculated result was quite consistent with the measured value at the operating currents of 6 A and 24 A. This agreement appears to be a direct result of accounting for both index resistance and joint resistance components.

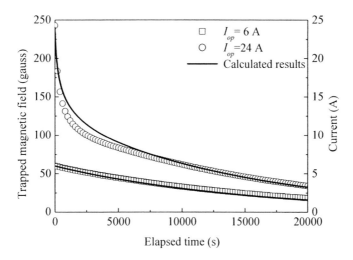

Fig. 12. Decay behavior of trapped magnetic field as a function of operating current in HTS magnet with persistent current switch

7   Conclusion

We fabricated Bi-2223/Ag jointed tapes and coils by the superconducting-joint and resistive-joint methods and evaluated the critical current ratio (CCR) of the jointed tape and the decay characteristics, joint resistance, and n-values of the closed coils. It was observed that the properties of the jointed tapes and closed coils were significantly dependent on the joining methods such as resistive- or superconducting-joint method. Specifically, the CCR was measured to be higher in the jointed tape made by the resistive-joint method than in that made by the superconducting-joint method. On the other hand, the measured value of the joint resistance was 4 orders of magnitude smaller in the superconducting-joint coil, in which approximately 40% of the critical current was retained in the persistent current mode and the joint resistance was 0.18×10^{-9} Ω.

We estimated the field decay behavior numerically and the calculated result was in good agreement with the measured one. In addition, it was observed that the joint resistance measured by the standard four-probe method was slightly higher than that obtained by the field decay technique. This difference is thought to have resulted from equipment noise and the limit of sensitivity being reached in the former method.

Finally, we fabricated a prototype HTS magnet with a persistent current switch and evaluated its decay characteristics in persistent mode. The relative contributions of both the index and joint resistances were highly dependent on the magnitude of the operating current. At an operating current of 6 A, the joint resistance component was dominant, and hence the magnetic field decayed linearly. In this case, the measured value of the total circuit resistance was approximately 10^{-8} Ω. At an operating current of 24 A, on the other hand, the index resistance component became dominant and the field began to decrease rapidly before decaying linearly.

References

[1]   Kiyoshi, T.; Inoue, K.; Kosuge, M.; Wada, H.; Maeda, H.; Current decay evaluation of closed HTS coil circuits, *IEEE Transaction on Applied Superconductivity*, 1997, vol. 7, pp. 870-880.

[2]   Iwasa, Y.; *Case Studies in Superconducting Magnets*; Plenum Press: New York and London, 1994. pp. 267-302.

[3]   Korpela, A.; Lehtonen, J.; Mikkonen, R.; Perälä, R.; Temperature dependent current–voltage characteristics of an HTS coil having a poor resistive joint, *Physica C*, 2003, vol. 386, pp.457-461.

[4]   Lee, H-G..; Kuk, I-H.; Hong, G-W.; Kim, E-A.; No, K-S.; Goldacker, W.; Filament-to-filament joining of multi-filamentary Ag/Bi-2223 superconducting tapes, *Physica C*, 1996, vol. 259, pp.69-74.

[5]   Sha, J.; Chen, X. J.; Wang, Z. B.; Ye-Jiao, Z. K.; A new method for joining of Bi(2223)/Ag tapes, *Physica C*, 1998, vol. 297, pp.91-94.

[6]   Vipulanandan, C.; Lu, W.; Iyer, A. N.; Factors affecting the joining of superconducting BSCCO tapes, *Physica C*, 2000, vol. 341-348, pp.2445-2446.

[7]   Wang, Y.; Luan, W. Z.; Lin, W.; Hua, P. W.; Li, Y. N.; Feng, R. B.; Zhou, Q.; Yao, Y. X.; Yuan, G. S.; Superconductive joint on Bi2223/Ag multifilamentary tape, *Physica C*, 2003, vol. 386, pp.174-178.

[8]   Kim, J. H.; Kim, K. T.; Joo, J.; Nah, W.; A study on joining method Bi-Pb-Sr-Ca-Cu-O multifilamentary tape, *Physica C*, 2002, vol 372-376, pp.909-912.

[9]   Kim, J. H.; Kim, K. T.; Jang, S. H.; Joo, J.; Choi, S.; Nah, W.; Kang, H.; Ko, T. K.; Ha, H-S, Oh, S-S.; Ryu, K-S.; Nash, P.; Measurement of joint properties of Bi(Pb)-Sr-Ca-

Cu-O (2223) tapes by field decay technique, *IEEE Transaction on Applied Superconductivity*, 2003, vol. 13, pp. 2992-2995.

[10] Kim, J. H.; Joo, J.; Fabrication and Characterization of the joining Bi-Pb-Sr-Ca-Cu-O superconductor tape, *Superconductor Science and Technology*, 2000, vol. 13, pp. 237-243.

[11] Hornung, F.; Kläser, M.; Leibrock, H.; Müller, H.; Schneider, T.; Suitability of Bi-HTS wires for high field magnets, *Physica C*, 2004, vol. 401, pp.218-221.

[12] Okada, M.; Tanaka, K.; Wakuda, T.; Ohata, K.; Sato, J.; Kiyoshi, T.; Kitaguchi, H.; Wada, H.; Bi-2212/Ag high-field magnets, *Physica C*, 2000, vol. 335, pp.61-64.

[13] Musenich, R.; Farinon, S.; Priano, C.; Fabbricatore, P.; Evolution of the ohmic voltage drop in connections of superconductors under time-varying current, *Cryogenics*, 2000, vol. 40, pp. 45-52.

[14] Kim, J. H.; Jang, S. H.; Kim, K. T.; Lim, J. H.; Joo, J.; Choi, S.; Nah, W.; Fabrication and characteristics of the joint properties in $(Bi, Pb)_2Sr_2Ca_2Cu_3O_x$ closed double pancake coil, *IEEE Transaction on Applied Superconductivity*, 2004, vol. 14, pp. 1094-1097.

[15] Leupold, M. J.; Iwasa, Y.; Superconducting joint between multifilamentary wires 1. joint-making and joint results, *Cryogenics*, 1976, pp. 215-216.

[16] Iwasa, Y.; Superconducting joint between multifilamentary wires 2. joint evaluation technique, *Cryogenics*, 1976, pp. 217-218.

[17] Tominaka, T.; Kakugawa, S.; Hara, N.; Maki, N.; Electrical properties of superconducting joint between composite conductors, *IEEE Transaction on Magnetics*, 1991, vol. 27, pp. 1946-1949.

[18] Tanaka, K.; Ninomiya, A.; Ishigohka, T.; Kurahashi, K.; Measurement of joint resistance of Bi-2223/Ag tapes using one-turn shorted coil, *IEEE Transaction on Applied Superconductivity*, 2001, vol. 11, pp. 3002-3005.

[19] Wilson, M. N.; *Superconducting Magnet*; Oxford University Press: New York: 1983; pp. 274-277.

[20] Kumakura, H.; Development of Bi-2212 conductors and magnets for high-magnetic-field generation, *Superconductor Science Technology*, 2000, vol. 13, pp. 34-42.

[21] Weijers, H. W.; Trociewitz, U. P.; Marken, K.; Meinesz, M.; Miao, H.; Schwartz, J.; The generation of 25.05 T using a 5.11 T $Bi_2Sr_2CaCu_2O_x$ superconducting insert magnet, *Superconductor Science Technology*, 2004, vol. 17, pp. 636-644.

[22] Kiyoshi, T.; Kosuge, M.; Yuyama, M.; Nagai, H.; Wada, H.; Kitaguchi, H.; Okada, M.; Tanaka, K.; Wakuda, T.; Ohata, K.; Sato, J.; Generation of 23.4 T using two Bi-2212 insert coils, *IEEE Transaction on Applied Superconductivity*, 2000, vol. 10, pp. 472-477.

[23] Wada, H.; Kiyoshi, T.; Development of 1 GHz class NMR magnets, *IEEE Transaction on Applied Superconductivity*, 2002, vol. 12, pp. 715-717.

[24] Kiyoshi, T.; Yoshikawa, M.; Sato, A.; Itoh, K.; Matsumoto, S.; Wada, H.; Ito, S.; Miki, S. T.; Miyazaki, T.; Kamikado, T.; Ozaki, O.; Hase, T.; Hamada, M.; Hayashi, S.; Kawate, Y.; Hirose, R.; Operation of a 920-MHz high-resolution NMR magnet at TML, *IEEE Transaction on Applied Superconductivity*, 2003, vol. 13, pp. 1391-1395.

[25] Rimikis, A.; Kimmich, R.; Schneider, T.; Investigation of *n*-values of composite superconductor, *IEEE Transaction on Applied Superconductivity*, 2000, vol. 10, pp. 1239-1242.

[26] Weijers, H. W.; Harken, B.; Kate, H. H.; Schwartz, J.; Critical current in Bi-Sr-Ca-Cu-O superconductor up to 33 T at 4.2 K, *IEEE Transaction on Applied Superconductivity*, 2001, vol. 11, pp. 3956-3959.

[27] Evetts, J. E.; Glowacki, B. A.; Relation of critical current irreversibility to trapped flux and microstructure in polycrystalline $YBa_2Cu_3O_7$, *Cryogenics*, 1988, vol. 28, pp. 641-649.

[28] Wesche, R.; Temperature dependence of critical currents in superconducting Bi-2212/Ag wires, *Physica C*, 1995, vol. 246, pp. 186-194.

[29] Kumakura, H.; Matsumoto, A.; Sung, Y. S.; Kitaguchi H, E-J characteristic of Bi-2212/Ag and Bi-2223/Ag tape conductors, *Physica C*, 2003, vol. 384, pp. 293-290.

[30] Yanagi, N.; Mito, T.; Morikawa, T. J.; Ogawa, Y.; Ohkuni, K.; Hori, D.; Yamakoshi, S.; Iwakuma, M.; Uede, T.; Itoh,, I.; Fukagawa, M.; Fukui, S.; Experiments of the HTS floating coil system in the mini-RT project, *IEEE Transaction on Applied Superconductivity*, 2004, vol. 14, pp. 1539-1542.

In: Recent Developments in Superconductivity Research
Editor: Barry P. Martins, pp. 85-101
ISBN 978-1-60021-462-2
© 2007 Nova Science Publishers, Inc.

Chapter 4

CHEMICAL SOLUTION DEPOSITION OF SrTiO₃-BASED BUFFER LAYERS FOR YBCO COATED CONDUCTORS

Xuebin Zhu and Yuping Sun

Key Laboratory of Materials Physics, Institute of Solid State Physics,
Chinese Academy of Sciences, Hefei 230031, P. R. China

Abstract

In this chapter, STO-based buffer layers will be prepared using chemical solution deposition (CSD) on Ni (200) tapes. First, the results show that seed layer is necessary for fabricating (200)-oriented STO buffer layers; second, the effects of addition of Ba element into seed layers are studied, the results show that when the seed layer is $Sr_{0.7}Ba_{0.3}TiO_3$ one can obtain highly (200)-oriented STO buffer layers when the annealing temperature is 900℃, which can be attributed to the orientation improvement of seed layers induced by the appearance of BaF_2 liquid phase; third, the conductive buffer layers $Sr_{1-x}La_xTiO_3$ (LSTO) ($0 \leq x \leq 0.4$) are fabricated using CSD method, the results show that it is possible to fabricate high-quality conductive LSTO buffer layers for YBCO coated conductors. Finally, thick STO buffer layers are fabricated using multi-seeded technique, the results show that the orientation of thick STO buffer layers can be improved using the so-called multi-seeded technique.

1   Introduction

In recently, the $YBa_2Cu_3O_7$ (YBCO)-based second generation high temperature superconducting (2nd HTS) tapes, YBCO coated conductors, are widely studied due to their high critical current density and strong flux pinning properties at liquid nitrogen temperature[1]. Among the several methods for preparing of YBCO coated conductors, the all chemistry-approach method based on rolling assisted biaxially textured substrates (RABiTS) technique is a non-vacuum route, with lower requirement to the equipments and is easy to be scaled up [2].

In the processing of YBCO coated conductors using all chemistry-approach method, one should construct the products with sandwich-like structure as YBCO/buffer layer(s)/metal substrate. The functions of buffer layers are twofold; one is to transmit the biaxial texture of the metal substrate to the YBCO film, the other is to block the chemical diffusion between the YBCO film and metal substrate [3].

As for the buffer layers, one ideal selection is to use a single buffer layer which can completely block the diffusion between the YBCO film and the metal substrate as well as transmit the biaxial texture. Among the numerous materials, $SrTiO_3$ (STO) is a feasible candidate. STO has the character of perovskite structure and is stable. In addition, the lattice parameter of 0.3905 nm of STO provides a very close lattice match with YBCO (2.3% mismatch) and Ni substrate (11.1% mismatch) [4]. Moreover, Siegal et al. [5] reported that using an all-chemistry method, YBCO coated conductors on STO-based materials buffered nickel tapes had critical current densities exceeding $10^6 A/cm^2$ at 77K under zero applied magnetic field.

In this chapter, STO-based buffer layers will be prepared using chemical solution deposition (CSD) on Ni (200) tapes. First, the results show that seed layer is necessary for fabricating (200)-oriented STO buffer layers; second, the effects of addition of Ba element into seed layers are studied, the results show that when the seed layer is $Sr_{0.7}Ba_{0.3}TiO_3$ one can obtain highly (200)-oriented STO buffer layers when the annealing temperature is 900□, which can be attributed to the orientation improvement of seed layers induced by the appearance of BaF_2 liquid phase; third, the conductive buffer layers $Sr_{1-x}La_xTiO_3$□LSTO□($0 \leq x \leq 0.4$) are fabricated using CSD method, the results show that it is possible to fabricate high-quality conductive LSTO buffer layers for YBCO coated conductors. Finally, thick STO buffer layers are fabricated using multi-seeded technique, the results show that the orientation of thick STO buffer layers can be improved using the so-called multi-seeded technique.

2   Effects of Seed Layers on the CSD-derived STO Buffer Layers

The STO solutions are produced by reacting titanium isopropoxide with acetylacetone before combining it with a solution of Sr acetate dissolved in trifluoroacetic acid (TFA) and diluted by acetylacetone to the desired concentration. In this work, the seed layer solution concentration is 0.04M, 0.065M, 0.08M and the solution concentration for the subsequent STO precursor layer is 0.25M [6].

A spinning rate of 4000 rpm and time of 60 s are used in the process of deposition both for the seed layers and the subsequent STO precursor layers on Ni (200) tapes, followed by a hot-plate treatment at 300□. These dried films are heated quickly to 900□ and kept at this temperature for two hours under 4% H_2/N_2 atmosphere. The treatment for each layer, seed layers as well as subsequent STO precursor layers, is repeated up to the numbers of layers.

In this section, all the samples are constructed with the structure of two layers STO /one STO seed layer/Ni. For the sake of description, we define here the seed layer samples fabricated using solution concentration of 0.04M, 0.065M, 0.08M and 0.25M as samples S_4, S_6, S_8 and S_{25}, respectively. The subsequent two layers STO are fabricated on the prepared STO seed layers using 0.25M solution concentration and defined as B_4, B_6, B_8 and B_{25}, respectively.

The θ-2θ x-ray diffraction (XRD) results of the sample B_4, B_6, B_8 and B_{25} are shown in Figs. 1(a) to (d), respectively. The peaks appearing at 32.1° and 46.2° belong to the (110) and (200) planes of STO buffer layers and the peak appearing at 44.3° belongs to the cubic Ni substrates. It can be seen that the strongest absolute intensity of STO (200) peak is attributed to the sample B_6. The STO (200) peaks of other samples are relatively weak. The results indicate that the crystalline qualities of STO buffer layers can be changed when the seed layers are different. The above difference of crystallization qualities of STO buffer layers can be uniquely attributed to the effect of seed layer solution concentrations because there are no other differences than the seed layer solution concentrations when we fabricate these STO buffer layers.

In order to check the microstructures of the different seed layers, experiments of field-emission scanning electronic microscopy (FE-SEM) are carried out, and the results are shown in Fig. 2(a) to (d). It can be seen that the samples S_4, S_6 and S_8 are composed of dispersed STO grains; however, the sample S_{25} is made up of connective STO grains. Moreover, the average distance between two STO grains of the sample S_6 is smaller than that of the sample S_4.

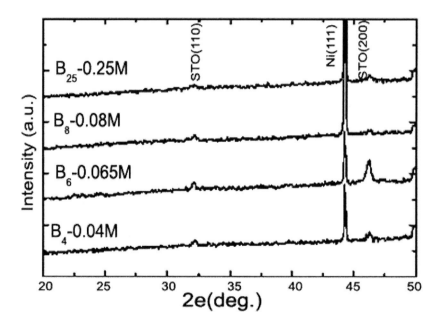

Fig.1 The θ-2θ XRD results of the sample B_4, B_6, B_8 and B_{25}

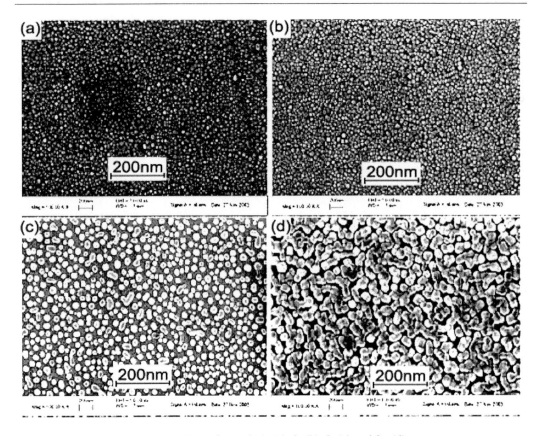

Fig.2 FE-SEM of sample S_4 (a), S_6 (b), S_8 (c) and S_{25} (d)

As discussed in Ref. [7], we think that the seed layers' grains are formed from the precursor continuous polycrystalline thin film and will be discussed in section 3. Additionally, as discussed by Schwartz et al. [8] the thickness of seed layer is very important for preparation of highly-oriented thin films. When the seed layer is too thick, the grains of seed layer will be nucleated both at the interface of film/substrate and within the bulk film, which will eliminate or depress the function of substrate resulting in randomly-oriented films.

It is well known that the subsequent precursor layers nucleate at the grains of seed layers. When the numbers of grains in the seed layer decrease, i.e., the nucleate centers for the subsequent STO film will decrease, the crystallization of the subsequent STO precursor layers will be degraded.

From the two foresaid viewpoints, it can be found some instructive hints as followings:

When the seed layer is very thick such as the sample S_{25} in our experiments, there have no plenteous nucleating centers for the subsequent films resulting in the crystallization of the subsequent film depressed; Moreover, the epitaxy or orientation of the grains in the seed layer are not very ideal due to the competition of nucleation between the film/substrate interface and within the bulk film, which will depress the orientation of the subsequent films too. In this case, the crystallization and the orientation or epitaxy of the films is not ideal.

On the other hand, when the seed layer is very thin, although the orientation or epitaxy is very good, the nucleating centers for the subsequent film are too sparse to induce all the grains of the subsequent film to nucleate on the prepared grains of the seed layers, which will also cause the crystallization and orientation of the subsequent film depressed.

The above description reminds of us that there maybe exist an optimal seed layer thicknesses or solution concentration for seed layers to fabricate highly-oriented or epitaxial films, which is realized in our experiments. That is to say, the best crystallization and orientation of the STO buffer layer is attributed to the sample B_6, whose seed layer is not the most thick or thin.

As for the orientation, the experiments of rocking curve or ω-scanning and the in-plane Phi scanning experiments are very useful methods. The full width at half maximum (FWHM) of the rocking curve result located at the STO (200)-plane for the sample B_6 is about $5°$, which indicates that our method is very useful to fabricate highly (200)-oriented STO buffer layers. However, the in-plane Phi scanning result of the sample B_6 shows that there have no distinguishable four peaks, indicating the in-plane orientation of the sample is not very good which maybe related to the appearance of stabilization agent (acetylacetone) in the STO solutions.

3   CSD Preparation of STO Buffer Layers on Ba-doped STO Seed Layers

In section 2, we have fabricated highly (200)-oriented STO buffer layers, but the results suggest that one cannot obtain ideal STO (200) buffer layers with high in-plane orientation when some stabilization agents, such as acetylacetone, are added into the precursor solutions. In this section, we change the preparation methods for STO solutions and the effects of Ba element into the STO seed layers are also studied. The motivation about using Ba-doped STO as seed layers are as followed:

In the brilliant paper, Dawley et al. [9] point out that the orientation of STO (200) can be improved when $Ba_xSr_{1-x}TiO_3$ (BST) is used as the seed layer, which can be attributed to the formation of a transient BaF_2-related liquid phase during processing. As far as we know, although the quantity of transient BaF_2-related liquid phase increases with the increase of Ba content in BST seed layer, which is favorable to fabricate (200)-oriented STO film, the lattice mismatches both for BST/Ni and STO/BST systems increase too, which is harmful to fabricate (200)-oriented STO film. Additionally, if one uses the BST as the whole buffer layers, the alternative current (AC) loss will be enhanced due to the larger permittivity of BST films. Therefore, we think there possible exists an optimal Ba content for STO seed layers to fabricate (200)-oriented STO films.

Briefly, the seed layer solution is produced by mixing the solution of the titanium isopropoxide dissolving in acetylacetone and the solution of Ba and Sr acetate dissolving in trifluoroacetic acid (TFA). The STO precursor layer solution is prepared by dissolving Sr acetate in heated glacial acetic acid followed by adding Ti butoxide and diluted by methanol to desired concentration. In this section, solution concentrations of 0.05M and 0.08M are used for the preparation of the seed layers and the solution concentration of 0.25M is used for the preparation of subsequent STO layer [10].

The procedures for heat treatment are as same as that of section 2. All the samples are constructed with the two layers of STO/one layer of BST seed layer/Ni.

For the sake of description, we define here the seed layer samples of SrTiO₃, $Ba_{0.3}Sr_{0.7}TiO_3$ (BST3/7) and $Ba_{0.5}Sr_{0.5}TiO_3$ (BST5/5) using the solution concentration of

0.05M and 0.08M as S_{05}, S_{08}, S_{35}, S_{38}, S_{55} and S_{58}, respectively. Correspondingly, STO samples deposited on these different seed layers are defined as the sample A_{05}, A_{08}, A_{35}, A_{38}, A_{55} and A_{58}, respectively. The first number of the subscript represents the Ba content in the seed layers; the second number of the subscript represents the solution concentration for the seed layers.

Fig. 3 is the standard θ-2θ XRD results of the sample of A_{05}, A_{08}, A_{35}, A_{38}, A_{55} and A_{58}, respectively. In Fig. 3(a), it can be seen that there have no obvious diffraction peaks about STO (110) except for the sample A_{05}, which indicates that the appropriate Ba addition into the STO seed layers can improve the (200) orientation of STO buffer layers. In Fig. 3(b), it can be seen that highly (200)-oriented STO buffer layers can be also prepared even when the seed layer is BST3/7 using the solution concentration of 0.08M. Although, the sample A_{08} is preponderant (200)-oriented, the obvious (110) peak attributed to STO can be also obviously observed. Moreover, when the seed layer is BST5/5 using solution concentration of 0.08M, the STO orientation is preponderant (110)-oriented, which further suggests that the orientation of STO buffer layers can be obviously affected by the seed layers.

In order to test the in-plane orientation of the STO buffer layers, the Phi scanning experiments of the highly (200)-oriented samples are carried out, the results are shown in fig.4. It can be seen that the best in-plane orientation is attributed to the sample A_{35}, then the sample A_{55}, the worst in-plane orientation is attributed to the sample A_{55}. From the results, it can be clearly seen that the addition of Ba into the STO seed layer can not only affect the out-of-plane orientation but also the in-plane orientation of the subsequent STO buffer layers. Moreover, there exists an optimal Ba concentration in the STO seed layers, in our results the Ba is 0.3, that is to say the seed layer is BST3/7.

Since there have no differences except for the seed layers in the processing of the above STO buffer layers, we carry out the experiments of field-emission scanning electronic microscopy (FE-SEM) and atomic force microscopy (AFM) for different seed layers, the FE-SEM results are shown in fig. 5. It can be seen that the sample S_{05}, S_{08} and S_{35} is composed of isolated grains, and the grain size of the sample S_{05} and S_{08} is larger than that of the sample S_{35}. The sample S_{38} and S_{55} is also dominatingly composed of isolated grains, but some continuous grains are synchronously appeared. The sample S_{58} is dominatingly composed of continuous grains; moreover, there exist some cracks in the grains of the sample S_{58}. Additionally, the AFM results (not shown here) reveal that the isolated grains grow in the form of island method.

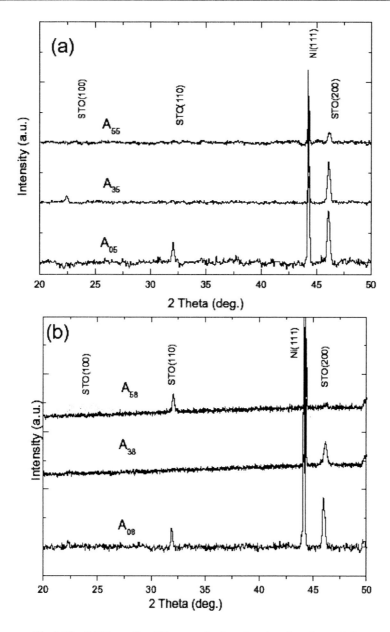

Fig.3 The XRD results of the sample A_{05}, A_{08}, A_{35}, A_{38}, A_{55} and A_{58}

As for the growth mechanism of the isolated island-like grains, two different mechanisms on single-crystal substrate were previously proposed. Miller et al. [7] proposed that isolated, highly oriented islands can be formed from a continuous polycrystalline thin film. Seifer et al. [11] proposed that isolated epitaxial islands can be formed from a continuous epitaxial single-crystal thin film. For our experiments, it can be found from the FE-SEM patterns of the seed layers that the isolated grains form from the continuous grains through shrinkage during annealing processing, which can be explained well by the model proposed by Miller et al. [7] indicating that the (200) orientations of continuous grains are not as good as these of isolated grains.

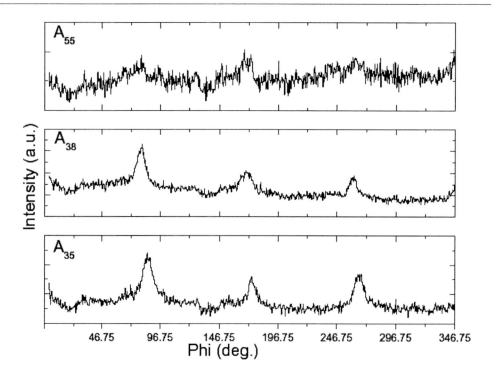

Fig. 4 Phi scanning results of A_{35}, A_{38} and A_{55}, respectively

As for the seed layers with same solution concentration but different Ba content, such as S_{05}, S_{35} and S_{55}, it can be seen that the S_{05} is composed of isolated grains, but large difference of the grain size is obviously observed and there exist many irregular grains which indicates that those grains are not highly (200)-oriented although further experiments are needed to prove this viewpoint. As for the S_{35}, the grain is also isolated; moreover, the grain shape is regular which means that the orientation of those grains is relatively higher (200)-oriented; For the S_{55}, although the majority of the grains are isolated and regular, but some irregular and continuous grains are also clearly observed.

It has been known that the subsequent STO films will inherit the orientation of the prepared seed layers. When the orientation of the seed layer is undesired, the orientation of the subsequent film will be not ideal, which causes the orientation of the sample A_{35} is the best in our experiments, then the sample A_{55}, and the worst orientation is attributed to the sample A_{05} in our experiments.

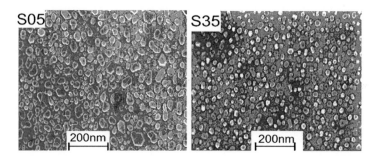

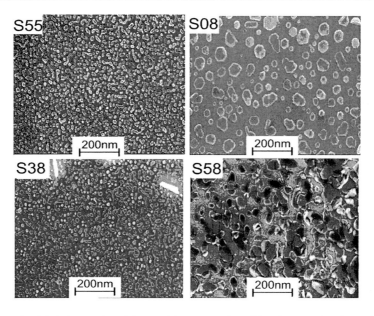

Fig. 5 The FE-SEM results of the seed layer samples of S_{05}, S_{08}, S_{35}, S_{38}, S_{55} and S_{58}

For different seed layers with same solution concentration, but different composition, the lattice parameter increases when the Ba content increases due to the larger ion radius of Ba than that of Sr ion. As discussed by Schwartz [12.], the contact angle of seed layer/substrate increases when the lattice mismatch of seed layer/substrate increases, which leads to the heterogeneous nucleation at interface depressed and results in the orientation of seed layer depressed too. For our experiments, the orientation of the seed layers is not linearly dependent on the lattice parameter, which maybe attributed to other factors. Different nucleation and growth mechanism is a possible reason.

When the Ba content in the seed layer is increased from 0 (viz. STO) to 0.3 (viz. BST3/7), although the mismatch between the seed layer/substrate is increased, but the appearance of the liquid BaF_2 phase may largely improve the orientation of the seed layer resulting in the well (200)-oriented BST3/7 seed layer; when the Ba content further increases, viz. from 0.3 to 0.5 (BST5/5), the amount of liquid BaF_2 phase is further increased as well as the improvement of the mismatch between the seed layer and substrate. In this condition, the adverse effects induced by the mismatch may exceed the favorable effect induced by the liquid BaF_2 phase resulting in the orientation of BST5/5 orientation depressed.

As for the seed layers with same content, but different solution concentration, viz. S_{35} and S_{38}, when the solution concentration is increased it can be found in the FE-SEM results that the orientation of the grains is suggested to be depressed, which maybe related to the competition between heterostructure nucleation at the interface of the seed layer/substrate and the homostructure nucleation within the film, as discussed by Schwartz et al. [12.]. As the solution concentration of the seed layer is increased, the probability of the homostructure nucleation increases, which results in the depression of the orientation of the seed layer.

As for BST seed layer with the highest Ba content and solution concentration in our experiments, viz. the sample S_{58}, it can be observed from the FE-SEM result that the seed layer grows in bulk method and there exist some cracks in the grains, which is suggested that the thickness of the sample S_{58} is larger than the critical thickness for appearance of cracking as discussed by for Zhao et al. [13].

For the differences of orientations of STO buffer layers deposited on different seed layers buffered Ni substrates, it is suggested that the orientation of the STO film maybe related to the competition of the seed layer orientation and the mismatch between the STO film / seed layer. As for the sample A_{05} and A_{35}, it is deemed that in the sample A_{35} the favorable effect for STO orientation induced by the seed layer orientation is exceeded the adverse effect induced by the mismatch effect resulting in the higher (200) orientation than that of the sample A_{05}. As for the sample A_{38} and A_{55}, although from the FE-SEM results one cannot observe any difference, but the mismatch between the STO/seed layer increases when the Ba content increases from 0.3 to 0.5 resulting in the orientation depression of the subsequent STO buffer layers. As for the sample A_{58}, the bulk growth of the seed layer causes the subsequent STO buffer layer inherit the growth method of the seed layer resulting in the bulk growth of the sample A_{58}; additionally, the cracks of the seed layer transmit to the subsequent STO buffer layer resulting in the appearance of cracks in the A_{58} too.

4   Annealing Efects on the Orientation of SrTiO₃/BST3/7 Seed Layer/NI Buffer Layers Derived by CSD Method

In section 3, we have known that when the seed layer is BST3/7 one can obtain highly (200)-oriented STO buffer layers. In this section, the annealing effects on the orientation of STO/BST3/7/Ni are studied. The results show that there exists an optimal annealing temperature, which may be related to the microstructure and orientation of the BST3/7 seed layers.

The processing procedures of the samples are as same as that of section 3. In particular, the annealing temperature is 850□, 900□ and 950□. The BST3/7 seed layers annealed at 850□, 900□ and 950□ are defined as the sample S_{85}, S_{90} and S_{95}; correspondingly, the STO buffer layers prepared on those different seed layers are defined as A_{85}, A_{90} and A_{95}.

Fig.6 is the XRD θ-2θ patterns of the sample A_{85}□A_{90} and A_{95}. It can be seen that all three samples are highly (200)-oriented without distinct differences. In order to check the in-plane orientation, the Phi scanning experiments are carried out, and the results are shown in fig.7. It can be seen from the Phi scanning results that the best in-plane orientation is attributed to the sample A_{90}, then the sample A_{95}, the worst in-plane orientation is attributed to the sample A_{95}. From the above results, it can be clearly seen that the orientation of STO buffer layers can be obviously affected by the annealing temperature of the seed layers.

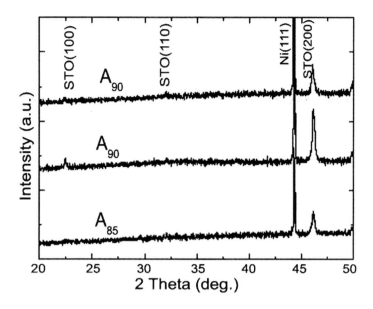

Fig.6 is the XRD θ-2θ patterns of the sample A₈₅ A₉₀ and A₉₅

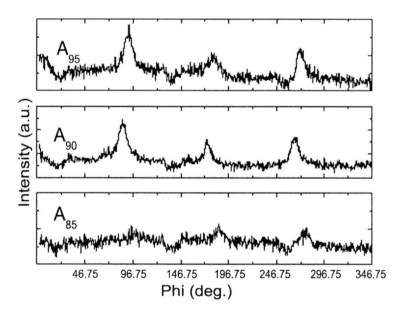

Fig.7 Phi scanning results of the sample A₈₅ A₉₀ and A₉₅

In order to study the annealing temperature effects on microstructure of the seed layers, the FE-SEM experiments are carried out and the results are shown in Fig.8. It can be seen that some abnormal grains exist in the sample S_{85}, which may be related to the low melting temperature materials. For the sample S_{90}, the grain is fine and dispersed uniformly. For the sample S_{95}, although the grain size is similar to that of the sample S_{90}, the grains do not disperse uniformly as that of the sample S_{90}. As discussed in Ref. [8□12] that the seed layer should be fine and dispersed uniformly to induce highly oriented subsequent film. In our experiments, it is found that at 900□ annealing temperature the grains of the BST3/7 seed layer is fine and rectangular-like and dispersed uniformly resulting in the best in-plane orientation of the subsequent STO buffer layers, which is similar with the results of Ong et al. [14].

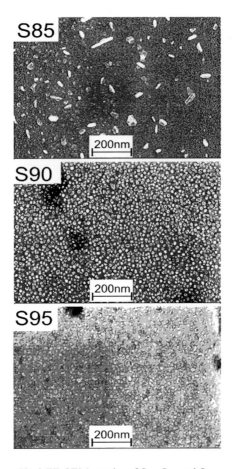

Fig.8 FE-SEM results of S_{85}, S_{90} and S_{95}

5 CSD Preparation of Conductive $Sr_{1-X}La_XTiO_3$ $(0 \leq X \leq 0.4)$ Buffer Layers on Ni Substrates

For the practical applications of YBCO coated conductors, the ability to shunt the current flow to metallic tapes in the event of a local supercurrent quench event is desirable. This requires the formation of conductive buffer layers for YBCO coated conductors. One candidate material system to use as a conductive buffer layer is $Sr_{1-x}La_xTiO_3$. $Sr_{1-x}La_xTiO_3$ has the structure of perovskite and the 3d electrons can be varied from 0 ($SrTiO_3$) to 1($LaTiO_3$). The end compounds $SrTiO_3$ and $LaTiO_3$ are band-gap and Mott-Hubbard insulators, respectively [15]. However, the insulate state can be easily broken and transformed to the metallic state with some La addition into $SrTiO_3$, or some Sr addition into $LaTiO_3$. In the previous study [16], it was reported that the $Sr_{1-x}La_xTiO_3$ shows metallic behavior in the range of $0.1 \leq x \leq 0.95$.

As for the fabrication of $Sr_{1-x}La_xTiO_3$ buffer layers on metallic tapes, Norton et al. [17□18] prepared epitaxial $LaTiO_3$ buffer layers using pulse laser deposition method. As far as we know, there has no related report about preparation of $Sr_{1-x}La_xTiO_3$ buffer layers on metallic tapes using CSD method. In this section, the conductive $Sr_{1-x}La_xTiO_3$ ($0 \leq x \leq 0.4$) buffer layers are fabricated using CSD method.

The solutions both for $Sr_{1-x}La_xTiO_3$ seed layers and subsequent layers are prepared by dissolving Sr acetate and La acetate in heated glacial acetic acid followed by the addition of Ti butoxide and diluted by methanol to the desired concentration. In this section, all the seed layer solution concentration is 0.05M, and the solution for the subsequent $Sr_{1-x}La_xTiO_3$ layers is 0.25M.

The heat treatment procedures are as same as that of section 3.

Fig.9 is the XRD patterns of the La-doped $SrTiO_3$ buffer layers. From the XRD results, it can be seen that all the samples are highly (h00)-oriented. Additionally, the intensity ratio of (200): (110) linearly decreases with the La content increasing. Moreover, the (200) peaks shift to the lower degree with increasing of La composition indicating the elongation of lattice parameter, which suggest that the La substitutes for the Sr-site and incorporates into lattice. The reason of linearly decreasing of intensity ratio of (200): (110) with the La content increasing is suggested that as the La content is increased the needed crystallization temperature is also improved. When the annealing temperature is fixed, viz. 900□ in our experiments, the crystallization of the $Sr_{1-x}La_xTiO_3$ will decrease resulting in the deteriorating of the orientation.

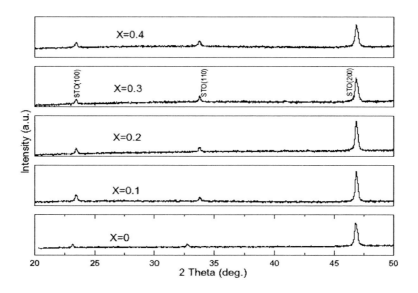

Fig.9 XRD patterns of the $Sr_{1-x}La_xTiO_3$ buffer layers

The FE-SEM results of the different $Sr_{1-x}La_xTiO_3$ buffer layers are shown in Fig. 10. It can be seen that the $SrTiO_3$ is rather smooth with some pinhole-like defaults, which maybe originate from the coalescence of STO grains. The $Sr_{0.9}La_{0.1}TiO_3$ and $Sr_{0.8}La_{0.2}TiO_3$ are consisted of granular grains, and the grains of $Sr_{0.8}La_{0.2}TiO_3$ are smaller than that of $Sr_{0.9}La_{0.1}TiO_3$, which indicates that the grain growth of $Sr_{0.8}La_{0.2}TiO_3$ is slower than that of

$Sr_{0.9}La_{0.1}TiO_3$. For the $Sr_{0.7}La_{0.3}TiO_3$ and $Sr_{0.6}La_{0.4}TiO_3$, the coalescence of grains appears again, which is obviously different with that of $Sr_{0.9}La_{0.1}TiO_3$ and $Sr_{0.8}La_{0.2}TiO_3$.

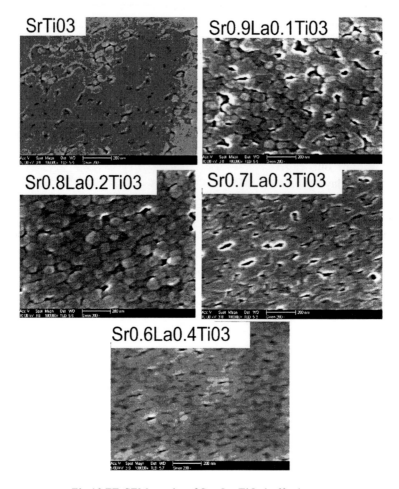

Fig.10 FE-SEM results of $Sr_{1-x}La_xTiO_3$ buffer layers

As for the differences of FE-SEM results, we think that it can be attributed to the competition of crystallization and orientation variation. The decreasing of crystallization will cause the grain size decrease; on the other hand, the deteriorating of orientation will lead to the grain size increase. When the La content is less than 0.3, the effect of the decreasing of crystallization exceeds the effects of the deteriorating of orientation, resulting in the grain size decreasing with the increasing of La content. When the La content is higher than 0.3, the adverse effects resulting in the grain size increasing with the improvement of La content.

6   Improving of the Orientation of SrTiO$_3$ Buffer Layers using Multi-seeded Technique in the CSD Processing

In the above sections, we have fabricated STO-based buffer layers with the structure of two layers of STO-based buffer layer/seed layer/Ni; the thickness of the buffer layers is about

200nm for this structure. In order to improve the blocking properties of STO buffer layers, one method is to enhance the thickness of the buffer layers.

In the processing of thick buffer layers, with the thickness increasing, the orientation will be decreased. As far as we know, there has no relative reports about increasing the orientation of thick buffer layers derived by CSD methods. In this section we attempt to improve the orientation of STO thick buffer layers using multi-seeded technique.

The preparing procedures of BST3/7 seed layer solution and STO solution and heat treatment procedures are as same as that of section 3. In this section, three samples with different structure are fabricated, one is 4 layers of STO/BST3/7 seed layer/Ni, the other is (STO/STO/BST3/7 seed layer)$_2$/Ni, and the last is (STO/BST3/7 seed layer)$_4$/Ni.

Fig.11 is the XRD θ-2θ results of the above three samples. It can be seen that the best (200) orientation in our experiments is attributed to the sample with the structure of (STO/BST3/7 seed layer)$_4$, then the (STO/STO/BST3/7 seed layer)$_2$ and the sample of 4 layers of STO/BST3/7 seed layer, which clearly indicates that the orientation of the thick buffer STO buffer layers can be largely improved using multi-seed layers, the so-called multi-seeded technique.

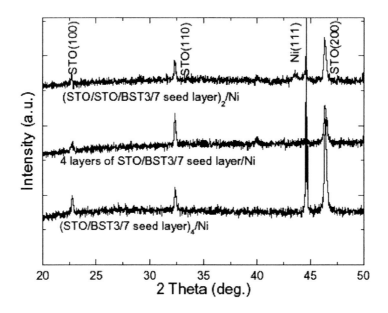

Fig.11 the XRD patterns of STO thick buffer layers with different structures

As for the origination of the orientation improving of the STO thick buffer layers, it is suggested to be related with the seed layers effects although further works is needed to clarify.

In the processing of films using CSD method, the orientation of the films will be deteriorated due to the improving of bulk nucleation within the films. One wants to obtain highly oriented thick films should suppress the bulk nucleation and induce the nucleation at the interface between the prepared template and the subsequent layer. In the section 2, we have demonstrated that seed layers can improve the nucleation at the interface and lead to the enhancement of the orientation. In this section, the seed layers may also play the same roles. When one directly deposits a layer of STO on the prepared STO template, such as

STO/BST3/7 seed layer, the heterostructure nucleation at the interface maybe decreased compared with the condition of BST3/7/STO/BST3/7 resulting in the depression of the orientation of STO layer. In brief, using multi-seeded technique, one can induce the STO layer to nucleate at the interface between the prepared template and the subsequent layer heterostructure, which leads to the orientation improving of thick STO buffer layers.

7 Conlusion

In summary, $SrTiO_3$ as an excellent buffer layer was prepared using chemical solution deposition method on Ni (200) tapes. First, the results showed that the seed layer was necessary for fabricating highly (200)-oriented $SrTiO_3$ buffer layers; second, the effects of addition of Ba element into seed layers were studied, the results showed that when the seed layer was $Sr_{0.7}Ba_{0.3}TiO_3$ one could obtain highly (200)-oriented $SrTiO_3$ buffer layers using the annealing temperature of 900□, which could be attributed to the orientation improvement of seed layers induced by the appearance of BaF_2 liquid phase; third, the annealing temperature on the orientation of $SrTiO_3/Sr_{0.7}Ba_{0.3}TiO_3$ seed layer/Ni were studied, the results showed that there existed an optimal annealing temperature to fabricate highly (200)-oriented $SrTiO_3$ buffer layer; fourth, the conductive buffer layers, $Sr_{1-x}La_xTiO_3$ ($0 \leq x \leq 0.4$), were fabricated using chemical solution deposition, and the results showed that with the increasing of La composition the crystallization of $Sr_{1-x}La_xTiO_3$ decreased when the annealing temperature was fixed, additionally, the observation of field-emission scanning electric microscopy showed that the grain size decreased with La composition increasing in the range of $0 \leq x \leq 0.2$, then the grains coalesced when x=0.3 and x=0.4, which indicated that it was possible to fabricate high-quality conductive $Sr_{1-x}La_xTiO_3$ buffer layers for YBCO coated conductors. Finally, multi-seeded technique was applied to improve the orientation of thick $SrTiO_3$ buffer layers, the results suggested that one could obtain highly oriented thick $SrTiO_3$ buffer layers using the so-called multi-seeded technique.

Acknowledgement

We would like to thank our close collaborators W.H. Song, J.M. Dai and J.J. Du for their contributions. This work was supported by the National Superconductivity 863 project under No.2002AA306211, No.2004AA306130, No. 2002AA306281, and the National Nature Science Foundation of China under contract No.10374033, 10474100 and the Fundamental Bureau, Chinese Academy of Sciences.

References

[1]   D. Larblestier et al. *Nature*. 2001, vol. 414, 368.

[2]   T. Araki et al. *Suprcond. Sci. Technol.* 2003, vol. 16, R71.

[3]   D.P. Norton et al. *Science*. 1996, vol. 274, 755.

[4]   Y.X. Zhou et al. *Supercond. Sci. Technol.* 2003, vol. 16, 901.

[5]   M.P. Siega et al. *Appl. Phys. Lett.* 2002, vol. 80, 2710.

[6]   X.B. Zhu et al. *Physica C* 2004, vol. 415, 57.

[7]   K.T. Miller et al. *J. Mater. Res.* 1996, vol. 5, 151.

[8]   R.W. Schwartz et al. *J. Am. Ceram. Soc.* 1999, vol. 83, 2359.

[9]   J.T. Dawley et al. *J. Mater. Res.* 2002, vol. 17, 1678.

[10] X.B. Zhu et al. *Physica C* 2004, vol. 411, 143.

[11] Seifert et al. *J. Mater. Res.* 1996, vol. 11, 1470.

[12] R.W. Shwartz. *Chem. Mater.* 1997, vol. 9, 2325.

[13] M.H. Zhao et al. *Acta Mater.* 2002, vol. 50, 4241.

[14] R.J. Ong et al. *J. Mater. Res.* 2003, vol. 18, 2310.

[15] Y. Tokura et al. *Phys. Rev. Lett.* 1993, vol. 70, 2126.

[16] C.C. Hays et al. *Phys. Rev. B* 1999, vol. 60, 10367.

[17] D.P. Norton et al. *Physica C* 2002, vol. 372-376, 818.

[18] K. Kim et al. *Solid State Electronics* 2000, vol. 47, 1277.

In: Recent Developments in Superconductivity Research   ISBN 978-1-60021-462-2
Editor: Barry P. Martins, pp. 103-132   © 2007 Nova Science Publishers, Inc.

Chapter 5

PREPARATION AND THERMOPOWER BEHAVIOR OF THE NEW COBALT OXYHYDRATES $(NA,K)_X(H_2O)_YCoO_2$ PREPARED BY THE AQUEOUS PERMANGANATE SOLUTION

Chia-Jyi Liu

Department of Physics, National Changhua University of Education,
Changhua 500, Taiwan, R. O. C.

Abstract

This chapter presents a novel route to prepare the superconductive sodium cobalt oxyhydrates using aqueous permanganate solution. Immersing γ-$Na_{0.7}CoO_2$ in $KMnO_{4\ (aq.)}$ leads to two distinct novel phases of $(Na,K)_x(H_2O)_yCoO_2$, depending on the concentration of the $KMnO_{4\ (aq.)}$ solution. These two phases differ in the *c*-axis but show little change from the parent compound in the *a*-axis. The phase transformation can be viewed as a topotactic process. The c $\approx$ 19.6 Å phase has bi-layers of water molecule inserted into the lattice, whereas the c $\approx$ 13.9 Å phase has mono-layers of water molecule inserted into the lattice essentially without changing the skeleton of the parent compound. Immersing γ-$Na_{0.7}CoO_2$ in $NaMnO_{4\ (aq.)}$ only leads to the c $\approx$ 19.6 Å phase of $Na_x(H_2O)_yCoO_2$. The formation mechanism and phase stability of $KMnO_4$-treated $(Na,K)_x(H_2O)_yCoO_2$ would be discussed. The hydration is a very slow process when using $KMnO_{4(aq.)}$ with a low molar ratio of $KMnO_4/Na$. Three related phases of c $\approx$ 19.6 Å, c $\approx$ 13.9 Å, and c $\approx$ 11.2 Å exist in the cobalt oxyhydrates, being associated with the water content in the lattice. The thermal stability of the c $\approx$ 19.6 Å phase of $(Na,K)_x(H_2O)_yCoO_2$ behaves differently from that of the c $\approx$ 13.9 Å phase, which is ascribed to the difference between the ion-dipole interaction of K^+-H_2O within the alkaline layers and that of Na^+-H_2O located between the CoO_2 layers and the alkaline layers. The thermopower behavior of three related phases of cobalt oxyhydrates will be presented. The generalized Heikes formula would be used to explain the variation of the thermopower between γ-$Na_{0.7}CoO_2$ and $Na_{0.33}K_{0.02}(H_2O)_{1.33}CoO_2$ arising from the variation of the concentration of Co^{4+}.

1   Introduction

The discovery of superconductivity of the cobalt oxyhydrate $Na_{0.35}(H_2O)_{1.3}CoO_2$ with $T_c \approx 4$-5 K by Takada et al. [1] is a surprise after several years of studies on its parent compounds of sodium cobalt oxides γ-Na_xCoO_2. The γ-$Na_{0.5}CoO_2$ is a potential candidate for thermoelectric applications due to its high electrical conductivity, large thermopower and low thermal conductivity. [2] It is an amazing and easy procedure to convert thermoelectric material to superconductor through virtually a one-pot reaction by immersing the parent material γ-$Na_{0.7}CoO_2$ in Br_2/CH_3CN solution for 5 days followed by washing with water. The structure of both the γ-$Na_{0.7}CoO_2$ and $Na_{0.35}(H_2O)_{1.3}CoO_2$ belongs to a hexagonal crystal system with the space group of $P6_3/mmc$ (No. 194). For γ-$Na_{0.7}CoO_2$, there are two CoO_2 layers of edge-sharing CoO_6 octahedra in the unit cell with the Na ions sandwiched between the two CoO_2 layers. The Na ions of γ-$Na_{0.7}CoO_2$ are located in two different crystallographic sites, i.e., *2b* and *2d*, and tend to disassociate from the crystal lattice when immersed in aqueous or acetonitrile solution. After intercalating water molecules into the lattice of parent material γ-$Na_{0.7}CoO_2$, the c-axis elongates from ca. 10.9 Å to ca. 19.6 Å, while the a-axis shows little change. For $Na_{0.35}(H_2O)_{1.3}CoO_2$, some of the water molecules are located between the CoO_2 layers and the Na ion layers, and others are located within the Na ion layers. The former tends to escape from the crystal lattice in the ambient environment accompanied by a shrinkage of the c-axis to ca. 13.8 Å with little change of the a-axis; the latter would dissociate from the lattice only upon heating due to a stronger bonding of water molecules with the Na ions within the Na ion layers as compared to the former. Complete removal of the water molecules would result in the de-hydrated form of $Na_{0.35}CoO_2$ with further shrinkage of the c-axis to ca. 11.2 Å, which is a little bit longer than c $\approx$ 10.9 Å for the parent material γ-$Na_{0.7}CoO_2$.

Due to the nature of the hexagonal lattice, the arrangement of Co ions forms a triangular lattice. As a result, it would frustrate the spin arrangement of electrons in such a triangular lattice if requiring the spin to align antiparallel to each other. It is a common way to ease up the electron-electron repulsion for an electron to align its spin antiparallel to its nearest neighbor for a system with strong electron correlations. Based on the heat capacity, [3] magnetic, [4] thermopower [5] and angle-resolved photoemission spectroscopy (ARPES) [6] measurements, it has been suggested that Na_xCoO_2 is a system with strong correlations of 3d electrons. It would be interesting to correlate the frustrated spin arrangements or the spin state with the superconductivity. $Na_{0.35}(H_2O)_{1.3}CoO_2$ is not only one of the few examples of layered transition metal oxide superconductors without containing copper, it is but also a particularly interesting system for comparison with the high-T_c cuprates in terms of the structure-electronic state correlations in light of the fact that both have 2D layers (triangular CoO_2 layers and square CuO_2 layers) in structure and have spin 1/2 ions (t_{2g}^5 for Co^{4+} in low spin state and $t_{2g}^6 e_g^3$ for Cu^{2+}) in electron configuration. In spite of their structural difference, mixed valency is a common feature to both systems. Besides, the family of sodium cobalt oxides Na_xCoO_2 is of particular interest because of their structural, magnetic and thermoelectric properties.

This chapter is organized as follows. Section 2 describes the crystal structures of the sodium cobalt oxides Na_xCoO_2. Section 3 discusses the preparation methods, phase formation mechanism of the superconducting cobalt oxyhydrates and related compounds using the aqueous permanganate solutions, and the kinetics of the topotactic transformation. Section 4 discusses the phase stability of the permanganate-treated cobalt oxyhydrates $(Na,K)_x(H_2O)_yCoO_2$. Section 5 describes the magnetization measurements on $(Na,K)_x(H_2O)_yCoO_2$ and discusses the thermopower behavior of three related phases of cobalt oxyhydrates. The generalized Heikes formula is used to explain the variation of the thermopower due to the variation of the spin and orbital degeneracy in the $3d$ electrons of the cobalt ions.

2 Crystal Structures of Sodium Cobalt Oxides Na_xCoO_2 with $x \leq 1$

According to Fouassier et al., there are four bronze type phases for Na_xCoO_2 with $x \leq 1$, where the phase formation is associated with the value x of the sodium content, i.e., (1) α-Na_xCoO_2 ($0.9 \leq x \leq 1$); (2) α'-$Na_{0.75}CoO_2$; (3) β-Na_xCoO_2 ($0.55 \leq x \leq 0.6$); and (4) γ-$Na_xCo_yO_2$ ($0.55 \leq x/y \leq 0.74$). [7] Fig. 1 shows the crystal structures of α-, β-, and γ-phase of Na_xCoO_2. For all the four phases, the volatile sodium ions are located between CoO_2 layers and are considered as a charge reservoir. Hence the valence state of Co ions can be tuned by varying the Na content in the materials, which would have significant effects on their structural, electrical, magnetic, and thermoelectric properties. The α-Na_xCoO_2 is isomorphous to α-$NaFeO_2$ and has a trigonal lattice with the space group $R\bar{3}m$ (No. 166). There are three crystallographically distinct positions: Co at $3b$ (0,0,0.5), Na at $3a$ (0,0,0), and O at $6c$ (0,0,z). [7,8] The α'-Na_xCoO_2 is isomorphous to α'-$NaMnO_2$ and has a trigonal lattice with the space group C2/m (No. 12). There are three crystallographically distinct positions: Co at $2a$ (0,0,0), Na at $2d$ (0,0.5,0.5), and O at $4i$ (x,y,0). [7,9] Fouassier et al. considered the crystal structure of β-Na_xCoO_2 to have a trigonal lattice with the space group of R3m (No. 160), but the recent studies show that the structure of β-Na_xCoO_2 should be assigned to have a monoclinic lattice with the space group C2/m (No. 12). There are three crystallographically distinct positions: Co at $2a$ (0, 0, 0), Na at $8j$ (0.810, 0.097, 0.495), and O at $4i$ (0.3898, 0, 0.1790). [7,10,11] The γ-Na_xCoO_2 is isomorphous to α-$NaMnO_2$ and has a hexagonal lattice with the space group of $P6_3/mmc$ (No. 194). There are four crystallographically distinct positions: Co at $2a$ (0,0,0), Na(1) at $2b$ (0,0,0.25), Na(2) at $2d$ (0.333,0.667,0.75), and O at $4f$ (0.333,0.667,z). [7,12]

(a)

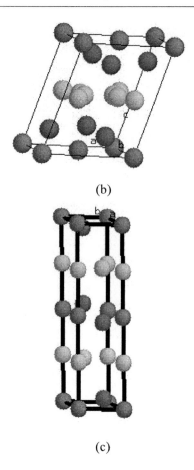

(b)

(c)

Figure 1. Crystal structures of (a) α phase; (b) β phase; (c) γ phase.(Na: ⬭ ; Co: ⬤ ; O: ⬤)

The material γ-$Na_{0.7}CoO_2$ belongs to a category of γ-$Na_xCo_yO_2$ phase with $0.55 \leq x/y \leq 0.74$. The Na ions in the gamma phase partially occupy the two crystallographically distinct sites *2b* and *2d*, and prefer occupying the *2d* site, since the *2b* site is directly between the Co ions when viewing along the *c* direction and results in a slightly higher energy state. [13,14] The calculations based on the density-functional theory within the local density approximation with the Ceperly-Alder exchange-correlation functional suggest that the arrangements of Na ions within a given plane of *2b* and *2d* tend to have ordering structures driven by the screened electrostatic interactions among the Na ions. [14] The Na ordering structures have been observed by electron diffraction experiments. [15] As a matter of fact, the phase formation of Na_xCoO_2 is not only associated with the sodium content but also with the firing temperature and partial pressure of oxygen. The γ-Na_xCoO_2 phase with $0.74 \leq x \leq 1$ can be readily synthesized by using appropriate firing temperature.

3   Materials Synthesis

Rational approaches to synthesize novel materials have been a great interest to the solid state chemists for producing solids with interesting properties and potential applications. Intercalation and ion exchange are the techniques among those low-temperature processes in the rational approaches. Intercalation refers to the insertion of a guest species into a crystalline host lattice. Ion exchange has been defined as a process in which ions are released from an insoluble permanent material in exchange for other ions in a surrounding solution. Both techniques involve a reversible reaction with the structural integrity of the host lattice essentially preserved throughout the process. In general, both types of reactions can occur at relatively low temperatures.

3.1 Synthesis Rationale

The superconducting phase of fully hydrated $Na_{0.35}(H_2O)_{1.3}CoO_2$ was obtained by immersing γ-$Na_{0.7}CoO_2$ powders in Br_2/CH_3CN solution followed by filtering and rinsing. Upon intercalation, the c-axis of the parent compound γ-$Na_{0.7}CoO_2$ expands from 10.9 Å to 19.6 Å to form the fully hydrated sodium cobalt oxyhydrates $Na_{0.35}(H_2O)_{1.3}CoO_2$ with little change in the a-axis. Since the fully hydrated phase preserves the structural integrity of the host lattice, the reaction can be viewed as a topotactic transformation process. Formation of superconducting $Na_{0.35}(H_2O)_{1.3}CoO_2$ is also generally considered as an oxidative deintercalation by removing Na^+ partially from the host lattice, followed by a hydration reaction. It is well-known that $KMnO_4$ is a strong oxidizing agent with a unique affinity for oxidizing organic compounds, which is particularly useful for in-situ remediation of organic compounds in ground water and subsurface soils when coupled with delivery techniques. Superconducting cobalt oxyhydrates can be obtained by immersing γ-$Na_{0.7}CoO_2$ in the aqueous $KMnO_4$ solution at a low molar ratio of $KMnO_4$ relative to Na in the parent compound without resort to the volatile and flammable Br_2/CH_3CN solution. [16] Since removing more Na^+ from the host lattice of γ-$Na_{0.7}CoO_2$ requires a higher oxidation potential, [17] one would expect that the sodium content in the cobalt oxyhydtates could be tuned by varying the concentration of aqueous $KMnO_4$ solution due to the fact that a higher concentration of aqueous $KMnO_4$ solution would have a higher redox potential according to the Nernst equation. Besides the potassium permanganate, sodium permanganate $NaMnO_4$ is another common form of permanganates. Both forms have similar chemical reactivity. Sodium permanganate has about 10 times the solubility in water that potassium permanganate has. This fact would be an advantage to produce superconducting potassium-free cobalt oxyhydrates in large quantities using $NaMnO_4$ as the de-intercalation and oxidation agent without resort to highly toxic Br_2/CH_3CN solution.

3.2 Preparation of $(Na,K)_x(H_2O)_yCoO_2$ and $Na_x(H_2O)_yCoO_2$

Using the conventional solid state reaction procedure or the sol gel route to prepare Na_xCoO_2 normally requires using excess Na to compensate for the loss of the Na during the gradual heating process. In addition, the loss of Na is accompanied by formation of the impurity phase of Co_3O_4. [18,19] In order to avoid the loss of Na, a rapid heat-up procedure [20] is adopted to prepare polycrystalline parent compounds of γ-$Na_{0.7}CoO_2$. High purity powders of Na_2CO_3 and CoO were thoroughly mixed and ground using a Retch MM2000 laboratory mixer mill and calcined in a preheated box furnace at 800°C for 12 h. In case of using aqueous $KMnO_4$ solution, the resulting powders (0.5 -1 g) were immersed and stirred in 50 -

680 ml of aqueous $KMnO_4$ solution with the molar ratio of $KMnO_4/Na$ between 0.05 and 40 at room temperature for 5 days. The concentration of $KMnO_4$ used to prepare $(Na,K)_x(H_2O)_yCoO_2$ depends on the molar ratio of $KMnO_4/Na$ and the amount of water used to dissolve $KMnO_4$. High molar ratio of $KMnO_4/Na$ requires a large amount of water to dissolve $KMnO_4$ due to the limited solubility of $KMnO_4$ in water. For the 40X sample, it needs 680 ml water to completely dissolve $KMnO_4$. The resulting products were filtered and washed by de-ionized water several times with the total volume of 150-200 cc. Note that the contents of potassium and sodium of the products depend slightly on the volume of water in the washing process because they would lose a small amount in the water as confirmed by ICP analyses. The powders were then stored in a wet chamber with relative humidity of 98% to avoid loss of the water content. In case of using aqueous $NaMnO_4$ solution, the molar ratio of $NaMnO_4/Na$ is between 2 and 40. The volume of water to rinse off $NaMnO_4$ is 350 cc.

3.3   Phase Formation of $(Na,K)_x(H_2O)_yCoO_2$ and $Na_x(H_2O)_yCoO_2$

Figs. 2 and 3 show the powder x-ray diffraction patterns of cobalt oxyhydrates obtained by immersing γ-$Na_{0.7}CoO_2$ in aqueous $KMnO_4$ solution having different molar ratios of $KMnO_4/Na$. It is clearly seen that different molar ratios of $KMnO_4/Na$ (different X)

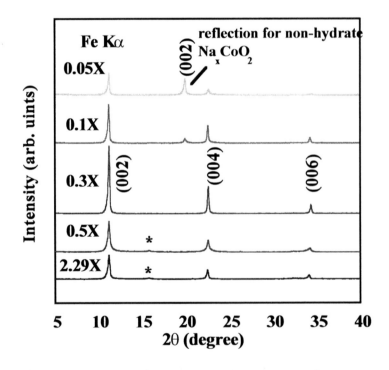

Figure 2. Powder x-ray diffraction patterns of $Na_{0.7}CoO_2$ and $(Na,K)_x(H_2O)_yCoO_2$ obtained by immersing $Na_{0.7}CoO_2$ in low molar ratio of aqueous $KMnO_4$ solution with respect to the Na content in the parent compound ($0.05 \leq KMnO_4/Na \leq 2.29$). The 0.5X represents that the molar ratio of $KMnO_4/Na$ is 0.5. The asterisk (*) indicates the (002) reflection of the $c \approx 13.9$ Å phase.

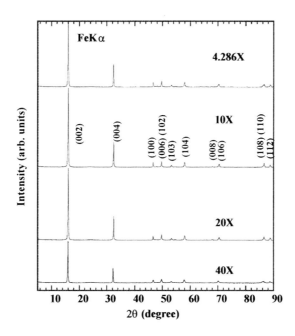

Figure 3. Powder x-ray diffraction patterns of $(Na,K)_x(H_2O)_yCoO_2$ obtained by immersing γ-$Na_{0.7}CoO_2$ in high molar ratio of aqueous $KMnO_4$ solution with respect to the Na content in the parent compound ($4.286 \leq KMnO_4/Na \leq 40$).

result in the variation of XRD patterns. The XRD patterns can be classified as four categories according to the molar ratio of $KMnO_4$/Na used to treat the parent material. These four categories are $KMnO_4/Na \leq 0.1$, $0.3 \leq KMnO_4/Na \leq 0.4$, $0.5 \leq KMnO_4/Na \leq 2.29$, and $4.286 \leq KMnO_4/Na \leq 40$. As shown in Fig. 2, the characteristic x-ray diffraction peak of the maximum intensity with the Miller index (002) shifts from $2\theta \approx 20.3°$ (d spacing ≈ 5.5 Å) for the parent compound irradiated by the Fe Kα radiation to $2\theta \approx 11.3°$ (d spacing ≈ 9.8 Å) for the samples between 0.3X and 2.29X, indicating that the c-axis expands from $c \approx 10.9$ Å to $c \approx 19.6$ Å in the unit cell, which is consistent with those for samples obtained by the Br_2/CH_3CN solution. Fig. 4 shows the XRD patterns of the $KMnO_4$-treated 0.05X sample, indicating the transformation of the non-hydrate phase with $c \approx 11.2$ Å to the fully-hydrated phase with $c \approx 19.6$ Å. For the 0.05X and 0.1X samples, the XRD patterns are a mixture of a fully hydrated phase ($c \approx 19.6$ Å) and a non-hydrate phase ($c \approx 11.2$ Å). However, it is found that when storing the as-prepared 0.05X and 0.1X samples in a wet chamber with a relative humidity of 98% for a long period of time, the fully hydrated phase grows at the expense of the non-hydrate phase, as evidenced by the fact that the (002) reflection peak ($2\theta \approx 11.4°$) of the fully hydrated phase grows, whereas the (002) peak ($2\theta \approx 20.1°$) of the non-hydrate phase nearly disappears. This result indicates that intercalation of the water molecules could undergo a gas-solid interaction in addition to that through a liquid–solid reaction.

As shown in Fig. 3, samples synthesized by the high concentration of $KMnO_4$ ($4.286 \leq KMnO_4/Na \leq 40$) surprisingly and clearly show distinct XRD patterns from low X samples. They have the characteristic peak of (002) reflection occurring at $2\theta \approx 16°$ (d spacing ≈ 6.95 Å), which is equivalent to have $c \approx 13.9$ Å. It has been shown that the fully hydrated phase obtained by the Br_2/CH_3CN solution is not stable at ambient conditions and tends to lose water becoming an intermediate hydrate phase with $c \approx 13.8$ Å. [21] Both the XRD pattern and the size of c-axis of our high X samples are very similar to the intermediate hydrate phase with $c \approx 13.8$ Å, which could be obtained by heating the sample at ca. 75□ or applying a hydraulic pressure at room temperature on the Br_2/CH_3CN-prepared fully hydrate phase. The $c \approx 19.6$ Å phase can be recovered by exposing the sample to sufficient

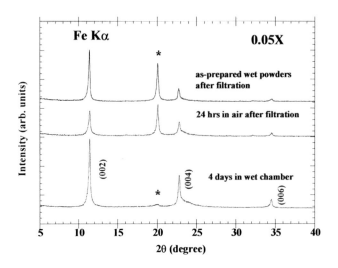

Figure 4. Powder x-ray diffraction patterns of the $KMnO_4$-treated 0.05X sample. The as-prepared wet powders are a mixture of $c \approx 19.6$ Å and $c \approx 11.2$ Å phases. The asterisk is the (002) reflection peak of $c \approx 11.2$ Å phase. After further exposing the as-prepared powders in the humid surroundings, the $c \approx 11.2$ Å phase transforms to $c \approx 19.6$ Å.

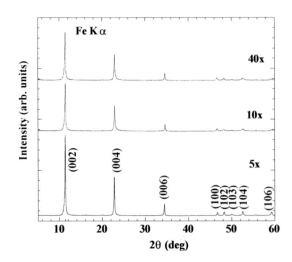

Figure 5. Powder x-ray diffraction patterns of $Na_x(H_2O)_yCoO_2$ prepared by immersing γ-$Na_{0.7}CoO_2$ in aqueous $NaMnO_4$ solution. The label 5X refers to that the molar ratio of $NaMnO_4$ relative to the Na in the parent compound is 5, i.e., 3.5 mole of $NaMnO_4$ is used for 1 mole of γ-$Na_{0.7}CoO_2$. All the samples indicate the fully hydrated phase with $c \approx 19.6$ Å.

humidity for a certain period of time as the Br_2/CH_3CN-prepared fully hydrate phase transforms to the $c \approx 13.8$ Å phase. Nevertheless, the $KMnO_4$-prepared $c \approx 13.9$ Å phase of the high X samples cannot be transformed to the $c \approx 19.6$ Å phase. Formation of this new phase is ascribed to the sodium site being replaced partially or mostly by the larger size of potassium, which will be discussed in the later section.

In case of using aqueous $NaMnO_4$ solution, the $c \approx 19.6$ Å phase can be readily obtained as well (see Fig. 5). Unlike the $KMnO_4$ case, the high molar ratio of $NaMnO_4/Na$ would not produce the $c \approx 13.9$ Å phase.

3.4   Formation Mechanism of $(Na,K)_x(H_2O)_yCoO_2$

As previously stated, formation of superconductive $Na_{0.35}(H_2O)_{1.3}CoO_2$ is generally considered as an oxidative de-intercalation process via the Br_2/CH_3CN route, followed by a hydration process. For using the $KMnO_4$ route, the direct formation of $c \approx 13.9$ Å phase involves not only the de-intercalation and hydration process but also an ion exchange reaction. Furthermore, we find that the oxidative de-intercalation can be achieved simply by immersed the parent compound in the water or hydrogen peroxide.

3.4.1   De-intercalation and Ion Exchange

Table I summarizes the chemical compositions obtained by the ICP-AES analyses and the lattice parameters of $KMnO_4$-prepared $(Na,K)_x(H_2O)_yCoO_2$. Chemical analyses show that the sodium content in $(Na,K)_x(H_2O)_yCoO_2$ decreases significantly at the molar ratio of $KMnO_4/Na$ equal to or greater than 4.286, whereas the potassium content increases with increasing molar ratio of $KMnO_4/Na$. The potassium almost replaces the sodium completely for the 40X sample. Besides, the sum of the sodium and potassium falls in the range of 0.28 and 0.38 in terms of the molar ratio with respect to cobalt. In an electrochemical study, [17] it is found that the chemical potential of the sample increases with the increasing amount of Na removed from the host lattice, meaning that to remove more Na from the host lattice requires a higher redox potential. According to Nernst equation,

Table I. Chemical analyses[a] of ICP-AES results and lattice constants[b] of $(Na,K)_x(H_2O)_yCoO_2$ synthesized using aqueous $KMnO_4$ solution.

Molar ratio of $KMnO_4/Na$	K	Na	x = Na+K	Mn[c]	Co	a (Å)	c (Å)
0.05X	0.00	0.38	0.38	0.04	1	2.8268(5)	19.626(2)
0.1X	0.01	0.34	0.35	0.07	1	2.8261(3)	19.573(1)
0.3X	0.02	0.33	0.35	0.08	1	2.8249(1)	19.669(1)
0.4X	0.02	0.34	0.36	0.08	1	2.8255(1)	19.686(1)
0.5X	0.02	0.34	0.36	0.08	1	2.8248(1)	19.679(1)
1X	0.03	0.35	0.38	0.08	1	2.8281(3)	19.685(2)
1.529X	0.05	0.31	0.36	0.07	1	2.8115(15)	19.750(5)
2.29X	0.08	0.28	0.36	0.07	1	2.8250(3)	19.735(2)
4.286X	0.18	0.15	0.33	0.07	1	2.8286(1)	13.917(1)

Molar ratio of KMnO$_4$/Na	K	Na	x = Na+K	Mn[c]	Co	a (Å)	c (Å)
10X	0.21	0.07	0.28	0.08	1	2.8261(2)	13.864(1)
20X	0.26	0.05	0.31	0.08	1	2.8252(12)	13.849(1)
40X	0.25	0.03	0.28	0.08	1	2.8251(2)	13.960(1)

[a]The error in weight % of each element in ICP-AES analysis is ±3 %, which corresponds to an estimated error of ±0.02 per formula unit for each element.

[b]The lattice constants are obtained using Rietica, a Rietveld structure refinement program, based on a hexagonal structure (P6$_3$/mmc).

[c]The manganese persistently exists in the sample after thoroughly washing procedure. The manganese might exist as the amorphous form of MnO$_2$ based on the reaction,

$$4MnO_4^- + (2+\frac{z}{2})H_2O + ze^- \underset{\leftarrow}{\rightarrow} 4MnO_2(s) + (3-\frac{z}{4})O_2(g) + (4+z)OH^-$$

; [38] however, MnO$_2$ can be partially removed by immersing the sample in alkaline solution to form brown Mn(OH)$_3$ precipitate.

$$E = E_0 + \frac{RT}{zF}\ln\left(\frac{a_{Ox}}{a_{Red}}\right),$$

(1)

where E is the potential for a particular half-cell reaction, a is the activity of oxidized or reduced species, and z is the number of electrons participating in the half-cell reaction. Therefore, a higher redox potential would be expected for higher concentration of KMnO$_4$ solution and consequently would result in de-intercalating more alkaline metal from the host lattice, which is confirmed by the smaller alkaline metal content (the sum of the sodium and potassium contents) in the higher X samples.

Fig. 6 shows the XRD patterns of Na$_{0.35}$CoO$_2$ obtained by immersing γ-Na$_{0.7}$CoO$_2$ in the 30% H$_2$O$_2$ for 5 days, followed by storage in a wet chamber for various periods of time. The (002) reflection is found to shift to a lower angle at 2θ = 20.12° − 20.08° (d = 5.542-5.553 Å) as compared to 2θ = 20.44° (d = 5.456 Å) for γ-Na$_{0.7}$CoO$_2$. The longer time the sample is stored in the wet chamber, the lower 2θ the (002) reflection would occur. According to the ICP-AES analysis, the sodium content is 0.35 which gives the evidence of de-intercalation from the host lattice after the treatment of hydrogen peroxide. The water content is estimated to be about 0.33 based on the TGA analysis shown in Fig. 7. As a result of de-intercalation of Na and hydration, the c-axis expands slightly. In fact, simply by immersing γ-Na$_{0.7}$CoO$_2$ in tap water, de-intercalation of Na can be achieved to have the composition of Na$_{0.30}$(H$_2$O)$_y$CoO$_2$ and the (002) reflection occurs at 2θ = 20.02° (d = 5.569 Å) and the water content of y ≈ 0.41 (Fig. 7).

3.4.2 Hydration

On contact with water vapor or with liquid water, A$_x$MS$_2$ (A = group 1A metal, M = Ti, Nb, or Ta) undergoes a topotactic process and forms hydrated compounds A$_x$(H$_2$O)$_y$MS$_2$ with water molecules intercalated into the host lattice A$_x$MS$_2$. As a result, the c-axis expands with little change in the a-axis. The size of the c-axis of is associated with the size of the alkaline metal in the A$_x$(H$_2$O)$_y$MS$_2$. [22] It has been found that the alkaline metal with the ionic radius >1 Å leads to bilayers of water intercalated between the MS$_2$ layers and hence a larger c-axis, whereas the alkaline metal with the ionic radius <1 Å leads to monolayer of water inserted within the alkaline metal layers and hence a smaller c-axis. The c-axis of $A_{0.3}(H_2O)_yTaS_2$ is 23.63 Å and 18.18 Å for A = Na^+ (ionic radius: ~ 1 Å) and K^+ (ionic radius: ~ 1.4 Å), respectively. The interlayer height between MS_2 layers is ca. 5.77 Å and 3.04 Å for A = Na^+ and K^+, respectively. They correspond well to once and twice the van der Waals diameter of a water molecule of ~ 2.8 Å. Similar relationships between the size of the alkaline metal and the c-axis could be found in the $KMnO_4$-prepared cobalt oxyhydrates, as shown in Figs. 2 and 3 and Table I.

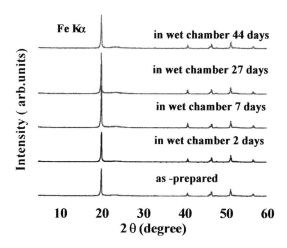

Figure 6. Powder x-ray diffraction patterns of $Na_x(H_2O)_yCoO_2$ prepared by immersing γ-$Na_{0.7}CoO_2$ in 30% H_2O_2 solution for 5 days and the resulting powders are stored in a wet chamber for different periods of time. The (002) reflection appears at $2\theta = 20.12° - 20.08°$ indicating slight elongation of the c-axis as compared to $2\theta = 20.44°$ for γ-$Na_{0.7}CoO_2$.

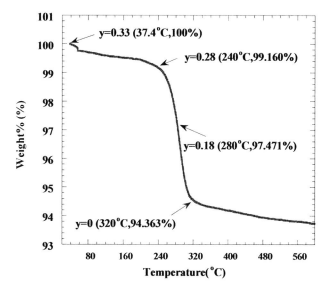

Figure 7. TGA curve of $Na_{0.35}(H_2O)_yCoO_2$ prepared by immersing γ-$Na_{0.7}CoO_2$ in 30% H_2O_2 solution for 5 days.

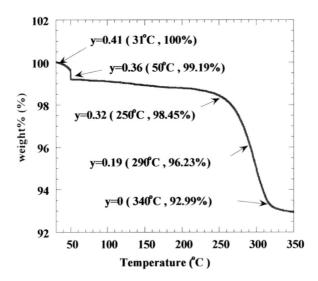

Figure 8. TGA curve of $Na_{0.30}(H_2O)_yCoO_2$ prepared by immersing γ-$Na_{0.7}CoO_2$ in tap water for 5 days.

Due to the presence of potassium, the following scheme is proposed to describe the formation of the new phase $(Na,K)_x(H_2O)_yCoO_2$ involving an ion exchange reaction in addition to the de-intercalation and hydration process.

Scheme 1[a]

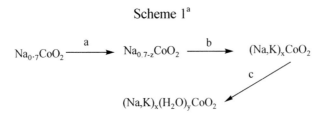

[a](a) oxidative de-intercalation process: MnO_4^- (aq) at room temperature; (b) ion exchange reaction: K^+ (aq) at room temperature and $x < 0.7$; (c) hydration process: H_2O at room temperature.

The whole process is completed as a one-pot reaction. Based on the XRD and ICP-AES results, the $c \approx 13.9$ Å phase having more K^+ content has an interlayer expansion of ca. 1.5 Å, whereas the $c \approx 19.6$ Å phase having less K^+ content has an interlayer expansion of ca. 4.35 Å. Direct formation of $c \approx 13.9$ Å phase of cobalt oxyhydrate could be attributed to the significant increase of K^+ in the $\geq 4.286X$ samples, since a larger size of alkaline metal would lead to a monolayer of water inserted within the alkaline metal layers, as discussed above in the case of $A_x(H_2O)_yMS_2$. It is therefore conceivable that formation of the two distinct c-axis phases of cobalt oxyhydrates is associated with the size of the alkaline metal ($K^+ \sim 1.4$ Å as compared to $Na^+ \sim 1$ Å). Further evidence can be seen in the TGA results, shown in Figs. 9 and 10. The $c \approx 13.9$ Å phase has the hydration water content of $y \approx 0.7$, while the $c \approx 19.6$ Å phase has $y \approx 1.33$. It should be noted that the water content was found to be $y \approx 0.5 - 0.7$ for A = K and $y \approx 1.6 - 1.8$ for A = Na in $A_x(H_2O)_yMS_2$. In addition, the TGA curves show the thermal stability of the water molecules in the lattice for the 0.3X sample is different from that of the 10X sample. This could be due to the variation of the hydration energy for the Na^+

and K^+ and the position of the water molecules in the lattice. The water molecules for the 0.3X sample are located both within the Na layers and between the Na layers and CoO_2 layers.

3.5 Kinetics of the Topotactic Transformation

The transformation from the non-hydrate phase (c ≈ 11.1 Å) to the fully hydrated phase ($c ≈$ 19.6 Å) in both of the 0.05X and 0.1X samples suggests that the whole process is slow at low concentration of $KMnO_4$ solution. It is known that the oxidation reaction follows the second order kinetics at low concentration of $KMnO_4$ solution, which means the reaction rate depends on both the concentration of the oxidized compound and the aqueous $KMnO_4$ solution. At low concentration of $KMnO_4$ solution, this could make the oxidative deintercalation slow, and in turn slow down completing the formation of fully hydrated phase. This is evidenced by the facts in the systematic study that $γ$-$Na_{0.7}CoO_2$ is immersed in different concentrations of aqueous $KMnO_4$ solution for different periods of time. As shown in Fig. 11, the $c ≈$ 19.6 Å phase has already appeared after one-day immersion for the 0.3X (the second category) and 1X sample (the third category). The 0.05 X sample (the first category) appears only as the non-hydrate phase ($c ≈$ 11.2 Å). Moreover, as shown in Fig. 12, the $c ≈$ 19.6 Å phase shows up after 2 days of immersion for the 0.05X sample. It can be readily seen that the amount of non-hydrate phase existing in the sample decreases with increasing X based on the relative reflection intensity of $I_{max}(c ≈$ 19.6 Å)/$I_{max}(c ≈$ 11.2 Å). Besides, for the 0.05X sample already immersed in the $KMnO_4$ solution for 5 days, a small amount of non-hydrate phase still remains after additional 4 days of exposure to the humidity. All these results suggest that the hydration reaction in intercalating water molecule into the host lattice is a slow process. For a higher X sample (≥ 0.3X), the non-hydrate phase would not be observed after 5 days of $KMnO_4$ treatment.

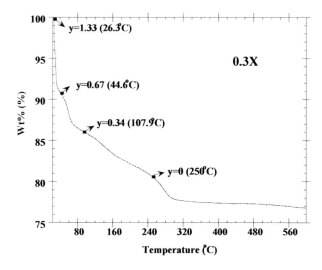

Figure 9. TGA for the 0.3X sample obtained by immersing $Na_{0.7}CoO_2$ in the aqueous $KMnO_4$ solution with the molar ratio of $KMnO_4$/Na = 0.3.

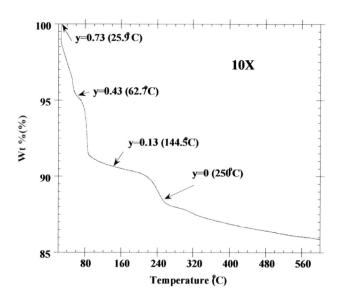

Figure 10. TGA for the 10X sample $Na_{0.07}K_{0.21}(H_2O)_yCoO_2$ obtained by immersing $Na_{0.7}CoO_2$ in aqueous $KMnO_4$ solution with the molar ratio of $KMnO_4/Na = 10$. The water content is determined by assuming the complete dehydration occurring at 250□.

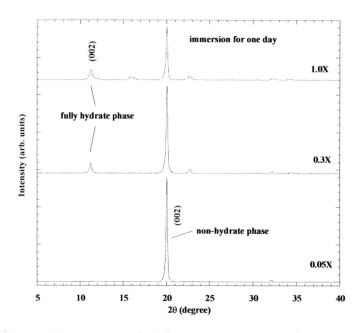

Figure 11. Powder x-ray diffraction patterns of the 0.05X, 0.3X and 1X samples which are obtained by immersing γ-$Na_{0.7}CoO_2$ in aqueous $KMnO_4$ solution for 1 day. The fully hydrate phase ($c \approx 19.6$ Å) is observed both in the 0.3X and 1.0X sample, indicated by the reflection peak at $2\theta \approx 11.3°$, but not for the 0.05X sample. It suggests that formation of the fully hydrate phase is a very slow process when using the low concentration of aqueous $KMnO_4$ solution.

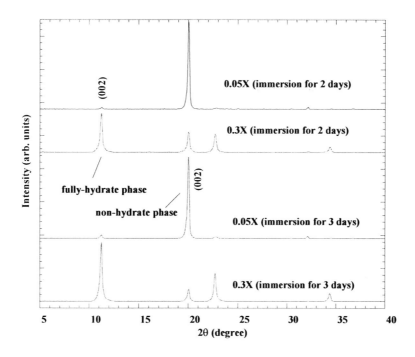

Figure 12. Powder x-ray diffraction patterns of the 0.05X and 0.3X samples obtained by immersing γ-$Na_{0.7}CoO_2$ in aqueous $KMnO_4$ solution for 2 days and 3 days, respectively. The fully hydrate phase ($c \approx$ 19.6 Å) appears after 2-day immersion in the aqueous $KMnO_4$ solution for the 0.05X sample. Formation of the $c \approx$ 19.6 Å phase for the 0.05X sample is much slower than that of the 0.3X sample.

4   Phase Stability

The fully hydrated phase is known to be very unstable in the ambient conditions. [21, 23] This situation would obviously affect the interpretation of transport property measurements. Even though with the nature of its instability, the $c \approx$ 19.6 Å phase of $KMnO_4$-prepared $(Na,K)_x(H_2O)_yCoO_2$ is found to be relatively more stable than those obtained by the Br_2/CH_3CN solution. The $c \approx$ 13.9 Å phase of $KMnO_4$-prepared $(Na,K)_x(H_2O)_yCoO_2$ is quite stable in the ambient conditions.

4.1   Phase Stability of the $c \approx$ 19.6 Å Phase of $(Na,K)_x(H_2O)_yCoO_2$

Fig. 13 shows the XRD pattern evolution of the 0.3X sample. It can be seen that the XRD peaks of the as-prepared wet powders, which are measured 40 min. after washing and filtration, are larger in intensity and sharper in peak width as compared to those of being stored in the ambient air for 24 h. This result indicates that the phase crystallinity is deteriorating due to gradual loss of water molecule from the host lattice when storing the sample in the ambient atmosphere. However, the fully hydrate phase can be brought back when stored in a wet chamber with relative humidity of 98 % for 4 days (Fig. 13c). After taken out from the wet chamber, the XRD peaks of the sample become smaller and broader in a relatively short time shown in Figs. 13 (d) and 13 (e). Fig. 14 shows that the $c \approx 19.6$ Å phase obtained by the Br$_2$/CH$_3$CN solution readily deteriorates in a shorter period of time as compared to that obtained by the aqueous KMnO$_4$ solution.

The $c \approx 19.6$ Å phase is also unstable when subjected to an evacuated environment. Figs.15 and 16 compare the phase stability of the KMnO$_4$-treated and Br$_2$/CH$_3$CN-treated sodium cobalt oxyhydrates in different evacuated conditions. The Br$_2$/CH$_3$CN-treated $c \approx$ 19.6 Å phase readily deteriorates in vacuum in a very short period of time. Knowing this fact

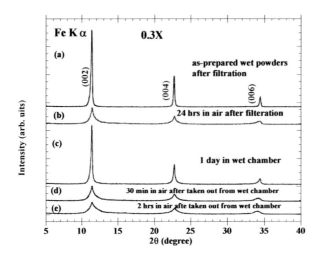

Figure 13. Evolution of XRD patterns for 0.3X sample obtained by immersing Na$_{0.7}$CoO$_2$ in aqueous KMnO$_4$ solution with the molar ratio of KMnO$_4$/Na = 0.3. The XRD pattern is obtained for (a) the as-prepared wet powders 35 min. after filtration; (b) the (a) powders stored in the ambient air for 24 h; (c) the (b) powders stored in a wet chamber with a relative humidity of 98 % for 1 day; (d) the (c) powders taken out from wet chamber and stored in ambient air for 30 min; (e) the (c) powders taken out from wet chamber and stored in ambient air for 2 h.

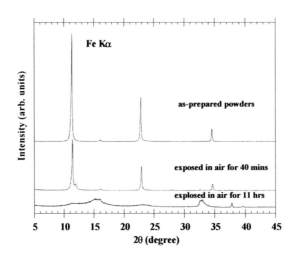

Figure 14. Evolution of the XRD patterns for the Br$_2$/CH$_3$CN-prepared (Na,K)$_x$(H$_2$O)$_y$CoO$_2$. The $c \approx$ 19.6 Å phase readily deteriorates evidenced by the (002) reflection disappearing in 11 hrs and a broad peak emerging at $2\theta \approx 16°$, where the (002) reflection of the $c \approx 13.9$ Å phase occurs.

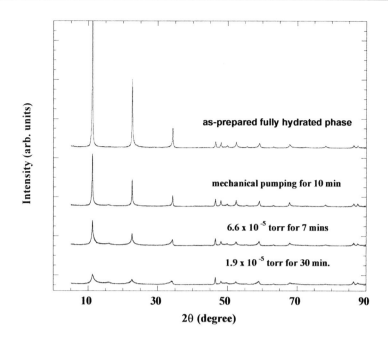

Figure 15. The evolution of XRD patterns of the $KMnO_4$-treated 0.4X sample in different evacuated conditions. The crystallinity of the $c \approx 19.6$ Å phase gradually becomes worse with higher vacuum but the $c \approx 19.6$ Å phase still remains.

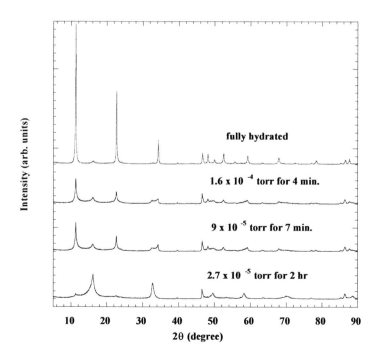

Figure 16. Evolution of the XRD patterns for the Br_2/CH_3CN-prepared $(Na,K)_x(H_2O)_yCoO_2$ in different evacuated conditions. The $c \approx 19.6$ Å phase readily deteriorates with evacuation. The (002) reflection disappears in 11 hrs and a broad peak emerges at $2\theta \approx 16°$, where the (002) reflection of the $c \approx 13.9$ Å phase occurs.

is important when characterizing the sample in a closed cycle refrigerator, which normally requires evacuating the sample chamber where the sample is located. It seems that the $KMnO_4$-treated $c \approx 19.6$ Å phase can sustain for a longer time in a similar vacuum condition and is more stable than the Br_2/CH_3CN-treated one.

As shown in Fig. 9, there is a multistage loss of weight as a result of loss of water upon heating, indicating thermally unstable nature for the 0.3X sample. The 0.3X full hydrate phase can be converted to the intermediate hydrate phase of c ≈ 13.9Å and the de-hydrate phase with c ≈ 11.2 Å upon heating to 90□ and 220□ for a couple of hours, as shown in Fig. 17.

4.2   Phase Stability of the $c \approx 13.9$ Å PHASE OF $(Na,K)_x(H_2O)_yCoO_2$

Fig. 18 shows the evolution of XRD patterns for the 10X samples. The XRD patterns remain unchanged after storing the samples in the wet chamber without any indication of changing the phase to the c ≈ 19.6 Å phase. After taken out of the wet chamber and stored in the ambient air for 2 days, the peak intensity has a slight decrease but remains unchanged after exposed to the ambient air for a long period of time, suggesting it is a stable phase. This situation is remarkably different from the cases in the low X samples. As shown in Fig. 19, the as-prepared wet powders contain the non-hydrate phase of $c \approx 11.2$ Å and convert to c ≈ 19.6 Å phase after storing the samples in the wet chamber for 4 days. The 10X sample also shows a multistage loss of water (Fig. 10). The thermally unstable nature still remains even with the Na mostly replaced by the K. As shown in Fig. 20, the XRD pattern indicates that the pure de-hydrate phase cannot be obtained for the 4.286X sample heated at the temperature as high as 600□. Above 300□, part of the sample has decomposed with the appearance of Co_3O_4. Note that for the 0.3X sample and samples obtained by the Br_2/CH_3CN solution, the de-hydrate phase can be readily obtained by heating at 220-250□. This result indicates the potassium-containing cobalt oxyhydrates exhibit different thermal behavior from the sodium cobalt oxyhydrates. This could be ascribed to the difference between the ion-dipole interaction of K^+-H_2O within the alkaline layers and that of Na^+-H_2O located between the CoO_2 layers and the alkaline layers.

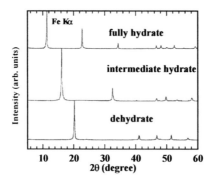

Figure 17. The XRD patterns of full hydrate (c ≈ 19.6 Å), intermediate hydrate (c ≈ 13.9 Å), and dehydrate (c ≈ 11.2 Å) of $Na_{0.33}K_{0.22}(H_2O)_yCoO_2$. The intermediate hydrate phase and dehydrate phase were obtained by heating the fully hydrate phase at 90□ and 220□, respectively.

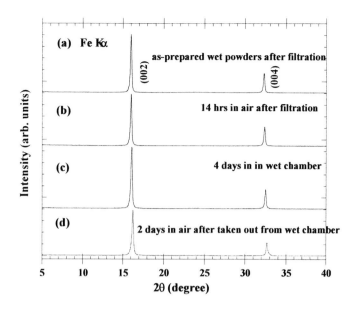

Figure 18. Evolution of XRD patterns for 10X sample obtained by immersing $Na_{0.7}CoO_2$ in aqueous $KMnO_4$ solution with the molar ratio of $KMnO_4/Na = 10$. The XRD pattern is obtained for (a) the as-prepared wet powders 40 min after filtration; (b) the (a) powders stored in the ambient air for 14 h; (c) the (b) powders stored in a wet chamber with relative humidity of 98 % for 4 days; (d) the (c) powders stored in the ambient air for 2 days after taken out from the wet chamber.

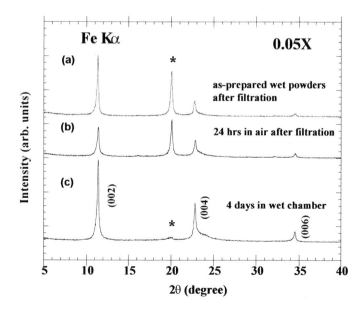

Figure 19. Evolution of XRD patterns for the 0.05X sample obtained by immersing γ-$Na_{0.7}CoO_2$ in the aqueous $KMnO_4$ solution with the molar ratio of $KMnO_4/Na = 0.05$. The XRD pattern is obtained for (a) the as-prepared wet powders 40 min after filtration; (b) the (a) powders stored in the ambient air for 24 h; (c) the (b) powders stored in a wet chamber with relative humidity of 98 % for 4 days. The asterisk indicate the (002) reflection of the de-hydrated $c \approx 11.2$ Å phase.

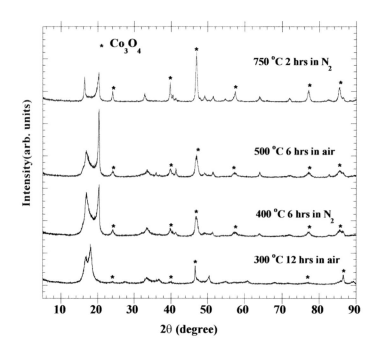

Figure 20. The XRD patterns for the 4.286X sample heated at various temperatures with attempt to obtain the de-hydrate phase. However, the pure de-hydrate phase cannot be obtained at different combinations of temperature and atmosphere. The $c \approx 13.9$Å phase persistently coexists with the dehydrate phase and the sample has decomposed partially accompanied by the formation of Co_3O_4, indicating different thermal stability from the low X samples.

5   Physical Characterization of $(Na,K)_x(H_2O)_yCoO_2$

5.1   Superconductivity

Long period of exposure to the water vapor for the 0.1X sample results in the growth of the $c \approx 19.6$ Å phase at the expense of the non-hydrate phase ($c \approx 11.2$ Å). This phase transformation also changes the onset superconducting transition temperature of the 0.1X sample from 3.4 K to 4.5 K, shown in Fig. 21. Apparently, a larger portion of $c \approx 19.6$ Å phase in the sample could result in a higher onset superconducting transition temperature. This could be caused by superimposition of the diamagnetic signal of a superconductor ($c \approx 19.6$ Å phase) and the paramagnetic signal of a non-superconductor ($c \approx 11.2$ Å). The positive

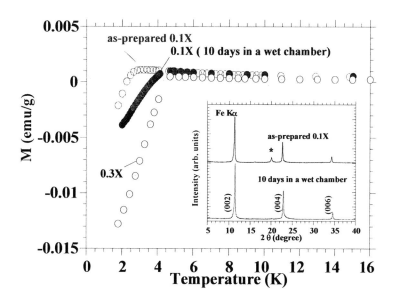

Figure 21. The zero-field cooled dc magnetization data of the as-prepared 0.1X sample, after-exposed-to-humidity 0.1X and the 0.3X samples. The applied field is 10 Oe and 20 Oe for 0.3X and 0.1X, respectively. The inset shows the growth of the c ≈ 19.6 Å phase of the 0.1X sample at the expense of the non-hydrate phase after 10-day exposure to the humidity, as evidenced by the disappearance of the (002) peak (*) of the non-hydrate phase (c ≈ 11.2 Å).

paramagnetic signal would shift the onset T_c to a lower temperature when the c ≈ 11.2 Å phase is present in the sample. It should be noted that the sodium content of the 0.1X sample is not expected to change after being exposed to the water vapor. This might pose a question on the dependence of superconducting transition temperature on the sodium content. [24] Fig. 22 shows the zero-field cooled magnetization data for the superconducting $KMnO_4$-treated samples containing the 19.6 Å phase. The onset superconducting transition, defined as the magnetization starting to decrease, is 3.2 K, 3.4 K, 4.6 K, 4.6.K, 4.5 K, and 3.5 K for the 0.05X, 0.1X, 0.3X, 0.5X, 1.529X, and 2.29X samples, respectively. The 2.29X sample shows an increasing magnetization with decreasing temperature before undergoing superconductive transition, which could be ascribed to the existence of the $c ≈ 13.9$ Å phase. Fig. 23 shows the zero-field cooled mass magnetization data obtained in an applied field of 20 Oe for the $KMnO_4$-treated samples having the pure 13.9 Å phase. It can be readily seen that the 13.9 Å phase of $(Na,K)_x(H_2O)_yCoO_2$ would not undergo superconductive transition down to 2 K.

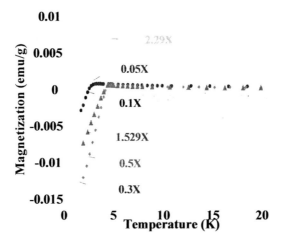

Figure 22. Zero-field cooled mass magnetization data for the low X samples containing the 19.6Å phase. The 0.05X and 0.1X are measured in a field of 20 Oe and the rest of the samples are measured in a field of 10 Oe. Note that only 0.3X sample is the pure 19.6Å phase. The 0.5X, 1.529X, and 2.29X sample are a mixture of 19.6Å and 13.9Å phase and the 0.05X and 0.1X samples are a mixture of 19.6Å and 11.2Å phase.

5.2 Thermopower

5.2.1 Introduction

Thermopower (TEP) measurements can provide both information of the type and the characteristic energy of charge carriers, which elucidate the conduction process and thermodynamics. Since TEP is a measure of the heat per carrier over temperature, we can thus view it as a measure of the entropy per carrier. For γ-Na$_x$CoO$_2$, the large thermopower has been ascribed to the spin and orbital degeneracy due to the competition of crystal field and Hund's rule coupling in cobalt ions. [25,26] Heat capacity, [27] magnetic, [28] thermopower, [29] angle-resolved photoemission spectroscopy (ARPES) [30], and

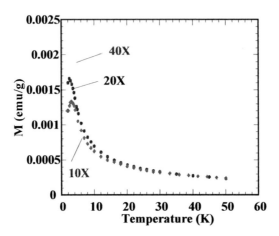

Figure 23. Zero-field cooled mass magnetization data obtained in an applied field of 20 Oe for the high X samples containing the pure 13.9Å phase.

polarization-dependent soft x-ray absorption [31] measurements have suggested a strong correlation of *3d* electrons in γ-Na$_x$CoO$_2$. In a strong correlated system based on the Hubbard model, the thermopower can be expressed as

$$S = -\frac{1}{T}\frac{S^{(2)}}{S^{(1)}} + \frac{\mu}{eT} \quad , \quad (2)$$

where *e* is the absolute value of the electron charge and

$$S^{(2)} = \frac{1}{2}\frac{e}{k_BT}\int_0^\infty \left\{ Tr\left[\exp\frac{1}{k_BT}\left(\mu\sum_i n_i - H \right) \right] \times \left[Qv(\tau) + v(\tau)Q \right] \right\} d\tau \quad ,$$

$$S^{(1)} = \frac{1}{2}\frac{e^2}{k_BT}\int_0^\infty \left\{ Tr\left[\exp\frac{1}{k_BT}\left(\mu\sum_i n_i - H \right) \right] \times \left[vv(\tau) + v(\tau)v \right] \right\} d\tau \quad , \quad (3)$$

where "Tr" denote the trace and is the summation over some complete set of states. H, μ, v and Q are the Hamiltonian, chemical potential, velocity and energy flux operators, respectively. In the high temperature limit, thermopower is dominated by the entropy term $\frac{\mu}{eT}$ in Eq. (2) since the energy-transport term $\frac{S^{(2)}}{TS^{(1)}}$ is small under the condition of strong Coulomb interaction with $\frac{U}{T} \to \infty$. [32,33] For the transition metal oxides, the room temperature thermopower can be considered close to the high temperature limit. [34, 35]

5.2.2 Thermopower Of (Na,K)$_x$(H$_2$O)$_y$CoO$_2$

As shown in Fig. 17, Na$_{0.33}$K$_{0.02}$(H$_2$O)$_y$CoO$_2$ can have three phases with the same metal composition and hexagonal structure except the variation in the *c* axis and the water content, i.e., the full hydrate phase (y ≈ 1.33) with *c* ≈ 19.6 Å, the intermediate hydrate phase (y ≈ 0.67) with *c* ≈ 13.9 Å and the dehydrate phase (y = 0) with *c* ≈ 11.2 Å. Since the alkali metal (Na,K) layer is considered as a charge reservoir, the average valence of Co is not expected to change by the variation of water content. Larger separation of trigonal CoO$_2$ layer is expected to enhance two-dimensional (2D) character. One should expect their physical properties will be affected by the anisotropy with respect to the bonding and structure.

As mentioned previously, the *c* ≈ 19.6 Å phase is unstable in the ambient environment. One has to take precaution when handling the sample for physical characterization. In particular, transport property measurements normally require pressing the sample into pellets. The *c* ≈ 19.6 Å phase could easily transform into the *c* ≈ 13.9 Å phase when applying too high a pressure on the sample. Besides, as shown in Fig. 24 the *c* ≈ 19.6 Å phase could transform into the *c* ≈ 13.9 Å phase during the lengthy measuring process. As shown in Fig. 25, a device is designed to overcome this problem by keeping the measuring chamber with

sufficient humidity and preventing the measured sample from losing water. Fig. 26 shows that
the $c \approx 19.6$ Å phase remains as it is after lengthy measurements.

In a strong correlated system with no double occupancy, namely, $t \ll k_B T \ll U$, where t is
the transfer integral of an electron between neighboring sites and U is on-site Coulomb
interaction, the thermopower in the high-temperature limit can be expressed by the
generalized Heikes formula as [32]

$$S = -\frac{k_B}{e} \ln\left(2\frac{1-\rho}{\rho}\right),$$
(4)

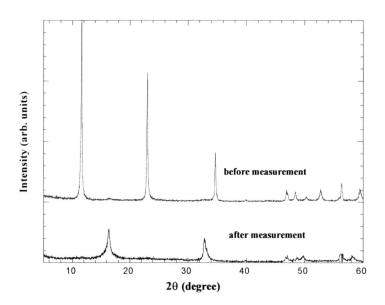

Figure 24. The $c \approx 19.6$ Å phase transforms into the $c \approx 13.9$ Å phase after thermopower measurements.

where $\rho = n/N$ is the ratio of particles to sites (n: the number of particles; N: the number of
available sites). For the strong correlated electron system of γ-Na$_x$CoO$_2$, considering the
effects of spin and orbital degeneracy, the thermopower can be expressed as [34,35]

$$S = -\frac{k_B}{e} \ln\left(\frac{g_3}{g_4}\frac{x}{1-x}\right),$$
(5)

where x is the concentration of Co^{4+}, g_3 and g_4 are the number of the configuration of Co^{3+}
and Co^{4+} ions, respectively. The term of k_B/e is 86.3 μV/K. Neglecting the possible non-
stoichiometry of the oxygen, the formal valence of the cobalt ion for the γ-Na$_{0.7}$CoO$_2$ and
Na$_{0.33}$K$_{0.02}$(H$_2$O)$_{1.33}$CoO$_2$ is +3.3 and +3.65, respectively, based on a simple calculation of
charge neutrality. Co$^{3.3+}$ corresponds to a mixed valence system comprising 70% of Co^{3+} and
30% Co^{4+} ($x = 0.3$); Co$^{3.65+}$ corresponds to a mixed valence system comprising 35% of Co^{3+}
and 65% Co^{4+} ($x = 0.65$). The spin state of cobalt ion could be low spin (LS), intermediate

spin (IS), or high spin (HS), being associated with the electronic state of cobalt ion. The number of the configuration of cobalt ion is determined by the competition between the crystal field splitting 10 Dq and the Hund's rule coupling as well as temperature. Table II lists the possible spin state of Co^{3+} and Co^{4+}, the ratio of g_3/g_4, and the calculated thermopower using Eq. (5) for γ-$Na_{0.7}CoO_2$ and $Na_{0.33}K_{0.02}(H_2O)_{1.33}CoO_2$. Fig. 27 shows the thermopower

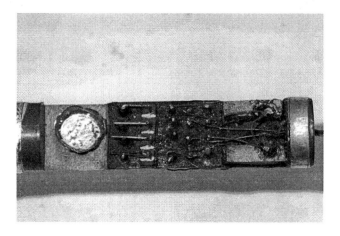

Figure 25. Sufficient humidity is maintained in the resistivity/thermopower measuring device by setting up a pan of water in the sample holder. The sample holder is then sealed by Teflon tape to prevent the water vapor from escaping from the holder.

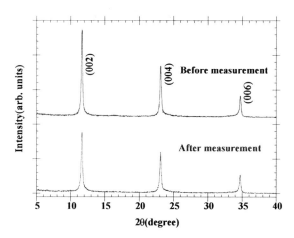

Figure 26. The $c \approx 19.6$ Å phase still remains as it is after thermopower measurements.

as a function of temperature for γ-$Na_{0.7}CoO_2$ and $Na_{0.33}K_{0.02}(H_2O)_{1.33}CoO_2$. When compared to the value of thermopower in Table II, both Co^{3+} and Co^{4+} of γ-$Na_{0.7}CoO_2$ and $Na_{0.33}K_{0.02}(H_2O)_{1.33}CoO_2$ seem to be a mixed spin system of coexisting HS, LS and IS with their states close in energy. However, based on magnetization measurements, both Co^{3+} and Co^{4+} of γ-$Na_{0.7}CoO_2$ should be in the LS state since the effective moment per cobalt in γ-$Na_{0.7}CoO_2$ is estimated to be ca. 1.2 μ_B and the theoretical effective moment for Co^{3+} and Co^{4+} in the LS state is 0 and 1.73 μ_B. [36] In both cases, the value of thermopower turns smaller as the Co^{4+} concentration increases, which is qualitatively consistent with the data in Fig. 27. Realizing that the valence state of cobalt ion for the $c \approx 13.9$ Å phase is not different from that for the $c \approx 13.9$ Å phase, [37] the smaller value, shown in Fig. 28, of the thermopower for the $c \approx 13.9$ Å phase as compared to the $c \approx 19.6$ Å phase should not be ascribed to the change of valence state of cobalt. Instead, it should be associated with the possible change of the electronic structure since thermopower is very sensitive to the topology of the Fermi surface. The nonlinear behavior of the thermopower in both the $c \approx$ 19.6 Å and the $c \approx 13.9$ Å phases is likely due to the enhancement of electron-electron correlation. If it is the case, the enhanced two-dimensionality in the $c \approx 19.6$ Å phase could play a role in affecting the electron-electron correlation and hence the extent of enhancement of thermopower. The thermopower of the $c \approx 11.2$ Å phase shows a distinct behavior from the other two, indicating a dramatic change of the electronic structure. In the wet-chemical analyses, the $c \approx 11.2$ Å phase shows a decrease of the valence state of cobalt ion as compared to the other two and has oxygen deficiency in the lattice caused by heating to obtain the dehydrate phase. [37] In particular, the sign change of thermopower suggests co-existence of electrons and holes, which could be associated with the oxygen deficiency. Fig. 29 shows the temperature-dependent thermopower of the $KMnO_4$-treated 10X sample of $Na_{0.07}K_{0.21}(H_2O)_{0.73}CoO_2$. Both the size and the temperature dependence of the thermopower bear a resemblance to the $c \approx 13.9$ Å phase of $Na_{0.33}K_{0.02}(H_2O)_yCoO_2$. Apparently, the potassium content would not affect the thermopower behavior; instead, the key factors rely on the concentration of Co^{4+} and the length of the c-axis (the dimensionality of the crystal structure). Note that the c axis of $Na_{0.07}K_{0.21}(H_2O)_{0.73}CoO_2$ phase is 13.864(1) Å.

Table II. The possible spin state of Co^{3+} and Co^{4+}, the ratio of g_3/g_4, and the calculated thermopower using Eq. (5) for γ-$Na_{0.7}CoO_2$ ($x = 0.3$)[a] and $Na_{0.33}K_{0.02}(H_2O)_{1.33}CoO_2$ ($x = 0.65$).

spin state of Co^{3+}	spin state of Co^{4+}	g_3/g_4	$S(\mu V/K)$ x=0.3	$S(\mu V/K)$ x=0.65
HS	HS	15/6	-6	-132
HS+LS	HS	16/6	-12	-138
HS	HS+LS	1/12	288	161
LS	LS	1/6	228	101
HS+LS+IS	HS+LS+IS	34/36	78	48

[a]The value of x is the concentration of Co^{4+}.

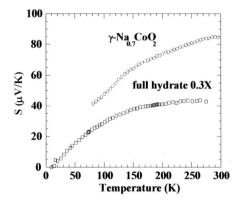

Figure 27. The temperature-dependent thermopower of γ-$Na_{0.7}CoO_2$ obtained and $Na_{0.33}K_{0.02}(H_2O)_{1.33}CoO_2$ (0.3X) with $c \approx 19.6$ Å. The γ-$Na_{0.7}CoO_2$ is prepared by the rapid-heat up procedure at 700°C, followed by sintering at 800°C in air.

Conclusion

Potassium sodium cobalt oxyhydrates $(Na,K)_x(H_2O)_yCoO_2$ can be synthesized using the aqueous $KMnO_4$ solution instead of using the toxic Br_2/CH_3CN route. By immersing the γ-$Na_{0.7}CoO_2$ in aqueous $KMnO_4$ solution with $0.05 \leq KMnO_4/Na \leq 2.29$, the superconductive $(Na,K)_x(H_2O)_yCoO_2$ phase with $c \approx 19.6$ Å can be obtained. For $KMnO_4/Na \geq 4.286$, the $c \approx$ 13.9 Å phase of $(Na,K)_x(H_2O)_yCoO_2$ do not undergo superconductive transition down to 2 K. Formation of the $(Na,K)_x(H_2O)_yCoO_2$ involves not only the oxidative deintercalation and hydration but also the ion exchange reaction. High molar ratio of $KMnO_4/Na$ (≥ 4.286) leads a pure $c \approx 13.9$ Å phase with a monolayer of water within the alkaline layer as a result of partial or almost complete substitution of K^+ for Na^+. The hydration is a very slow process particularly when using the aqueous $KMnO_4$ solution with a low molar ratio of $KMnO_4/Na$. By using the aqueous $NaMnO_4$ solution, the potassium-free superconductive $Na_x(H_2O)_yCoO_2$ phase can be also obtained. By simply immersing γ-$Na_{0.7}CoO_2$ in tap water or H_2O_2, de-intercalation of sodium would occur with the sodium content of 0.3-0.4, which is about the same as using the $KMnO_4$ or the Br_2 routes. As a result, it only leads to a slight expansion of the c-axis (ca. 11.2 Å), which is very close to that of the de-hydrate phase. However, there is

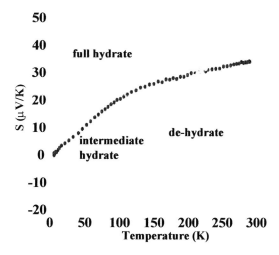

Figure 28. The temperature-dependent thermopower of three $Na_{0.33}K_{0.02}(H_2O)_yCoO_2$ phases with $c \approx$ 19.6 Å (full hydrate), $c \approx 13.9$ Å (intermediate hydrate), and $c \approx 11.2$ Å (de-hydrate), respectively.

a sign for the appearance of the $c \approx 19.6$ Å phase by simply storing the γ-Na_xCoO_2 in a wet chamber in our recent results.

The $c \approx 19.6$ Å phase of $(Na,K)_x(H_2O)_yCoO_2$ is unstable in the ambient air, evacuated chamber, and the hydraulic pressure and can transform into the $c \approx 13.9$ Å phase. The $KMnO_4$-treated $(Na,K)_x(H_2O)_yCoO_2$ seems to be more robust than the Br_2-treated one. The $c \approx 13.9$ Å phase of $(Na,K)_x(H_2O)_yCoO_2$ is quite stable when subjected to the above conditions. The $c \approx 19.6$ Å phase of $(Na,K)_x(H_2O)_yCoO_2$ is also unstable and would lose water upon heating accompanied by converting to the $c \approx 13.9$ Å phase and $c \approx 11.2$ Å phase at 90°C and 220°C, respectively.

For transport property measurements, one should check the XRD after measurements to make sure the phase being the same as before measurements. The effects of the spin and orbital degeneracy seem to play a role in the cobalt oxide system with strong electron-electron correlation. The size of thermopower at room temperature for $Na_{0.33}K_{0.02}(H_2O)_{1.33}CoO_2$ decreases as compared to that for γ-$Na_{0.7}CoO_2$ due to the increasing

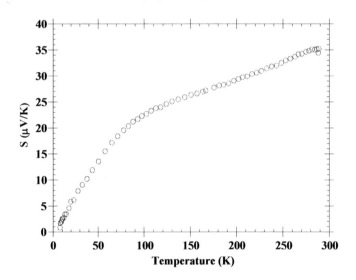

Figure 29. The temperature-dependent thermopower of the $Na_{0.07}K_{0.21}(H_2O)_{0.73}CoO_2$ phase with $c \approx$ 13.864(1) Å.

concentration of Co^{4+}, which is qualitatively consistent with the calculations based on the generalized Heikes formula. In spite of the fact that both the $c \approx 19.6$ Å and $c \approx 13.9$ Å phases have the same concentration of Co^{4+}, the variation of the thermopower could be arising from the enhanced two-dimensionality of the $c \approx 19.6$ Å phase, which affects the thermopower enhancement of the electron-electron correlation. The $Na_{0.07}K_{0.21}(H_2O)_{0.73}CoO_2$ phase with $c \approx 13.864(1)$ Å exhibits similar thermopower behavior to the $c \approx 13.9$ Å phase of $Na_{0.33}K_{0.02}(H_2O)_yCoO_2$, suggesting that the content of potassium plays insignificant role in the transport property and that the length of the c-axis, i.e. the dimensionality of the crystal structure, has a role on the parameters such as electron-electron correlation, Hund's rule coupling or energy level splitting between orbitals, affecting the transport properties.

Acknowledgment

This work is supported by the National Science Council of ROC, grant Nos. NSC 92-2112-M-018-005 and NSC 93-2112-M-018-003. I would like to thank C.-Y. Liao, W.-C. Hung, J.-S. Wang for the experimental works and acknowledge fruitful collaboration with Y.-Y. Chen, S. Neeleshwar, H.-S. Sheu, and J.-Y. Lin.

References

[1] Takada, K,; Sakurai, H.; Takayama-Muromachi, E.; Izumi, F.; Dilanian, R. A.; Sasaki, T. *Nature* 2003, 422, 53.

[2] Terasaki, I.; Sasago, Y.; Uchinokura, K. *Phys. Rev. B* 1997, 56, R12685.

[3] Ando, Y.; Miyamoto N.; Segawa, K.; Kawata, T.; Terasaki, I. *Phys. Rev. B* 1999, 60, 10580.

[4] Ray, R.; Ghoshray, A.; Ghoshray, K.; Nakamura, S. *Phys. Rev. B* 1999, 59, 9454.

[5] Wang, Y.; Rogado, N. S.; Cava, R. J.; Ong, N. P. *Nature* 1999, 423, 425.

[6] Hasan, M. Z.; Chuang, Y.-D.; Kuprin, A. P.; Kong, Y.; Qian, D.; Li, Y. W.; Mesler, B. L.; Hussain, Z.; Fedorov, A. V.; Kimmerling, R.; Rotenberg, E.; Rossnagel, K.; Koh, H.; Rogado, N. S.; Foo, M. L.; Cava, R. J. (2003). Fermi surface and quasiparticle dynamics of Na_xCoO_2 (x=0.7) investigated by angle-resolved photoemission spectroscopy. cond-mat/0308438.

[7] Fouassier, C.; Matejka, G.; Reau, J.-M.; Hagenmuller, P. *J. Solid State Chem.* 1973, 6, 532.

[8] Wycoff, Ralph W. G. Crystal Structures; Inorganic Compounds RX_n, R_nMX_2, R_nMX_3; Robert E. Krieger Publishing Company: Malabar, FL, 1986; Vol. 2, pp 291-296.

[9] Parant, J.-P.; Glazcuaga, R.; Devalette, M.; Fouassier, C.; Hagenmuller, P. J. *Solid State Chem.* 1971, 3, 1.

[10] Cushing, B. L.; Wiley, J. B. *J. Solid State Chem.* 1998, 141, 385.

[11] Ono, Y.; Ishikawa, R.; Miyazaki, Y.; Ishii, Y.; Morii, Y.; Kajotani, T. *J. Solid State Chem.* 2002, 166, 177.

[12] Hahn, T. International Tables for Crystallography; Space-Group Symmetry; D. Reidel Publishing Company: Dordrecht, Holland, 1983; Vol. A, pp 590-591.

[13] Zhang, P.; Luo, W.; Crespi, V. H.; Cohen, M. L.; Loui, S. G. *Phys. Rev. B* 2004, 70, 085108.

[14] Zhang P.; Capaz, R. B.; Cohen, M. L.; Louie, S. G. (2005) Theory of Sodium Ordering in Na_xCoO_2. cond-mat/0502072.

[15] Zandbergen, H. W.; Foo, M.; Xu, Q.; Kumar, V.; Cava, R. *J. Phys. Rev. B* 2004, 70, 085108.

[16] Liu, C.-J.; Liao, C.-Y.; Huang, L.-C.; Su, C.-H.; Neeleshwar, S.; Chen, Y.-Y.; Liu, C.-J. C. *Physica C* 2004, 416, 43.

[17] Chou, F.C.; Cho, J. H.; Lee, P. A.; Abel, E. T.; Matan, K.; Lee Y. S. *Phys. Rev. Lett.* 2004, 92, 157004.

[18] Kawata, T.; Iguchi, Y.; Ito, T.; Takahata, K.; Terasaki, I. *Phys. Rev. B* 1999, 60, 10584.

[19] Liu, C.-J.; Liao, J.-Y.; Wu, T.-W.; Jen, B.-Y. *J. Mater. Sci.* 2004, 39, 4569.

[20] Motohashi, T; Naujalis, E.; Ueda, R.; Isawa, K; Karppinen, M.; Yamauchi, H. *Appl. Phys. Lett.* 2001, 79, 1480.

[21] Foo, M. L.; Schaak, R. E.; Miller, V. L.; Klimczuk, T.; Rogado, N. S.; Wang, Y.; Lau, G. C.; Craley, C.; Zandbergen, H. W.; Png, N. P.; Cava, R. J. *Solid State Commun.* 2003, 127, 33.

[22] Lerf, A and Schöllhorn, R. *Inorg. Chem.* 1977, 16, 2950.

[23] Chen, D. P.; Chen, H. C.; Maljuk, A.; Kulakov, A.; Zhang, H.; Lemmens, P.; Lin, C. T. *Phys. Rev. B* 2004, 70, 024506.

[24] Schaak, R. E.; Klimczuk, T.; Foo, M. L.; Cava, R. J. *Nature* 2003, 424, 527.

[25] Koshibae, W.; Tsutsui, K.; Maekawa, S. *Phys. Rev. B* 2000, 62, 6869.

[26] Koshibae, W.; Maekawa, S. *Phys. Rev. Lett.* 2001, 87, 236603.

[27] Ando, Y.; Miyamoto, N.; Segawa, K.; Kawata, T.; Terasaki, I. *Phys. Rev. B* 1999, 60, 10580.

[28] Ray, R.; Ghoshray, A.; Ghoshray, K.; Nakamura, S. *Phys. Rev. B* 1999, 59, 9454.

[29] Wang, Y.; Rogado, N. S.; Cava, R. J.; Ong, N. P. *Nature* 1999, 423, 425.

[30] Hasan, M. Z. Chunag, Y.-D.; Qian, D.; Li, Y. W.; Kong, Y.; Kuprin, A.; Fedorov, A. V.; Kimmerling, R.; Rotenberg, E.; Rossnagel, K.; Hussain, Z.; Koh, H.; Rogado, N. S.; Foo, M. L.; Cava, R. J. *Phys. Rev. Lett.* 2004, 92, 246402.

[31] Wu, W. B.; Huang, D. J.; Okamoto, J.; Tanaka, A.; Lin, H.-J.; Chou, F. C.; Fujimori, A.; Chen, C. T. *Phys. Rev. Lett.* 2005, 94, 146402.

[32] Chaikin, P. M.; Beni, G. 1976, 13, 647.

[33] Beni, G. *Phys. Rev. B* 1974, 10, 2186.

[34] Oguri A.; Maekawa, S. *Phys. Rev. B* 1990, 41, 6977.

[35] Pálsson G.; Kotliar, G. *Phys. Rev. Lett.* 1998, 80, 4775.

[36] Wu, T.-W. (2003) Studies on the structure and thermoelectric properties of bronze type cobalt oxides. *Master thesis of National Changhua University of Education*, Taiwan.

[37] Karppinen, M.; Asako, I.; Motohashi, T.; Yamauchi, H. *Chem. Mater.* 2004, 16, 1693.

[38] Skoog, D. A.; West, D. M., *Fundamental of Analytical Chemistry*, 3rd ed. Holt, Rinehart and Winston, Inc.: New York, 1976; p. 341.

In: Recent Developments in Superconductivity Research
Editor: Barry P. Martins, pp. 133-199

ISBN 978-1-60021-462-2
© 2007 Nova Science Publishers, Inc.

Chapter 6

MOCVD "DIGITAL" GROWTH OF HIGH-T$_C$ SUPERCONDUCTORS, RELATED HETEROSTRUCTURES AND SUPERLATTICES

Adulfas Abrutis[1] and Carmen Jiménez[2]***

[1] Vilnius University, Dept. of General and Inorganic Chemistry,
Naugarduko 24, 03225 Vilnius, Lithuania
[2] Laboratoire des Matériaux et du Génie Physique (LMGP), INPG-ENSPG,
UMR 5628, BP 46, 38402 St Martin d'Hères, France.

Abstract

The large possibilities and advantages of pulsed injection metal organic vapour deposition technique (PI-MOCVD) for the growth of high-T$_c$ superconductors and related multilayered heterostructures are demonstrated. Accurate precursors micro-dosing by computer controlled electromagnetic injectors and easy variation of injections parameters allow a flexible control of the MOCVD process concerning the film growth rate, thickness and composition. Film growth can be controlled reproducibly at a level of angstrom per pulse and the film thickness is directly related with pulses number; consequently, the growth can be named as "digital." Such a feature is very favourable for the growth of sophisticated multilayered heterostructures like superlattices, where precise thickness and composition control is a key factor. In this case multiple injectors are used.

High quality superconducting films suitable for various applications can be grown by PI-MOCVD. A number of various heteroepitaxial structures in which the YBCO layer is combined with dielectric, conducting or ferromagnetic layers were deposited, including superconductor/dielectric superlattices and polarized-spin injection devices. Their properties and possible applications are discussed. The recent achievements of the PI-MOCVD technique in the field of YBCO coated conductors are also presented. The results on the ex-situ or in-situ growth and properties of various buffer layers-YBCO architectures on metallic substrates and tapes are discussed in the context of state-of-art in this field. Finally, the industrial developments of PI-MOCVD technology are reviewed.

---

* E-mail address: adulfas.abrutis@chf.vu.lt, http://www.chf.vu.lt/MOCVD/index.htm
** E-mail address: Carmen.Jimenez@inpg.fr, http://www.lmgp.inpg.fr

1   Introduction

The perspectives of practical applications of high-T_c superconductors (HTSC), and in particular $YBa_2Cu_3O_{7-x}$ (YBCO), are highly interconnected with their use in the form of thin films. HTSC thin films as well as their synthesis methods must meet high requirements which can vary according to a particular application. In general, the main properties to be achieved are epitaxy and related high transition temperature T_c and high critical current density j_c, uniformity of high critical parameters overall (and often large) deposited surface, smooth surface and low surface resistance (for microwave applications). Synthesis methods must ensure, not only high critical parameters of superconducting films, but also high reproducibility. The last feature is not a simple task because there are many factors influencing film properties: film composition, oxygenation, defects, growth rate, crystallite size and orientation, etc., which have to be reproduced in order to reproduce the same film properties.

The other problem which needs to be addressed is the formation of integrated multilayer structures combining various oxide layers with different physical properties, including HTSC. These complex materials offer the possibility to tailor magnetic, ferroelectric and superconducting properties by changing the coupling between layers. The possibility of artificial control of the crystal structure combined with the growth of superconducting superlattice materials is being extensively studied in order to find new functions in ceramic systems for new applications. The growth of such multilayered structures and superlattices presenting a high and reproducible quality is still a challenge for materials scientists. In this area, the features of the deposition technique, such as its versatility, play a very important role.

For the deposition of HTSC oxide layers various physical (PVD - physical vapor deposition: laser ablation, on-axis and off-axis sputtering, co-evaporation, etc.) and chemical methods are used. The PVD methods require solid sources from metal or oxide. The main difficulty for PVD is the study of the effect of subtle variation of the stoichiometry (or doping), which may have a drastic effect on the properties of the deposited layer. Among the chemical methods, the metal-organic chemical vapor deposition (MOCVD) technique allows to obtain the best quality HTSC thin films. It is practically the only chemical method which allows *in-situ* deposition of HTSC films [1,2] exhibiting properties close to those obtained by PVD. Moreover, MOCVD offers several potential advantages over PVD: more versatile composition control, higher film growth rate, ability to coat complex shapes and large areas. However, despite these apparent advantages, the classical MOCVD technique has been less frequently used than PVD techniques for the deposition of complex oxides such as HTSC for a long time. It has taken much longer for the first reports on the successful growth of superconducting $YBa_2Cu_3O_{7-x}$ (YBCO) layers by MOCVD to appear in the literature than for PVD techniques [1]. The main reason for such a delay was the fact that in the "classical" multi-source MOCVD version used for the first attempts to grow YBCO layers, there were chemical problems related to the nature of MO precursors, the source systems and the vapour decomposition on substrate. The solutions to these problems are not simple, even for chemists, while the first attempts to obtain YBCO layers by MOCVD was made mainly by physicists.

The principle of YBCO deposition by conventional MOCVD is the following: volatile Y, Ba and Cu metal-organic compounds (usually solids) are placed into special separate evaporators (sources), where they are evaporated under vacuum by heating. The vapors are carried by flows of inert gas into a collector, then mixed with oxygen and transported into the reaction chamber. The vapor decomposition and film growth occur on a heated substrate. The process taking place near the hot substrate contains several stages (Fig. 1): a) diffusion of precursor molecules from vapour flow, b) adsorption of species on the substrate surface, c) decomposition, d) desorption and evacuation of organic by-products, e) migration and aggregation of the atoms on substrate surface and f) growth of oxide layer.

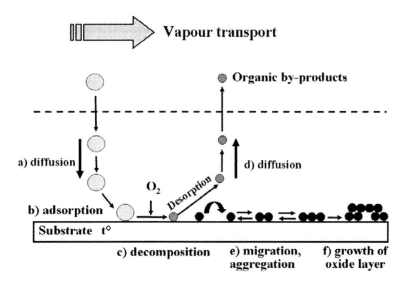

Fig. 1. Schematic presentation of the stages of the MOCVD process on the substrate surface.

The properties of YBCO films are very dependent on the film composition. The film composition is mainly defined by the composition of the vapor phase, so it is very important to achieve a precise and reproducible control of the vapor composition. This is not an easy task in the conventional MOCVD, because the precursors used for YBCO have usually very different and not thermally stable properties. By accurate choosing and precise control of the evaporation temperatures and the gas flow rates, it is possible to obtain a composition of the vapor phase which leads to YBCO films exhibiting good or even excellent properties. However, the main problems are to maintain stable vapor composition during the long process and to reproduce the vapour composition and film properties in different deposition runs. The main factors contributing to the complexity of MOCVD process for multi-component oxides and influencing the reproducibility are:

- Low and very different vapor pressure of solid MO precursors requiring high evaporation temperature, which should be accurately chosen and precisely controlled for each precursor.
- Different thermal stability of precursors in vapor phase leading to large differences between vapor and film compositions.

- Thermal and chemical instability of solid precursors during a prolonged heating in the evaporators (decomposition in solid phase, polymerization, hydrolysis) leading to time dependent changes in vapor phase (and film) composition.
- "Ageing" problem of solid precursors during their storage, related with degradation reactions and leading to changes of volatility.
- Difference in purity and properties of different batches of the same commercial precursor, especially if produced by different companies. As a result it is difficult to reproduce the same evaporation rate under the same conditions.
- Bad reproducibility of the granularity of solid precursors leading to the irreproducibility of the free evaporation surface and thus evaporation rate. Moreover, the evaporation rate fluctuates with the amount of precursor remaining in the source. So, controlling of vaporisation rate by heating under low pressure is a real challenge even in the case of thermally stable solid precursor.

The most complicated precursors in regard of their application in MOCVD are those of barium, strontium and Rare Earth elements. So far metal β-diketonates, 2,2,6,6-tetramethyl-3,5-heptanedionates (thd) in particular, have been used for the MOCVD deposition of oxides films containing these metals, including HTSC. In the case of MOCVD of YBCO films, the main problems of reproducibility are related with available Ba precursors. Barium β-diketonates are predominantly oligomers, display medium volatility and low thermal and chemical stability in solid phase during heating; moreover, they exhibit "ageing" phenomenon and different volatility between different batchs of commercial precursors. A large number of chemistry laboratories works intensively in this area and offer some potential ways of molecular chemistry to improve the properties of precursors (adducts of β-diketonates with Lewis bases in solid and in vapour phase, new chelating ligands, some success in the synthesis of liquid MO precursors) [2-8]. However, at the moment there is no fully satisfactory precursor for the conventional MOCVD of barium, strontium, calcium and Rare Earth elements. Despite several promising developments in the chemistry of new precursors, this problem remains unsolved and impedes the technological application of conventional MOCVD for multi-component oxides including HTSC. It is worth noting that irreproducibility of film properties is a common problem for all deposition techniques.

The problem of irreproducibility had to be solved or reduced in order to make a profit on many advantages of MOCVD technique for large-scale applications. Two main ways - "chemical" and "technical" - have been explored for improving of the conventional MOCVD method. The "chemical way", based on the improving of the properties of MO precursors, has been shortly explained above. The "technical way" has been mainly focused on the elaboration of new technical principles and systems for the generation of vapour phases. The invention of new vapour source systems for MOCVD has led to modifications of the MOCVD technique. These developments have been based on the following ideas:

- Long heating of precursors should be avoided. How to do this?

  o the precursor should be kept in a source at room temperature under inert gas
  o only small amount of precursor should be steadily introduced into an evaporator

o   evaporator temperature should be sufficiently high to ensure a flash evaporation of the introduced precursor.

- Rapid vapour transport into the deposition chamber towards a hot substrate
- A mixture of precursors should be used as a single vapour source for the deposition of a multi-component oxide film.

The realisation of these ideas provided distinctive advantages over conventional MOCVD:

- Decomposition and other chemical changes of precursors in the source can be avoided
- Flash evaporation and rapid vapour transport permits overheating of the evaporator without significant decomposition of precursor, what allows to use very high precursors' feeding rate and to achieve, if necessary, very high vapour pressure and thus very high growth rates.
- Less volatile and less thermally stable precursors can be used.
- The evaporator temperature does not require a very high stabilisation as in the conventional MOCVD.
- The main advantage is the possibility to use mixtures of precursors (i.e., Y, Ba, Cu thd) as single vapour sources for the deposition of multi-component oxide films. Flash evaporation ensures that all precursors of the introduced mixture volatilise completely. As the result, the composition of vapour phase is the same as of the initial mixture. This allows the easier control of vapour phase and film composition and to simplify the MOCVD process.

The main problem is the accuracy of control of the feeding rate of the precursor into the evaporator. Several systems have been proposed for solid precursors or mixtures of solid precursors: i) direct introduction of the powder using a vibrating feeder [9,10] and ii) gradual introduction into the evaporator of a compact road made with a mixture of precursors [11], at a very step thermal gradient. However, is not a simple task to made homogenous roads from sticky (if wetted) or electrostatic (if dried) powders. Also it is difficult to handle directly such powders with a high accuracy of feed rate and homogeneity. The realisation of such systems for large scale application is still more complicated as compared to the laboratory scale use. The high degree of control of the precursors feeding rate as well as the uniformity of composition of precursors' mixture is key factors in the synthesis of multi-component oxides, such as YBCO, from a single source.

The control of the feeding rate is considerably easier for a liquid than for a solid. For example, MO precursor or a weighted mixture of precursors can be dissolved in a suitable organic solvent. Several ways have been proposed for the introduction of solution in the evaporator. In the first way precursor solution is introduced in the form of an aerosol generated ultrasonically or pneumatically and carried by a vector gas [12-15]. The main problem is the stabilisation of the flow of the aerosol: the evaporation of the solvent occurs prematurely due to the solution heating and to the direct exposition to the low pressure zone. The second way is based on the use of a micro pump, a syringe driven by a motor or a liquid mass flow controller [16,17]. In these systems the solution transfer from the zone at room temperature to the evaporator held at high temperature and under low pressure is usually made through a capillary tube which reduces the direct contact of the solution with the evaporation zone. This capillary tube is the main weak point, since it can become clogged-up by the precipitation of solid precursor due to solvent evaporation. All these systems for liquid delivery can be named as systems with "direct contact", because they do not prevent completely the contact of solution with the evaporator.

In order to overcome the mentioned problems, a very simple and versatile liquid delivery system of precursors has been conceived and developed in collaboration between LMGP-Grenoble and Vilnius University [18,19]. This system of vapour generation for MOCVD process is based on the principles similar to those used for the fuel injection in thermal motors. It became the basis of a variation of the MOCVD process called Pulsed Injection MOCVD (PI-MOCVD).

In next sections we describe the principles and advantages of this technique as well as the most interesting results of its successful application for the deposition of HTSC materials.

2   Pulsed Liquid Injection MOCVD

In our source system the solution of precursors is maintained at room temperature in a container connected to a high speed precision micro electro-valve. The pressure (0.5-1.5 Bar of argon) is applied to the liquid inside the container to push the liquid toward the valve. When valve opens, a precise microdose of the liquid is ejected (Fig. 2) at high speed into an evaporator held at high temperature (150-300 °C) and low pressure ($\sim$ 1-10 Torr), where the flash evaporation occurs. The resulting vapour from the precursors mixture and from the solvent is transported rapidly by a high flow of a carrier gas (usually Ar + O_2) into the reaction chamber towards the heated substrate. The solution feeding rate can be simply controlled by the control of injection parameters, as in the fuel injection system. The parameters which control the precursors feeding rate are: a) the opening time of the valve (typically 1-10 ms), b) the injection frequency (1-100 Hz, typically 0.5-3 Hz), c) the concentration of solution, d) the pressure in the solution container. The first two parameters are defined by electrical pulses driven by a computer, which provides a high degree of flexibility to the deposition process. Varying the last two parameters gives additional possibilities to vary precursor feeding rate and growth rate. The mass of injected micro-doses of solution can range a large scale from 0.2-3 mg per pulse, depending on the injector type, up to continuous injection. The composition of the vapour phase and of the film can be controlled by the composition of the initial solution.

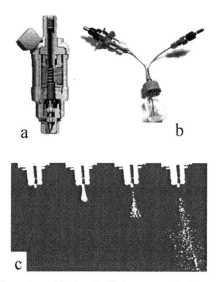

Fig. 2. Injector: a) cross-section view, b) simple "injector - solution container" system, c) different stages of solution injection (into vacuum, without heating).

Our precursor delivery system has some distinctive advantages over other solid and liquid sources:

- High reproducibility of the system (< 5 %).
- Very large range of precursor flow rates is available by adjusting the mentioned four parameters. So the growth rate can be easy varied from very slow to very fast growth.
- The use of several consequently or simultaneously working injectors and the possibility of easy change of injection parameters during the process allows a very flexible growth of various sophisticated multilayers and superlattices.
- Process works "digitally": the thickness may be simply controlled by the number of drops injected, what is very interesting for the growth of multilayers and superlattices. By adjusting of the drop size and solution concentration, the thickness can be controlled at the level of angstroms per pulse (Fig. 3).

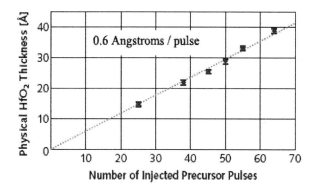

Fig. 3. Atomic precision thickness control for HfO$_2$ films grown by PI-MOCVD. By courtesy of Aixtron AG, Germany.

The main peculiarities of our source system are the pulsed character of the process and the "digital growth". Each injection and following flash evaporation of the solvent and precursors leads to a slight pressure pulsation in the reactor (Fig.4). Pressure change depends on the size of injected microdose, but the high injection frequency flattens the pressure variations. The direct measurement of the pressure pulses may be used for in-situ control of the size of injected drops. The high quality YBCO films obtained by this technique shows that the pulsed character of the process has not any negative influence. Moreover, pulsed character with a possibility of the precise control of precursor dose size and related thickness is a desirable feature of a process for the preparation of artificial structures.

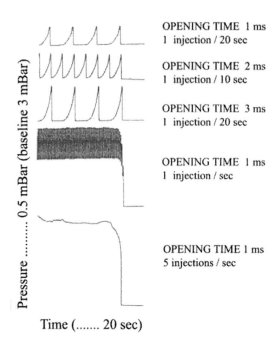

Fig. 4. Pressure pulsation during injections.

Two types of MOCVD reactors based on our precursor delivery system have been built and used for oxides deposition. In the reactors of the first type, a mixture of precursors and solvent vapors formed after solution injection and flash evaporation is directly transported to the reaction chamber (Fig. 5). In the reactors of the second type, the solvent vapour is eliminated from vapour phase and only the precursors vapour is transported to the reaction zone (Fig. 6). The presence of high amount of vapour from organic solvent increases the total carbon concentration in the reaction zone. However, we have not found any significant difference in the carbon contamination of the oxide layers grown using either conventional or liquid-injection MOCVD sources.

In general, we have not found any detrimental influence of the presence of solvent vapour on the properties of YBCO films. The solvent vapours become a problem when very high growth rates are required working at high temperature. In this case, high feeding rate of precursor solution has to be used, what is related with the significant decrease of the oxygen activity due to high concentration of solvent vapour in the reaction zone. Very low oxygen activity in MOCVD process may be detrimental for the stability of oxide materials, including YBCO. This can be compensated by increasing the oxygen partial pressure, but high oxygen pressure in a gas phase containing a large amount of organic solvent at high temperature (~800 °C) leads to an explosive situation. So, in the case when high growth rates needs to be achieved at high temperature, the system of solvent elimination from the reaction zone is preferable.

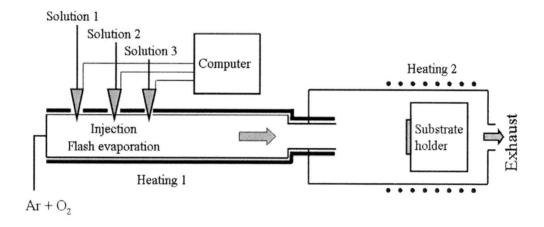

Fig. 5. Schematic representation of a PI-MOCVD reactor with three injectors. In real reactors the vertical configuration of the reaction chamber with an upward vapour flow is used.

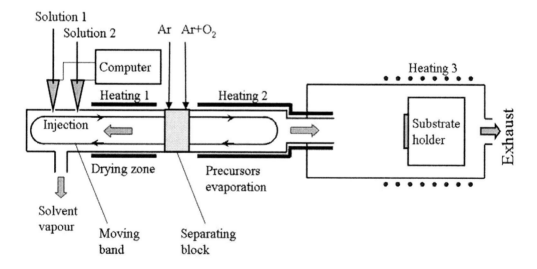

Fig. 6. Schematic representation of a PI-MOCVD reactor with *in-situ* elimination of the solvent. The drops are injected under low pressure on a moving tape, in a slightly heated zone. The solvent volatilises and the Ar gas flowing towards the right carries away its vapor. The moving tape carries the dried precursors into the heated evaporation zone, and the formed vapors are carried by the Ar+O_2 gas flowing toward the substrate.

A general view of some PI-MOCVD reactors installed at Vilnius University is presented in Fig. 7.

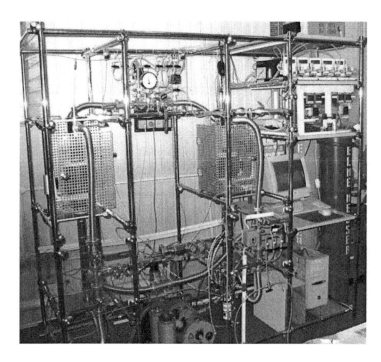

Fig. 7. A system of two laboratory-scale PI-MOCVD reactors in Vilnius University, with a common process control block. Left reactor is for 2 inch substrates, right reactor - for 1 inch substrates.

3   Applications of PI-MOCVD Technique for High-T$_c$ Superconductors

3.1   Introduction

The main factors acting in MOCVD process and influencing properties of the YBCO films are deposition temperature, vapour phase composition, partial oxygen and overall gas pressures, gas flow rates. In PI-MOCVD process where a single liquid source is used and works in the pulsed regime, some additional factors may be used for process optimization, such as solution concentration and injection parameters (opening time and injection frequency) determining the precursor feeding rate. Due to flash evaporation of micro-doses of the injected solution, the vapour phase composition can be easy modified by the control of the solution composition. However, the different precursors generally have different level of thermal stability leading to different yields of decomposition reactions on the substrate. So, the deposition of YBCO films from precursors mixture is very far from congruent. Moreover, the difference between the film and vapour phase (solution) compositions is highly influenced by other process parameters (temperature, oxygen partial pressure, etc.). The properties of YBCO films are very sensible to film composition, so the optimization of solution composition is very important in the MOCVD process based on single liquid source.

The general deposition conditions under which high quality YBCO layers were reproducibly grown, are given in Table 1. Under such conditions the YBCO growth rate is several μm/h, i.e., about 5-10 times higher than that achieved in most classical MOCVD reactors.

Table 1. General deposition conditions for YBCO films by PI-MOCVD

Precursors	M(thd)$_n$
Substrate temperature	Mainly 825 °C
Evaporator temperature	280-300 °C
Total pressure	6.6 hPa (5 Torr)
Oxygen partial pressure	2.6 hPa (2 Torr)
Solvent	Monoglyme
Solution concentration	0.02-0.03 M (Y)
Pulse time	2-4 ms
Injection frequency	1-2 Hz
Mass at each injection	3-6 mg

The influence of the main deposition conditions (substrate temperature and vapour composition) as well as of the nature of the substrate material on YBCO films properties was studied. This systematic study allowed us to gain a better control of the quality of YBCO films. The structure and properties of the substrate material is important factor determining the microstructure and properties of the film, so much attention has been paid to the comparison of films grown on various substrates.

Only metal – 2,2,6,6-tetramethyl-3,5-hepanedionates have been used as precursors for the deposition of HTSC films by PI-MOCVD in this study. Precursors were synthesized and purified at Vilnius University. Monoglyme (1,2-dimethoxiethane) has been mainly used as a solvent. This solvent was chosen due to high solubility of precursors and due to chemical stability of solutions, moreover, the medium volatility of this solvent is convenient for liquid injection and flash evaporation processes.

The critical temperature (T$_c$) of HTSC layers corresponding to the transition onset was determined by contactless AC susceptibility measurements or by resistance vs. temperature measurements using the standard four-probes technique. Critical current density (J$_c$) was determined from AC susceptibility third harmonic measurements or by voltamperometric measurements in a narrow bridge, using short (10-15 ns) current pulses.

Surface morphology was studied by optical microscopy, scanning electron microscopy (SEM) or by atomic force microscopy (AFM). The film composition was determined by energy-dispersive X-ray analysis (EDX) operating with SEM. The calibration of the EDX has been made by RBS on some selected samples. X-Ray photoelectron spectroscopy (XPS) and secondary ion mass spectrometry (SIMS) have been also used for the analysis of the composition homogeneity of the layers and of the inter diffusion at the substrate-film interface. Films microstructure was investigated by X-ray diffraction (XRD) in Bragg-Brentano and Schulz geometries, transmission electron microscopy (TEM), high resolution transmission electron microscopy (HRTEM) and BKD (Backscattered Kikuchi Diffraction). Film thickness was measured by profilometry, optical interference microscopy, RBS, ellipsometry and optical (laser) reflectometry (for *in-situ* measurements).

3.2   HTSC Films on Monocrystalline Substrates

3.2.1   YBCO Films on Perovskite Substrates

Three perovskite substrates - LaAlO$_3$ (001), NdGaO$_3$ (001) and YAlO$_3$ (001) - have been studied in details for YBCO depositions by PI-MOCVD [20-23]. LaAlO$_3$ has a rhombohedral slightly distorted perovskite structure (a=0.379 nm, α=90.5 °). It can be considered as a pseudo-cubic. NdGaO$_3$ (a=0.543 nm, b=0.550 nm, c=0.770 nm) and YAlO$_3$ (a=0.518 nm, b=0.533 nm, c=0.737 nm) are orthorhombic with a higher orthorombocity in YAlO$_3$. The main advantage of these perovskite substrates is the good match of their lattice parameters and thermal expansion coefficients to those of orthorhombic YBCO (a=0.382 nm, b=0.388 nm, c=1.168 nm), therefore, it is quite easy to obtain epitaxial YBCO films. The best much is in the case of NdGaO$_3$. Other perovskite substrate often used for YBCO films growth – cubic SrTiO$_3$ (a=0.390) - has been sometimes used in our work. The details on the YBCO films grown by PI-MOCVD on SrTiO$_3$ are not discussed in this work, because their microstructure and properties were similar to those of films grown on LaAlO$_3$.

YBCO on LaAlO$_3$ substrates.

In order to establish the correlation between solution and film compositions, YBCO films (0.2-0.3 μm) were grown at 825 °C from vapour phases with various compositions. This relation is presented in Fig. 8a. In all cases, the films had relatively less yttrium and barium, but more copper than in the vapour phase. On the other hand, all films were enriched in yttrium and poor in barium, compared with the 123 composition, while the copper proportion varied on both sides of the stoichiometric quantity (50 %).

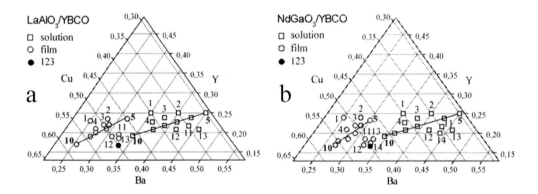

Fig. 8. Composition (metal atom %) of both the solutions and the YBCO films on a) LaAlO$_3$ and b) NdGaO$_3$ substrates. Note the clear shift related to the different decomposition reaction yields of the precursors.

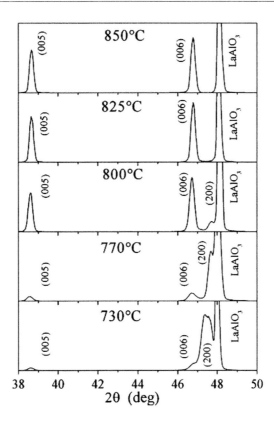

Fig. 9. XRD patterns for YBCO films grown on LaAlO₃ (001) at various temperatures.

Most of the YBCO films grown on LaAlO₃ at 825 °C were purely c-axis textured (c⊥) and had a perfect 0° in-plane orientation on a (001) plane of a pseudo-cubic LaAlO₃ structure. Small values (0.25-0.45°) of the FWHM of the (005) rocking curves and a clearly observed YBCO twinning in two perpendicular directions ([110], [1-10]) demonstrated the very high epitaxial quality of the films. However, a small number (<10 %) of a-axis textured (a⊥) crystallites appeared in the films with the highest barium amount (films 11-13 in Fig. 8a). These a-axis crystallites could be observed in XRD θ/2θ and φ-scans as well as in the SEM pictures of the surface of these films. The surface of the films rich in yttrium and without excess of copper was rather smooth, but some pinholes were visible. In the films with an excess of copper a great number of Cu-rich precipitates appeared, and they became larger and denser when the Cu percentage increased.

The vapour phase composition in the studied range had no noticeable influence on critical temperature of films: T_c varied in the range 90-91 K while the full transition width changed between 0.3 and 1.0 K. The majority of films had high values of critical current density (4-6 MA/cm² at 77 K). Only film No 5 in Fig.8a, having the lowest Cu percentage exhibited clearly lover $J_c = 1$ MA/cm².

In order to establish the influence of the deposition temperature a series of YBCO films (~ 0,4 µm) was deposited varying the temperature between 730 and 850 °C. A pure YBCO phase was present in all films on LaAlO₃ of this series. However, only films grown at high temperature (825-850 °C) were purely c-axis textured, while other contained a mixture of c-axis and a-axis textured crystallites. The amount of a⊥ crystallites increased when the temperature decreased up to almost full prevailing at 730 °C (Fig. 9). Such an effect of deposition temperature on the ratio of c-axis and a-axis textures is well known for YBCO films deposited by various techniques. In our case the transition temperature from a-axis to c-axis growth is about 150 °C higher than in films deposited by magnetron sputtering [24]. Such a difference is determined by different deposition techniques based on very different deposition conditions. All films grown at various temperatures demonstrated "cube-on-cube" epitaxial relations with the substrate lattice. The twinning of c-axis crystallites was obvious in films deposited on LaAlO₃ at 825 °C. Twinning was less pronounced in films grown at 800 °C and it was not found in films deposited at lower temperatures.

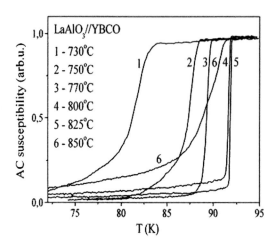

Fig. 10. Suprconducting transition of YBCO films grown at different temperature on LaAlO₃ (001) substrates.

The effect of deposition temperature on superconducting transition of YBCO films grown on LaAlO₃ substrates is presented in Fig. 10. The figure demonstrates that all films deposited in the range 730-850 °C are superconducting, but the critical temperature and transition width strongly depended on deposition temperature. The best films in this series were obtained at 825 °C: T_c ~92 K, ΔT ~0.2 K. Their J_c was ~4 - 5 MA/cm^2 at 77 K. Quite similar properties exhibited films grown at 800 °C, but the transition was not so sharp (ΔT ~0.4 K) and J_c values were lower (~2-3 MA/cm^2). The properties of films prepared at other studied temperatures were clearly worse.

It was found that the T_c - onset values of films on LaAlO₃ substrates only slightly decrease with the decrease of film thickness from 300 nm to 35 nm, while the superconducting transition becomes clearly wider for the thinnest films (Fig. 11). The thickness influenced the critical current density of YBCO films as well. The highest J_c values were usually obtained for films having the thickness in the range 100-400 nm, while the further increase in film thickness followed by a decrease in J_c (Fig. 12), most probably due to accumulation of defects, a-axis textured crystallites or microcracks in thick films. The accumulation of a-axis crystallites was observed in thick (~1 μm and more) YBCO films on LaAlO₃.

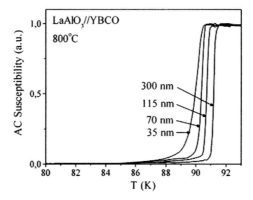

Fig. 11. Suprconducting transitions of YBCO films of different thickness grown on LaAlO₃ (001) substrates.

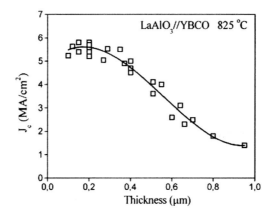

Fig. 12. Critical current density vs. thickness of YBCO films on LaAlO₃ substrates.

YBCO on NdGaO₃ substrates

YBCO films on NdGaO₃ substrates were grown at 825 °C from vapour phase of different composition. The relation between vapour phase and film composition (Fig. 8b) shows the same trends as in the case of films on LaAlO₃. Films on NdGaO₃ were highly textured but a mixed c-axis and a-axis texture are present. This result correlates with the fact that both orientations have excellent lattice match with NdGaO₃ (001) plane. The ratio of the c-axis (c⊥) and a-axis (a⊥) textured crystallites varied in a large game (~5-70 % of a⊥) depending on the vapour phase composition. The higher amount of Ba in the vapour and film usually led to the higher percentage of a-axis textured crystallites (e.g. films 11-14 in Fig. 8b). Furthermore, the proportion of a⊥ crystallites was higher in thicker films grown under the same conditions. The morphology of films surface was influenced by the ratio of a⊥ and c⊥ crystallites. For the films with a dominant c⊥ texture a small quantity of prolonged shape a⊥-crystallites is visible on the surface. With the increasing proportion of a⊥-oriented crystallites their density on the surface increased until they almost totally cover the film surface. XRD study in Schulz geometry shoved that both a-axis and c-axis textured crystallites had a perfect

45° in-plane orientation. This is in agreement with the best match in the case of such oriented crystal lattices of YBCO film and substrate. We observed that a-axis textured crystallites were oriented in two mirror directions turned 45° and -45° in respect of the substrate lattice, but one orientation is clearly dominant. It was found that both (110) and (1-10) twin planes coexist in the $c\perp$ domains of films on $NdGaO_2$ (001), similarly to the case of films on $LaAlO_3$ substrates, however, one twinning direction is slightly prevailing. This can be explained by the orthorombocity of the substrate structure in which a- and b-axis lengths are slightly different. It is worth noting that a detailed XRD study in Schulz geometry revealed the twinning not only in $c\perp$ but also in $a\perp$-oriented domains of our YBCO films on $NdGaO_3$ [20]. On the other hand, contradictory data have been reported for PLD grown YBCO films on $NdGaO_3$, where unidirectional twinning [25,26] or even untwined structure [27] were found. Such opposite results can probably be explained by very different deposition condition, especially different oxygen partial pressure.

Vapour phase composition almost did not influenced T_c values of YBCO films on $NdGaO_3$. All films exhibited a high $T_c \sim 90$ K and a sharp superconducting transition (full $\Delta T_c = 0.3$-0.7 K). However, vapour phase composition strongly influenced the J_c of the films due to variable ratio of $a\perp$ and $c\perp$ orientations. J_c values in these films varied from $4 - 5$ MA/cm^2 (at 77 K) in films with low quantity of $a\perp$ crystallites to 0.2-0.5 MA/cm^2 in films in which these crystallites were dominating. In general, the increase of the quantity of $a\perp$ crystallites in films led to a significant decrease of J_c values as can be seen from Fig. 13.

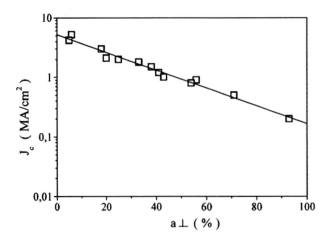

Fig. 13. J_c dependence on the quantity of a-axis textured crystallites in YBCO films on $NdGaO_3$ substrates.

YBCO on $YAlO_3$ substrates

The orthorhombic $YAlO_3$ (001) substrate has higher mismatch of its lattice with YBCO than in the case of $NdGaO_3$ or other perovskite substrates, so the deposition of epitaxial YBCO films on it is more difficult and need the careful choice of deposition conditions and vapour phase composition. The best properties of YBCO films on $YAlO_3$ (001) were obtained when deposited at 800 °C. Films of about 300 nm thick were purely or almost purely c-axis textured and had 45° in plane orientation, as in the case of NdGaO$_3$ substrates. For both YAlO$_3$ and NdGaO$_3$ substrates the 45° orientation is defined by the match of a/b axis of the substrate lattice with a diagonal in a-b plane of YBCO. Higher degree of the ortorombocity of YAlO$_3$ compared to NdGaO$_3$ determines the different twinning behaviour of YBCO on these substrates. The b-axis (0.533 nm) of YAlO$_3$ is noticeably longer than the a-axis (0.518 nm) and matches better the diagonal in the a-b plane of YBCO (d=(a^2$_{YBCO}$ + b^2$_{YBCO}$)$^{1/2}$ = 0,544 nm). So in this case we have a clearly preferable twinning direction corresponding to the axis b of the substrate (Fig. 14) what explains the clear unidirectional twinning found in our YBCO films on YAlO$_3$ (001) substrates. In these films the propagation of micro-cracks aligned normally to the main direction of twin boundaries was found by TEM (Fig. 15), and this had already been observed in YBCO films grown by PLD on NdGaO$_3$ [28]. The release of the stress that accumulates at the tetragonal-orthorombic transition usually is brought by bidirectional twinning. It seems that in the case of the unidirectional twinning the stress release can be taken over by the cracking aligned along the second possible twinning direction. However, it is not very clear whether the unidirectional twinning cause the cracking or the cracking quenches the alignment of the twin boundaries along a single direction. The second presumption might be supported by the fact that thinner films (75-150 nm) had no cracks and exhibited bidirectional twinning.

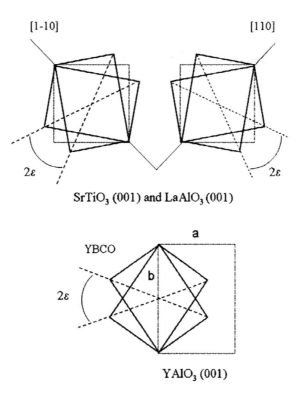

Fig. 14. Schematic representation of the YBCO twinning on cubic (SrTiO$_3$) or pseudo-cubic (LaAlO$_3$), and orthorhombic (YAlO$_3$) substrates. The orthoromocity of lattices is exaggerated for clearness. ε is YBCO twinning angle (~0.9 °).

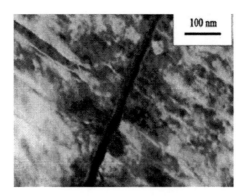

Fig. 15. TEM picture of a YBCO film on $YAlO_3$ showing a crack across twin boundaries.

The superconducting properties of YBCO samples on $YAlO_3$ substrates were in good agreement with their microstructure. T_c for all grown samples was ~90 K and ΔT_c varied in the range 0.3-0.5 K, what means that superconducting transition was not affected by twin boundaries and micro-cracks. However, J_c values for thicker samples (~300 nm and more) was low (~0.2-0.4 MA/cm^2 at 77 K) due to unidirectional twinning and micro-cracking, while thinner films (75-150 nm) exhibited clearly higher J_c (~1.1-1.5 MA/cm^2).

3.2.2   YBCO Films on Non-perovskite Substrates

The possibility to grow high quality YBCO layers on non-perovskite MgO (001), YSZ (001) (yttrium stabilized zirconium oxide) and sapphire (R-plane) substrates has also been studied [29-32]. MgO (a=0.421 nm) and YSZ (a=0.516 nm) are cubic, while sapphire (α-Al_2O_3) is hexagonal (a=0.476 nm, c=1.299 nm). Two planes of sapphire substrates, namely the planes (10-10) (plane M) and (1-102) (plane R) have been studied by various authors for YBCO growth. These two planes may be considered as having a pseudo-square structure with parameters 0.433 nm, 0.476 nm for M-plane and 0.476 nm, 0.512 nm for R-plane. The M-plane has a bigger mismatch with YBCO lattice compared to the R-plane, for this reason only sapphire substrates having the R-plane on the surface were used in our work.

All these non-perovskite substrates have noticeably bigger mismatch of their lattices with the YBCO lattice than perovskite ones, so the growth of epitaxial YBCO films exhibiting good superconducting properties is more complicated. Moreover, sapphire interacts with YBCO at high deposition temperature and this interaction has detrimental consequences for films properties. In general, it is difficult to obtain a pure single orientation of YBCO crystallites in films on these substrates, as a result films contains high angle grain boundaries decreasing J_c values. In the next sections we present the properties of YBCO films grown by PI-MOCVD on non-perovskite substrates and discuss about some promising ways to improve in-plane orientation and superconducting properties of YBCO films using vicinally polished substrates or buffer layers.

YBCO on MgO substrates

MgO (001) substrates has rather big lattice mismatch with the YBCO lattice (~9 %), so the growth of epitaxial YBCO films with high J_c is problematic. The MgO sensitivity to moisture and the poor reproducibility of the surface quality make things still worse. On the other hand, MgO is one of the cheapest substrate and has dielectric properties suitable for use in high frequency devices.

Depositing of YBCO layers on standard (commercial) MgO substrates by PI-MOCVD gave only mediocre results. Even the best films grown at 825 °C did not exhibit high critical parameters: T$_c$ ~ 86-88 K, ΔT$_c$ ~2-3 K, J$_c$ ~ 0.1 MA/cm^2 at 77 K. The disappointing J$_c$ values are due to the microstructure of the films which are strongly c-axis textured, however in the substrate plane YBCO crystallites are oriented in several different directions with 0° predominating. The presence of high angle inter-grain boundaries causes low critical current densities. These current densities are lower than the best values reported for YBCO films obtained by conventional MOCVD. We have no explanation on the poor performance of PI-MOCVD in the case of MgO substrates, but it is possible that the presence of the solvent vapour increases the concentration of CO, CO$_2$, H$_2$O in the reaction zone, which reacts with the MgO surface. On the other hand, the poor reproducibility of the conventional MOCVD for YBCO films leads, in ~ 90 % of the samples, to much lower performances. PI-MOCVD layers on MgO are at least reproducible.

The attempts to obtain pure in-plane orientation in YBCO layers on standard MgO substrates by varying deposition conditions (temperature, vapour phase composition), was unsuccessful. Furthermore, we have made attempts to improve microstructure and critical current density by introducing buffer layers between MgO substrate and YBCO film. We expected to reduce the effect of the big lattices mismatch between MgO and YBCO by depositing YSZ or CeO$_2$ buffer layers [33], because these materials have better lattices match with YBCO. However, this way also did not allow improving of the in-plane orientation and J$_c$ of YBCO films on MgO.

Many efforts have been made by other groups to improve microstructure and critical current density of YBCO films on MgO applying various methods for modification of substrate surface (mechanical polishing, chemical treatment, annealing, surface milling by low energy Ar-ions, vicinal polishing). We expected to improve YBCO properties by low angle vicinal polishing of MgO substrates. The substrate surface was polished with an inclination of 1.4 – 1.9° from the (001) plane along [100] direction. Using these substrates, the influence of the vapour phase composition on the composition, surface morphology, in-plane orientation and critical parameters of YBCO layers was investigated. Layers of the thickness of ~0.2-03 µm were deposited at 825°C. The solution composition was varied in a wide range as for perovskite substrates. The relationship between vapour phase (solution) and film composition revealed a trend consistent with that established for perovskite substrates (see Fig. 8).

XRD study revealed that all films on vicinally-polished MgO substrates had a pure c-axis texture while the in-plane orientation of YBCO crystallites was sensitive to the vapour phase composition (Fig. 16). All films on vicinally-polished MgO had a predominant 45° in-plane orientation, but a small number (5-30 %) of crystallites with other orientations (0°, ~15°, ~40°) were present in most films.

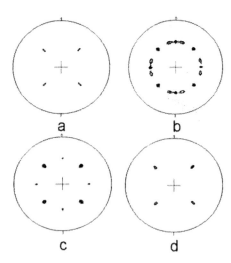

Fig. 16. Pole figures of MgO (111) (a) and YBCO (102) (b-d) planes for films on vicinally-polished MgO substrates. Y:Ba:Cu ratio in films determined by EDX was b) 1:1.45:2.15, c) 1:1.6:3.25, d) 1:1.85:2.85

Only films with average composition close to the stoichiometric 123 had pure 45° orientation, which is very different from that generally observed on MgO where the 45° orientation did not exist or appeared as a minor one. These films had rather smooth surface (Fig. 17) without a noticeable number of inclusions of the crystallites of secondary phases. Films with compositions very rich in yttrium had many inclusions of various Y-rich phases phases, which can be associated with Y_2BaCuO_5, $Y_2Cu_2O_5$, while films rich in Cu had Cu-rich grains on the surface.

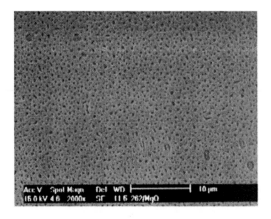

Fig. 17. SEM image of YBCO film on vicinally-polished MgO substrate. Film composition is close to 123.

Based on the results of this study, vicinal polishing of MgO substrates can be considered to be the most important factor for making the 45° in-plane orientation in YBCO films more preferential compared to the 0 ° orientation. However, it is likely that the pure 45° in-plane orientation on these substrates may only be obtained for films with average composition close to 1:2:3, i.e. exhibiting low precipitate density and absence of secondary phases. The presence of misoriented secondary phases in some other films may probably influence the growth and nucleation mechanism leading to the appearance of different orientations of YBCO crystals in the films.

The critical temperatures of the films on vicinal MgO were not noticeably dependent on the vapour phase composition (highest T$_c$ ~ 89 K, ΔT$_c$ ~ 1 K). However, the critical current densities varied largely depending on the in-plane orientation of the films. As expected, the highest J$_c$ values (>1 MA/cm^2 at 77 K) were found in the films with the pure 45° orientation. Significantly lower J$_c$ values for films with double or multiple in-plane orientation may be easy to understand in terms of high-angle grain boundaries.

YBCO on YSZ substrates

YBCO and YSZ have different crystal structures and a rather large mismatch (~6 %) of the lattices, so it is difficult to deposit epitaxial YBCO layers with high critical parameters. Moreover, chemical reaction between YSZ and YBCO with the formation of a BaZrO$_3$ interlayer can occur at high deposition temperature and this interlayer may also be responsible for the poor epitaxial quality and low critical parameters usually found for YBCO films on YSZ. The properties of films on YSZ depend strongly on the deposition technique used and on the deposition conditions. There are several reasons for interest in the study of YBCO depositions on YSZ: a) this substrate is cheaper than perovskite substrates, b) YSZ is widely studied as a potential buffer layer for YBCO deposition on sapphire, silicon, or metallic substrates, and these studies may be partly based on the data obtained from YBCO deposition on YSZ single crystal substrates.

YBCO films (0.3-0.4 μm) were grown by PI-MOCVD at various temperatures and in a wide range of vapour phase compositions. Layers with the best critical parameters were obtained at the solution composition Y:Ba:Cu = 1:2:2.5, so this composition was used for the study of the influence of deposition temperature. YBCO films grown at 850-875 °C were strongly c-axis textured, while films grown at 825 °C already contained a small amount of a-axis textured crystallites. The quantity of a⊥ crystallites increased with the decrease of the deposition temperature, and at 750 °C they become dominant.

The critical temperature of YBCO films clearly dependeds on the substrate temperature. At 775°C deposited films exhibited semiconductor-type resistivity vs. temperature dependence. The best films were obtained at 825°C (T$_c$ = 89 - 90 K, ΔT$_c$ ~ 1 K) and the properties are slightly worse at 850°C. Films deposited at 875°C had considerably lower T$_c$~82 K and those grown at 900°C do not exhibit transition to superconducting state over 77 K. The degradation of the critical temperature with the increase in growth temperature might be caused by the interaction between YSZ substrate and YBCO film. Evidence for Zr diffusion into YBCO layer was established by SIMS. The SIMS dept profiles reveal rather profound Zr diffusion into YBCO layer at 850-875 °C (Fig. 18). Additionally, XRD θ/2θ spectra for these films showed weak BaZrO$_3$ reflections.

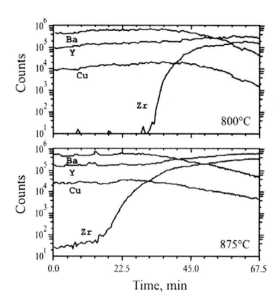

Fig. 18. SIMS concentration depth profiles for YBCO films grown at 800 °C and 875 °C on YSZ.

The composition of the vapour phase strongly influenced the in-plane orientation of YBCO films on YSZ (this study was made at 825 °C). YBCO c-axis textured crystallites in the films have several different in-plane orientations but the 45° orientation dominates in agreement with the best coincidence of two lattices. Additional 0°, 9° and 36° orientations was found in films and their relative quantity depended on the vapour phase composition. In the best layers, the 45° orientation dominated and a very small amount of 0° oriented crystallites was found. However, the full widths at half maximum (FWHM) of φ-scan peaks and rocking curves were rather large (4-5° and ~0.9° respectively), what shows that even the best films were not completely epitaxial. This may be caused not only by the rather large lattices mismatch, but also by the formation of non-oriented intermediate $BaZrO_3$ layer at high deposition temperature. The not high epitaxial quality of YBCO films leads to a rather low critical current density: even the best films have J_c values only ~0.4 MA/cm^2 at 77 K. These values are rather high compared with the layers obtained by MOCVD on YSZ, but lower than those grown by PVD.

YBCO on CeO$_2$ buffered YSZ substrates

We expected to improve the YBCO epitaxy on YSZ by modifying the YSZ surface with a thin layer of simple oxides, such as CeO_2. Although cubic CeO_2 (a = 0.541 nm) does not have a very good lattice match with YSZ (mismatch is ~5 %), the epitaxial "cube on cube" growth is possible. On the other hand, CeO_2 has a very good lattice match with YBCO, and epitaxial YBCO films are expected to be easily deposited on epitaxial CeO_2 layer.

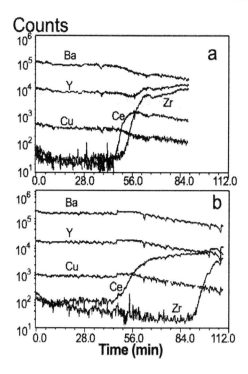

Fig. 19. SIMS concentration profiles for two YSZ//CeO$_2$/YBCO heterostructures. CeO$_2$ thickness: a) ~30 nm, b) ~160 nm.

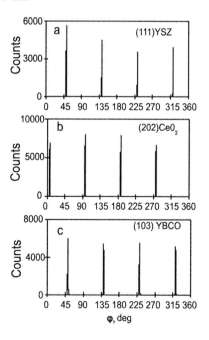

Fig. 20. Typical XRD φ-scans for YSZ//CeO$_2$/YBCO heterostructures: a) YSZ (111), b) CeO$_2$ (202), c) YBCO (103).

CeO$_2$/YBCO structures with a variable thickness of CeO$_2$ layer (15-160 nm) and a thickness of YBCO layer between 300 and 400 nm were deposited in situ, using two independently operating injectors. Ce(thd)$_4$ solution in monoglyme was used for the deposition of CeO$_2$ films. This precursor was synthesized in our laboratory. Thin CeO$_2$ layers were grown at 850 °C and YBCO layers at 825 °C. The first positive effect found in these heterostructures as compared to YBCO films grown directly on YSZ was the fact, that CeO$_2$ buffer layer stops Zr diffusion to YBCO. SIMS concentration profiles presented in Fig. 19 shows rather sharp interfaces in the heterostructures. It seems that even very thin CeO$_2$ layer is able to stop Zr diffusion.

Table 2. Properties of CeO$_2$/YBCO heterostructures on YSZ substrates.

CeO$_2$ thickness, nm	FWHM of CeO$_2$ (200) rocking curve, deg	FWHM of YBCO (005) rocking curve, deg	T$_c$, K	ΔT$_c$, K	J$_c$, MA/cm^2 at 77 K
15	0.34	0.30	90.5	0.8	3.6
30	0.51	0.46	90.6	0.3	4.8
80	0.44	0.47	91.0	0.4	4.0
110	0.50	0.34	90.4.	0.3	4.2
160	0.45	0.48	90.3	0.4	2.4

The main positive result was substantial improvement of YBCO epitaxy and critical current density. Single CeO$_2$ layers on YSZ as well as CeO$_2$ layers in heterostructures were strongly (001) textured and perfectly in-plane oriented according to the "cube on cube" growth mode (Fig. 20b). YBCO layers in heterostructures were also strongly c-axis textured and had a perfect 45° in-plane orientation in agreement with the best match between so oriented CeO$_2$ and YBCO lattices (Fig. 19c). In contrast to the layers grown directly on YSZ, in heterostructures an evident YBCO bidirectional twinning was found, and this confirm the excellent epitaxial quality of the heterostructures. The deposited YSZ/CeO$_2$/YBCO heterostructures had high critical parameters (Table 2) in accord to their high epitaxial quality. The J$_c$ values were improved by an order and became comparable to those obtained on perovskite substrates. It seems that the most suitable thickness of CeO$_2$ buffer layer is between 30 nm and 100 nm (Table 2).

YBCO on buffered sapphire substrates

Sapphire is a substrate widely used in the semiconductors technology. Monocrystals of very large dimensions and of very high purity are commercially available. Suitable dielectric properties of sapphire make this substrate very attractive for YBCO deposition for microwave applications. We used R-plane sapphire substrates having pseudo-square structure on the surface (parameters 0.433 and 0.476 nm). The mismatch between this cell and YBCO cell is quite important (~6 %). In addition, taking into account the important Al diffusion into the YBCO film at high deposition temperature, the direct deposition onto this substrate by MOCVD is prospectless. In general, only the using of convenient buffer layers offers the possibility to grow superconducting YBCO layers on sapphire by MOCVD. In this section we present the promising results on the in-situ growth of YSZ or CeO$_2$ buffer layers and

YBCO films on sapphire substrates by PI-MOCVD. Solutions of Zr(thd)$_4$, Y(thd)3, and Ce(thd)$_4$ in monoglyme were used for the deposition of buffer layers.

YSZ layers (~100 nm) were deposited at different temperatures in the range of 600-800 °C. All YSZ films contained a mixture of (111) and (001) textured domains, but in most cases the (001) texture was more or less predominant. Moreover, (001) textured crystallites were in-plane oriented. YBCO layer grown at 825 °C over YSZ buffer layer was c-axis textured and had a dominating 45 ° in-plane orientation with respect to YSZ (001) textured domains, but some YBCO crystallites with other orientations were also found. Not completely epitaxial growth of YSZ/YBCO heterostructures on sapphire caused rather low critical parameters of YBCO layers. Even for the best structures the superconducting transitions (occurring at ~90 K) were quite large (~2 K), and critical current densities did not exceed 0.1 MA/cm^2 at 77 K.

Table 3. The properties of the best films grown by PI-MOCVD on various monocrystalline substrates

Substrate	YBCO texture	FWHM rocking (005)YBCO	YBCO in-plane orient.	YBCO twinning[a]	T$_c$, K	ΔT[b]	J$_c$, MA/cm^2 At 77 K
LaAlO$_3$ (001)	c⊥ pure	~0.3	0°	Strong 2d	~92	~0.2	5 − 6
SrTiO$_3$ (001)	c⊥ pure	~0.5	0°	Strong 2d	~90	~0.5	2 − 3
NdGaO$_3$ (001)	c⊥ >>a⊥	~0.3	45°	Strong 2d	~90	~0.3	4 - 5
YAlO$_3$ (001)	c⊥ pure	~0.4	45°	Weak 2d (thin film) / Strong 1d (thick film)	~90 / ~90	~0.3 / ~0.4	1 − 1.5 / 0.2 - 0.4 (cracks)
MgO (001) standard	c⊥ pure	~1	0°>45°	Not found	~88	2-3	~0.1
MgO (001) vicinal	c⊥ pure	~0.5	45°	Weak 2d	~89	~1	1-2
YSZ (001)	c⊥ pure	~0.9	45°>>0°	Not found	~90	~0.6	~0.4
YSZ (001)/ CeO$_2$ (001)	c⊥ pure	~0.4	45°	Strong 2d	~90	~0.3	~4
Sapphire-R	-	-	-	-	~86	~5	~0.0002
Sapphire-R/ YSZ	c⊥ pure	~1	45°>>0°	Not found	~90	~2	~0.1
Sapphire-R/ CeO$_2$ (001)	c⊥ pure	~0.5	45°	Weak 2d	~90	~0.8	~1

[a] 1d – unidirectional twinning, 2d – bidirectional twinning, [b] full transition width

Much better results were obtained when CeO$_2$ was used as buffer layer on sapphire. CeO$_2$ (15, 30, 100 and 200 nm) and YBCO (0.4 and 0.6-0.7 μm) films were deposited at 850 °C and 825 °C respectively. Thin CeO$_2$ layers (15 and 30 nm) on sapphire were grown epitaxially, while thicker ones (100 and 200 nm) contained a very small amount of (110) textured crystallites in the matrix of (001) texture. Thinner (~0.4 μm) YBCO films on a thin CeO$_2$ layer were also grown epitaxially (c-axis texture and 45 ° in-plane orientation). Thicker films had some a-axis textured crystallites, moreover some signs of micro-cracking were observed in these films. The best superconducting properties exhibited thinner YBCO films grown on sapphire substrates buffered by ~30 nm-thick CeO_2 layer. Their T_c was ~90 K, ΔT_c ~0.8 K and J_c ~1 MA/cm^2 at 77 K. Despite the nearly equal epitaxial quality of heterostructures with 30 nm and 15 nm buffer layers, the later exhibited lower J_c (~0.3 MA/cm^2). Probably, a very thin CeO_2 layer was not sufficient to stop completely the Al diffusion into YBCO.The best results obtained in YBCO deposition by PI-MOCVD on various monocrystalline substrates are summarised in Table 3. From this table one can conclude that the most promising results concerning the critical current density were obtained on $LaAlO_3$ (001) substrates.

3.2.3   Large Area Deposition of YBCO Films

Electrical power engineering and microwave device applications require large area homogeneous $YBa_2Cu_3O_{7-x}$ (YBCO) films with high critical parameters and low surface resistance. MOCVD is considered as a technique suitable for large area deposition of various materials. However, homogeneous large area deposition of high quality YBCO films is rather problematic, and only few MOCVD groups working in industrial companies (Hewlett Packard Co [11], Emcore Corp [34], Conducts Inc [35], Aixtron GmbH [36]) reported about such deposition.

Here we demonstrate the ability of PI-MOCVD technique to deposit reproducibly high quality YBCO films on large area substrates [37]. A modified pulsed injection MOCVD reactor assigned for large area deposition was built. For the homogenization of the vapour flow, a special shower-type diffuser was installed at the inlet of the reactor to spread the gas flow symmetrically towards the substrate. A stainless steel substrate holder (8 cm in diameter) was centered in a quartz reactor chamber of 10 cm internal diameter and heated by an external electrical furnace.

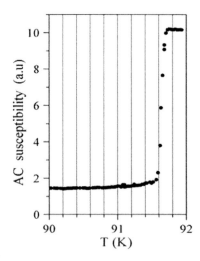

Fig. 21. Sharp superconducting transition of an YBCO film on $LaAlO_3$ substrate. Note that full ΔT_c is only ~ 0,1 K.

The optimisation of various deposition parameters influencing the film uniformity was performed by depositing ZrO_2 (Y) and Al_2O_3 films (~0.5 μm) on large area (3 inches in diameter) Si (100) substrates. Under optimal conditions oxide layers of uniform colour on the surface of ~70 mm in diameter were obtained. Only edges of the films had different colour (edge effect). Ellipsometry measurements showed that in the area of uniform colour the difference between the maximal and minimal values of the thickness and refractive index did not exceed 10-12 % of the mean value. In the second step, the reactor was optimised for the deposition of high quality YBCO films on small LaAlO$_3$ (001) substrates (~ 1 cm^2). The optimal conditions were found to be similar to those presented in Table 1, but precursors feeding rate was increased due to a larger volume of the reaction chamber. YBCO films (~0.2-0.3 µm) grown under optimised conditions exhibited high superconducting properties. Standard values of T$_c$, ΔT$_c$ and J$_c$ were ~91.5-92 K, 0.2 K and 5-6 MA/cm^2 (at 77 K), respectively. Some films showed very sharp superconducting transition equal to ~0.1 K (Fig. 21). In the next step YBCO films were grown on 2-inch and 3-inch LaAlO$_3$ (001) substrates. Two examples of the distribution of J$_c$ values in such films, measured by SIEMENS and THEVA are presented in Fig. 22. One can see that high J$_c$ values 4-6 MA/cm^2 (77 K) were found over large area and only the edges had clearly lover J$_c$.

The surface resistance (R$_s$) was measured for an YBCO film on 3-inch LaAlO$_3$ substrate. The maximum field levels (B$_s$) were excited on a circle (about 8 mm in diameter) around the center of the film. The film was measured against another YBCO film of known quality. The absolute level of the low-field R$_s$ was rather good. Low-field R$_s$ (not corrected for the thickness effect) was found to be 0.35 mΩ (8.5 GHz and 77 K) and R$_s$-B$_s$ dependence remained flat up to 4 mT.

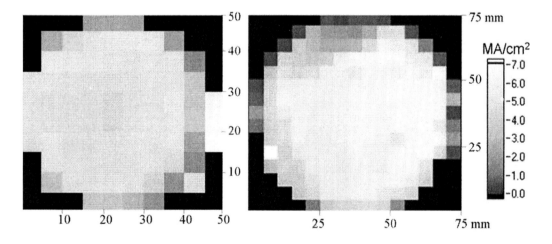

Fig. 22. Distribution of the critical current density in ~0,2 µm thick YBCO films on 2-inch (left) and 3-inch (right) LaAlO$_3$ substrates. Left - measured by SIEMENS, right – by THEVA.

3.2.4   SmBCO Films

SmBa$_2$Cu$_3$O$_{7-x}$ (SmBCO) seems to be an attractive material for coated conductors applications because it exhibits higher transition temperature (T$_c$) than YBCO and has higher critical current density (J$_c$) in strong magnetic fields. Due to the comparable ionic radius size of the Sm and Ba, the substitution of Sm^{3+} into the Ba^{2+} site takes place, changing the superconducting properties. Some authors underline that films of Sm-rich solid solution, Sm$_{1+x}$Ba$_{2-x}$Cu$_3$O$_{7-\delta}$, exhibits even better superconducting properties and has smoother surface as compared to stoichiometric composition. On the other hand, the ratio of substitution is difficult to control and triggers irreproducibility in the film properties.

Thin SmBCO films on various single crystal substrates are commonly prepared by physical vapor deposition (PVD) techniques such as PLD, dc sputtering, co-evaporation. There are only few reports concerning attempts to grow SmBCO thin films by MOCVD on single crystals substrates [37,38].

We studied the possibility to grow high quality thick (>2 μm) SmBCO films by PI-MOCVD, on LaAlO$_3$ (001) substrates [39]. The solution of precursors used for SmBCO film deposition consisted of a mixture of Sm, Ba and Cu 2,2,6,6-tetramethyl-3,5-heptanedionates (thd) dissolved in monoglyme with appropriate molar ratios. The precursor materials were synthesized in our laboratory. After the deposition, the reactor was filled with oxygen to atmospheric pressure and the films were slowly cool down.

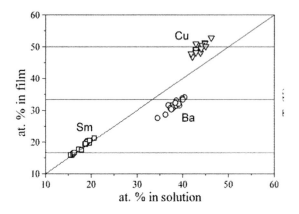

Fig. 23. Relationship between the percentage of elements in deposited SmBCO films and in initial solution. Vertical and horizontal lines correspond to stoichiometric quantity of elements.

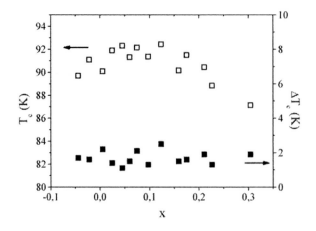

Fig. 24. Influence of Sm/Ba substitution ratio x in Sm$_{1+x}$Ba$_{2-x}$Cu$_3$O$_{7-\delta}$ films to their superconducting properties (T$_c$ and ΔT$_c$). Substrate – LaAlO$_3$, film thickness is ~0,5 μm.

The composition of the MOCVD films always differs from the initial solution composition. This effect is especially important in the case of solid solution of Sm$_{1+x}$Ba$_{2-}$

$_x$Cu$_3$O$_{7-\delta}$, which properties depend on x-value, and consequently on film composition. Firstly, we deposited at 825°C a series of thin (0.5 μm) SmBCO films varying the composition of solutions, and established the correlation curve between the composition in solution and in films (Fig. 23). We can conclude that Sm content in the films is slightly higher than in solution, while Ba and Cu percentage in films are quite different from those in solutions (Ba-lower, Cu-higher). Most of the films deposited in this series contained an excess of Sm and a lack of Ba (x>0). In Fig. 24 we present critical temperature T$_c$ (onset) and ΔT$_c$ values of these films in relation with Sm-Ba substitution ratio (x). Figure shows that T$_c$ values are high (>90 K) in a large range of x-values up to ~0.2, what is larger than reported in [37] for SmBCO films grown by MOCVD. Transition width for our films varied from ~1 K to ~2 K. The reproducibility of T$_c$ and ΔT$_c$ values for SmBCO is clearly lower than for YBCO films deposited in the same PI-MOCVD process. This fact suggests that superconducting properties of SmBCO films are defined not only by Sm-Ba substitution ratio but also by other factors occurring in the vapour phase which are difficult to control. Nevertheless, the reproducibility of the superconducting properties of our films is quite better than those presented in [37] what might be attributed to the advantages of PI-MOCVD technique.

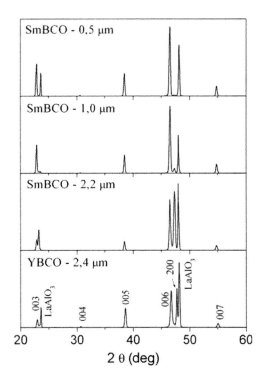

Fig. 25. XRD for SmBCO films of various thicknesses grown on LaAlO$_3$ substrates. For comparison, a thick YBCO film is presented as well.

In the second step, SmBCO films of various thicknesses (from 1 to 2.2 μm) were deposited at 825°C on LaAlO$_3$ substrates. Sm/Ba substitution ratio (x) in these thick films was about 0.05. For comparison, a 2.4 μm-thick YBCO film was deposited as well. The main properties of deposited films are summarized in Table 4. XRD characterizations of SmBCO and YBCO films on LaAlO₃ substrates are reported in Fig. 25. We can notice that almost purely c-axis textured (c⊥) SmBCO phase is present in thin (0,5 µm) film, while in thick films the percentage of a-axis textured (a⊥) crystallites becomes prevailing (~80 % in SmBCO and ~70 % in YBCO films). XRD study in Schultz geometry (φ-scans) showed that both c⊥ and a⊥ crystallites are perfectly in-plane oriented. Epitaxial relations between c⊥ crystallites and pseudo-cubic LaAlO₃ substrate can be presented as SmBCO $<100>\|$ LaAlO₃$<100>$. In the case of a⊥ crystallites, four peaks are obtained (instead of two) in the φ-scans of (102) reflection having twofold symmetry. This corresponds to two different in-plane orientations of a-axis textured crystallites (SmBCO $<001>\|$ LaAlO₃$<100>$ and SmBCO $<001>\|$ LaAlO₃$<010>$). The same epitaxial relations were found in thick YBCO films on LaAlO₃ substrates.

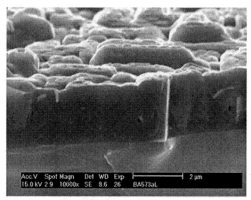 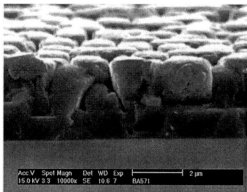

SmBCO, ~2,2 µm   YBCO, ~2,4 µm

Fig. 26. SEM cross-section views for thick SmBCO (left) and YBCO (right) films.

Table 4. Properties of SmBCO films

Layer	Thickness, µm	Rocking curve (005), FWHM, deg	a⊥ crystallites, %	T_c, K	ΔT_c, K	J_c, MA/cm² at 77 K
SmBCO	0,5	0,39	8	92,4	1,6	3,5
SmBCO	1,0	0,38	26	91,7	1,5	2,1
SmBCO	2,2	0,53	79	92,6	1,2	0,8
YBCO	2,4	0,52	71	90,4	0,2	1,4

No clear dependence of T_c and ΔT_c on film thickness was found for SmBCO films (T_c varied around 92 K and ΔT_c was about 1.5 K). The 2.4 µm-thick YBCO film had very narrow superconducting transition (ΔT_c=0.2 K) comparable to thin YBCO films. J_c values of thin SmBCO (0.5 µm) films on LaAlO₃ substrates were about 3.5 MA/cm² at 77 K but follow a drastic decay when increasing the thickness (Table 4). The reason of such a J_c-thickness dependence might be the increase of the amount of a-axis textured crystallites. Although we did not observe by SEM any evidence of cracks in thick SmBCO films, the presence of micro-cracks is not excluded. Thick YBCO films (~2.4 µm) on LaAlO₃ substrates exhibit higher J_c value than thick SmBCO films despite similar a-axis crystallites ratio. SEM surface and cross section views for thick SmBCO and YBCO films are shown in Fig. 26. We can see two different behaviors for these materials during thick film growth. It seems that a-axis textured crystallites of SmBCO start to grow from the substrate surface while YBCO grows predominantly in c-axis texture up to ~ 1 μm before changing to a-axis growth. This phenomenon could explain the higher J_c values of thick YBCO films as compared to thick SmBCO films.

3.3   Multilayered Structures Containing HTSC

3.3.1   Introduction

Over the last few years much attention has been focused on deposition and investigation of various multilayered structures combining perovskite oxides exhibiting different specific properties. This attention has been justified by possible applications in novel hybrid electronic devices and sensors. High temperature superconductors have also considerable potential for novel electronic applications. They can be associated with other multicomponent oxides with perovskite and perovskite like crystallographic structure, which have electrical properties varying in a wide range including insulating, metallic ($LaNiO_3$, $SrRuO_3$, $La_{1-x}Sr_xCoO_3$,...), ferroelectric ($BaTiO_3$, $SrTiO_3$, $Ba_{1-x}Sr_xTiO_3$, ...), ferromagnetic and colossal magnetoresistive ($La_{1-x}M_xMnO_3$,...) behaviours. Number of important device applications are related to the above mentioned properties combined with high temperature superconductivity: Josephson effect devices (HTSC and insulating oxides), interconnection and passive low resistance components, microstrip lines for high speed electronics (insulating, metallic and HTSC), microwave components, tunable filters (HTSC and ferroelectrics), spin-polarised quasiparticle injection devices (ferromagnetic, dielectric and HTSC). Moreover, the oxide superlattice materials with an artificial control of the crystal structure are actually studied in order to find new and peculiar functions from ceramic systems, leading to many potential applications. Superconducting superlattices, formed by artificial stacking of superconducting and insulating components of various thicknesses can contribute to the understanding of high-T_c superconductivity and of the basic coupling between anisotropy and physical properties.

The present technology of metal oxide heterostructures relies mainly on physical PVD techniques, which in most cases are very difficult to implement for industrial purposes. The use of cost-effective deposition techniques such as MOCVD would contribute to an industrial-scale development of related technologies, as was already proven for semiconducting-related technologies. On the other hand, the difficulty in obtaining multilayered oxide structures by classical MOCVD is the major obstacle for the use of this technique in industry. Classical MOCVD of multicomponent metal-oxide films is based on the evaporation of solid MO precursors and meets the problem of irreproducibility of multichannel vapour generation and transport, thus making the growth process and consequently the film properties less reproducible. This drawback is still more unfavourable for devices based on oxide multilayers, whose physical properties have proven to be much dependent not only on the quality of single layers, but also on the interface characteristics. Due to these reasons new single liquid source MOCVD based on flash evaporation principle is more suitable for chemical vapour deposition of metal-oxide multilayers.

In present section we demonstrate the ability of our PI-MOCVD technique to deposit high quality various metal oxide multilayers and superlattices containing HTSC materials.

3.3.2   General Investigations on Perovskite Stacking

Various series of multilayers containing $SrTiO_3$ (STO), $La_{1-x}Sr_xMnO_3$ (LSMO), $Ba_{1-x}Sr_xTiO_3$ (BSTO), $SrRuO_3$ (SRO) and YBCO have been in-situ or ex-situ deposited by PI-MOCVD on $LaAlO_3$ (001) and $NdGaO_3$ (001) substrates: STO/YBCO, BSTO/YBCO, LSMO/YBCO, STO/LSMO/YBCO, LSMO/STO/YBCO, STO/LSMO/STO/YBCO, SRO/YBCO [40-43]. Mainly conventional $M(thd)_n$ precursors synthesized in our laboratory have been used for depositions except the mixed thd-alkoxide precursors for Ti ($Ti(OR)_2(thd)_2$). The principal deposition conditions of the layers in heterostructures are presented in Table 5. The number of injectors used by deposition corresponds to the number of different materials to be incorporated into the multilayer. Linear dependence of the thickness on the number of injections for each material allows the construction of multilayered heterostructures with various thicknesses of components. A description of some representative deposited heterostructures as well as the properties of the incorporated YBCO layer can be found in Table 6.

Table 5. Principal deposition conditions of various materials in perovskite multilayers

Parameter	$La_{1-x}Sr_xMnO_3$	$SrTiO_3$, $Ba_{1-x}Sr_xTiO_3$	$SrRuO_3$	$YBa_2Cu_3O_{7-x}$
T deposition (°C)	750-825	825-850	825	825
T evaporator (°C)	290	290	280	290
Total flow rate ($Ar+O_2$) (l/h)	108	108	108	108
Total pressure (Torr)	5	5	5	5
O_2 partial pressure (Torr)	1.6	1.6	2.1	1.6-2.1
Solvent	Monoglyme	Monoglyme	Monoglyme	Monoglyme
Precursors	$M(thd)_n$	$M(thd)_2$, $Ti(O^iPr)_2(thd)_2$	$Sr(thd)_2$ $Ru(acac)_3$	$M(thd)_n$
Injection frequency (Hz)	2	2	2	1-2
Growth rate (μm/h)	~0.6	~0.5	~0.4	~1-2

Ferroelectric materials like STO and BSTO are structurally compatible with HTSC regarding low surface resistance and losses at microwave frequencies. Integration of ferroelectric thin films with high temperature superconductor materials can be implemented with bilayer structures for coplanar wave-guides and trilayer structures for microstrip wave-guides (Fig. 27). $Ba_xSr_{1-x}TiO_3$ is well suited for the development of ferroelectric based microwave electronics. The Curie temperature of bulk BST ranges from 30 to 400K for x between 0 and 1, respectively. The ability to control the dielectric properties in a simple way will allow optimising device structures for maximum tunability and minimum loss at the desired frequency.

The composition of precursors' solution was adjusted for the growth of pure phase and high epitaxial quality STO and BSTO layers on $LaAlO_3$ substrates. An AFM study revealed quite a flat surface with a surface roughness of ~1-2 nm in the case of ~100 nm thick films. Finally, YBCO films were deposited in-situ or ex-situ on $SrTiO_3$ or $Ba_xSr_{1-x}TiO_3$ (x=0.1) layers of various thickness at 825 °C or 850 °C. X-ray diffraction study for heterostructures showed a perfect alignment between the different layers. Bidirectional twinning of YBCO crystallites observed in the heterostructures confirmed they perfect epitaxial quality. Superconducting properties of these structures (Table 6) were almost as good as for YBCO layers grown directly on LaAlO$_3$ substrates.

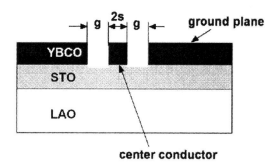

Fig. 27. Scheme of a wave-guide formed from a bilayer SrTiO$_3$/YBCO.

Table 6. Properties of YBCO layers in heterostructures with other perovskite materials

Substrate (001)	Heterostructure (thickness of layers, nm)	Texture of YBCO layer	T$_{co}$,[*] K	ΔT$_c$ (full), K
LaAlO$_3$	STO(~5)/YBCO(~300)	c⊥	~92	~0.3
LaAlO$_3$	STO(~15)/YBCO(~300)	c⊥	~92	~0.2
LaAlO$_3$	STO(~25)/YBCO(~300)	c⊥	~91.5	~0.3
LaAlO$_3$	STO(~35)/YBCO(~300)	c⊥	~91.5	~0.3
LaAlO$_3$	STO(~55)/YBCO(~300)	c⊥	~92	~0.5
LaAlO$_3$	STO(~100)/YBCO(~300)	c⊥	~92	~0.5
LaAlO$_3$	Ba$_{0.1}$Sr$_{0.9}$TiO$_3$ (~100)/YBCO(~300)	c⊥	~91.5	~0.7
NdGaO$_3$	STO(~5)/YBCO(~300)	c⊥	~92	~0.2
NdGaO$_3$	STO(~35)/YBCO(~300)	c⊥>>a⊥	~92	~0.3
LaAlO$_3$	LSMO(~55)/YBCO(~300)	c⊥	~91.5	~0.3
LaAlO$_3$	LSMO(~120)/YBCO(~300)	c⊥	~91.5	~0.2
LaAlO$_3$	STO(~15)/LSMO(~120)/YBCO(~300)	c⊥	~91.5	~0.2
LaAlO$_3$	LSMO(~140)/STO(~5)/YBCO(~300)	c⊥	~92	~0.2
LaAlO$_3$	LSMO(~110)/STO(~15)/YBCO(~300)	c⊥	~91.5	~0.3
LaAlO$_3$	LSMO(~140)/STO(~25)/YBCO(~300)	c⊥	~91.5	~0.3
LaAlO$_3$	STO(~15)LSMO(~120)/STO(~5)/YBCO(~300)	c⊥	~91	~0.4

[*] Critical current density of the heterostructures on LaAlO$_3$ waried in the range of 2-4 MA/cm^2 at 77 K

In the field of magnetoelectronic applications, in recent years, a series of spin electronic devices made from oxide multilayers have been proposed or realized as a challenge to conventional electronics. Device concepts currently investigated are spin valves [44-46] and spin polarized quasi-particle injection device (SPQID) [47-52], including ferromagnetic (FM) layers of mixed valence manganites or other FM materials. SPQID are based on a ferromagnetic layer and superconducting layer electrically isolated by a thin dielectric layer (FM/I/HTSC). PI-MOCVD technique was optimized for the growth of LSMO layers and multilayers with YBCO, as a result high epitaxial quality heterostructures exhibiting good superconducting properties were grown on $LaAlO_3$ substrates (Table 6). The formation and characterization of spin injection devices from such heterostructures are described in a further section.

Ruthenates, especially strontium ones (SRO), during recent years are of great interest because of their unique electrical and magnetic properties. $SrRuO_3$ possesses high chemical stability and metallic conductivity. These properties combined with the possibility of epitaxial growth on various perovskite substrates (determined by similar crystal structure) make it attractive as electrode in possible multilayer device applications, as well as buffer layer for the fabrication of thin films of high-T_c superconductors. Several works reported on the deposition of $SrRuO_3$ films by classical Metal-Organic Chemical Vapour Deposition (MOCVD) [53,54]. In these works a separate evaporation of solid Sr and Ru MO precursors were used to generate a vapor phase. Only physical vapor deposition techniques are usually used for the deposition of $YBa_2Cu_3O_{7-x}/SrRuO_3$ heterostructures, moreover, the reduction of the critical temperature of superconductor in presence of $SrRuO_3$ layer was observed [55].

SRO films (~0.1 μm) were grown by PI-MOCVD on perovskite substrates $LaAlO_3$ (100), $SrTiO_3$ (100) and $NdGaO_3$ (100) [41]. Deposition temperature was higher (825 °C) than in classical MOCVD of ruthenates using solid MO precursors (500-750 °C). [53,54]. Such a high temperature was chosen because it was the optimum temperature for YBCO deposition. Similarity of conditions for SRO and YBCO deposition (Table 5) allows depositing of SRO/YBCO heterostructures in the same deposition run (*in-situ*) by a simple off/on operation of two injectors connected to the reactor. Films on perovskite substrates contained pure (100) textured and perfectly in-plane oriented cubic $SrRuO_3$ phase, when deposited from solutions with Ru/(Ru+Sr) ratio in the range 0.23-0.33. The film resistivity (at RT) versus vapor composition curves showed a minimum of resistivity around Ru/(Ru+Sr)=0.27, so this ratio was considered as optimum. In the case of Ru/(Ru+Sr)=0.17 the presense of additional Sr_2RuO_4 and other phases was found. All SRO films on perovskite substrates exhibited metallic behavior. The resistivity vs temperature curves have a clear break at about 160 K, which corresponds to the Curie temperature of paramagnetic to ferromagnetic transition typical for this oxide. The $SrRuO_3$ films had low resistivity values at room temperature (about 0.2 mΩ cm), comparable to the best values reported in literature.

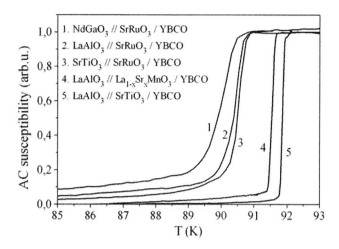

Fig. 28. Superconducting transitions of SRO/YBCO heterostructures on various perovskite substrates. Compared with similar STO/YBCO and LSMO/YBCO structures.

YBCO/SrRuO$_3$ heterostructures (~0.3 μm / ~0.1 μm) were grown *ex-situ* or *in-situ*. No great difference was found comparing the properties of *ex-situ* and *in-situ* deposited heterostructures. YBCO layers in heterostructures on perovskite substrates were c-axis textured and had in-plane orientation corresponding to the "cube on cube" growth mode (with respect to cubic or pseudocubic substrate lattices). However, small amount of a-axis textured YBCO crystallites were also observed in XRD spectra of these films. The best superconducting properties had YBCO/SrRuO$_3$ heterostructures on SrTiO$_3$ and LaAlO$_3$ substrates – T$_c$ close to 91 K, ΔT$_c$ < 1 K (Fig. 28) and J$_c$ ~1.5 MA/cm^2 at 77 K. These results are among the best results reported for such heterostructures. However, the superconducting properties of YBCO/SRO structures were slightly lower as compared to similar YBCO/STO or YBCO/LSMO structures grown by PI-MOCVD (Fig. 28).

3.3.3   Thick SmBCO/YBCO Heterostructures

To obtain high current transport I$_c$ in superconducting films, one way is to fabricate thick films. However, the critical current density J$_c$ of YBCO was shown to decrease with increasing film thickness. The reason of the J$_c$ drop in thick films might be due to cracks, a-axis growth or formation of a non-superconducting top layer, when the thickness of YBCO films exceeds 1 to 2 μm. A possible way to solve this problem could be the use of other REBCO phases (RE= Sm, Nd,...) [56] or to grow multilayers of thick structures, in which SmBCO would play the role of interlayer between YBCO films [57]. In this configuration SmBCO was shown to act as a remedy of failure in the c axis growth.

YBCO/SmBCO (1.3 μm/1.1 μm) and YBCO/SmBCO/YBCO (0.9 μ/0.5 μm/0.8 μm) heterostructures were deposited in-situ at 825°C on LaAlO$_3$ (001) substrates using two separately functioning injectors [39]. Both heterostructures on LaAlO$_3$ were grown epitaxially ("cube on cube orientation", FWHM$_{Roc(005)}$ =0.38 ° and 0.67 °, respectively) and had less a-axis textured crystallites (12% and 40% for double and triple structure) as compared with thick single SmBCO (79 %) or YBCO (71 %) films. Superconducting transitions of heterostructures in comparison with thick single films on LaAlO$_3$ substrates are presented in Fig. 29. One can see that the transition temperatures of the heterostructures are intermediate between those of thick YBCO and SmBCO films. It is surprising that despite the perfect epitaxial quality and the lower quantity of a-axis textured crystallites, heterostructures exhibited slightly lover J_c (0.6-0.7 MA/cm^2) than single thick layers (Table 4). It can be attributed to morphological imperfections in heterostructures which can be visible in SEM cross- section views in Fig. 30.

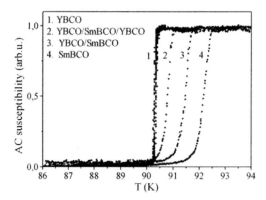

Fig. 29. Superconducting transitions of thick SmBCO and YBCO layers and their thick heterostructures.

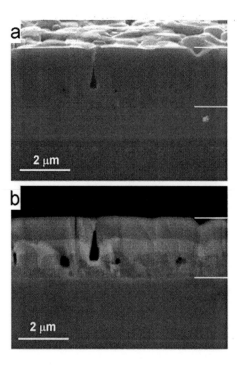

Fig. 30. SEM cross-sections views of YBCO/SmBCO/YBCO heterostructure grown on LaAlO$_3$ substrates. Images were obtained using SE (a) or BSE (b) detectors.

3.3.4   Spin-Injection Devices

Combination of HTS and colossal magnetoresistance manganites such as La$_{1-x}$Sr$_x$MnO$_3$ in the heterostructures provides great interest both for basic research of nonequilibrium superconductivity and novel applications. CMR manganites exhibiting almost 100% spin-polarized carriers in the ferromagnetic (FM) state offer promising possibility to inject spin-polarized quasiparticles into a superconductor in order to suppress superconductivity by braking Cooper pairs. Significant pair braking effect has been demonstrated by injecting spin-polarized electrons into the YBCO superconductor through a thin (d = 1-5 nm) intermediate insulating (I) layer inhibiting proximity effect between HTS and FM [47-52].

A key issue for spin-polarized quasiparticle injection devices (SPQID) based on HTS/I/FM heterostructures is perfect crystalline structure of all constituent layers ensuring certain electrical and magnetic properties. In addition, high quality (flat and sharp) interfaces are highly appreciated. The HTS/I/FM heterostructures reported up to now have been prepared by PVD techniques, mainly by pulsed laser deposition. Here we demonstrate similar SPQID based on high quality YBCO/I/FM heterostructures prepared through fully MOCVD way using our PI-MOCVD technique [58-60]. This work was carried out in a close collaboration between Vilnius University and Vilnius Semiconductors Physics Institute.

The possibility to grow high crystalline quality single phase LSMO, STO and YBCO films and their heterostructures have been demonstrated in previous sections. Here we present the formation and characterization of spin injection devices from LSMO/STO/YBCO trilayers. The thicknesses of layers in heterostructures were 150-200 nm, 5-10 nm and 200-250 nm, respectively. The La$_{1-x}$Sr$_x$MnO$_3$ (x ~ 0.3) ground layer was grown on the whole surface of LaAlO$_3$ (001) substrate. In following, it was masked partly by thin MgO diaphragms to grow the SrTiO$_3$ and YBCO overlayers on an unmasked central part of the manganite film. The grown heterostructures were cooled down slowly to a room temperature in an oxygen atmosphere. Epitaxial quality with "cube on cube" orientation has been found for all the constituent layers. SEM investigations revealed smooth surface of the LSMO and STO layers. Cu-rich precipitates were observed on a surface of the YBCO top layers, however their presence on a top layer is not detrimental for device properties. Thin Ag coatings (d ~ 0.3 μm) used as electrodes for electrical measurements were magnetron sputtered at T~300 K. Finally, lithographic process and wet etching were applied to form a tape-like YBCO layers for investigation of longitudinal and transverse electrical transport in the heterostructures.

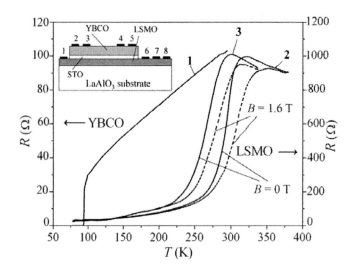

Fig. 31. Resistance vs. temperature of the YBCO top layer (1), LSMO bottom layer (2) and barrier resistance of the YBCO/STO/LSMO heterostructure multiplied by a factor of 1.2×10^4 (3).

Electrical resistance measurements of individual YBCO and LSMO layers were carried out in a conventional four-probe configuration using AC current $(I_\sim)$ of about 1 μA. The critical temperature (T_c) and critical current density (J_c) of the YBCO layers were evaluated from electrical transport measurements. Resistance $(R_{YBCO}=U_{12}/I_{34})$ versus temperature dependence of the patterned tape-like YBCO layer of the dimensions 5 mm × 0.5 mm × 0.25 μm showed zero resistance state at $T \cong 92$ K $(\Delta T_c < 0.5$ K) (see inset and curve 1 in Fig. 31) in a good accordance with the susceptibility data for similar structures. Temperature dependent resistance of the LSMO layers (curve 2 in Fig. 31) showed ferromagnetic transition at T~320 K. Resistivity of the the LSMO layers was found to be ~1000 μΩcm and about 20 μΩcm, at 300 and 100 K respectively. Magnetoresistance of LSMO layers reached value of ~40 % at ~300 K in the field of 1.6 T. The transverse electrical transport in the heterostructure is characterized by the effective barrier resistance R_{bar} defined by the relationship $R_{bar} \approx U_{36}/I_{12}$ where U_{36} is the voltage drop between points 3 and 6 if passing electrical current through 1 and 2 (see curve 3 in Fig. 31). The area resistivity of the STO barrier layer (d = 5 nm) of about 5×10^{-5} Ωcm^2 at 100 K has been estimated. Following the measured $R_{bar}(T)$ dependence we point out lower peak temperature of the barrier resistance compared to that of the LSMO layer. Taking into account that tunneling current should depend on density of states of the adjacent YBCO and LSMO layers we conclude that the LSMO layer exhibits depressed FM properties in a close vicinity to the LSMO/STO boundary.

The critical current density (Fig. 32) of the patterned tape-like YBCO overlayer was measured by passing simultaneously DC (I = 0-100 mA) and small AC $(I_\sim \sim 1$ μA) current through the same pair of Ag electrodes. The AC current was used in this case to test the resistive transition in the superconductive film induced either by the transport (I) or by the injection (I_{inj}) current. The as measured J_c values were lower compared to those determined from the AC susceptibility measurements before wet etching $(>10^6 A/cm^2$ at 77 K).

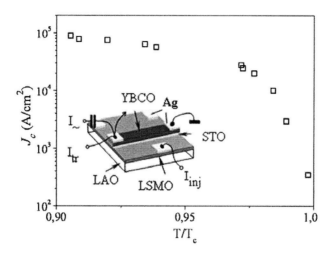

Fig.32. Critical current density vs. temperature of tape-like YBCO top layer. Inset shows the measurement scheme.

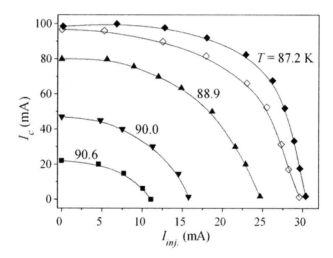

Fig. 33. Critical current of YBCO layer vs. injection current at various temperatures. Full and open points show different I_{inj} polarities.

Fig. 33 shows the I_c versus I_{inj} dependencies for the prepared three terminal device (see inset to Fig.32) with the dielectric layer thickness of 5 nm measured at various temperatures in a close vicinity to T$_c$. Figure demonstrate significant supercurrent suppression effect for the prepared three terminal device by injecting spin-polarized current from the underlying LSMO layer at different temperatures. However, the use of a thicker isolator layer (d$_I$ = 10 nm) in heterostructure resulted in reduced effect with significantly lower current gain (G = $\Delta I_c/I_{inj}$) of less than unity.

MOCVD grown LSMO/YBCO bilayers without intermediate insulating layer have also been investigated as well as possible structures for spin injection devices. Clear indication of the spin-polarized carrier injection effect has been earlier reported in PVD grown HTS/FM heterostructures [61-63]. La$_{1-x}$Sr$_x$MnO$_3$ (x ~ 0.2, 150-200 nm) / YBCO (200-250 nm) bilayers were grown ex-situ at 825 °C on LaAlO$_3$ (001) substrates. The LSMO tape-like layer (width ~3 mm) was formed using thin MgO sheets to mask partially substrate during film deposition. YBCO films were deposited subsequently on the whole substrate surface. XRD study revealed that the both constituent layers in heterostructures were grown epitaxialy. Conventional optical lithography and wet etching were applied to form stripe (3.5 mm x 0.1 mm) crossing the underlying LSMO tape to investigate longitudinal and transverse electrical transport in the heterostructures. Ag contact pads were magnetron sputtered at room temperature through a mask onto YBCO and LSMO film surface (see inset in Fig. 34).

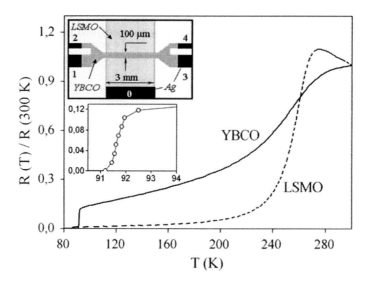

Fig. 34. Resistance vs. temperature measured for YBCO stripe and LSMO layer in LSMO/YBCO heterostructure. Insets shows the scheme of patterned heterostructure and the enlarged view of superconducting transition

The effective stripe resistance of heterostructure and temperature dependent resistance of the LSMO layer are displayed in Fig. 34. Sharp superconducting transition with a zero resistance at ~ 91 K is in agreement with our previous results on similar hetereostructures. Clearly defined resistance peak and a very sharp ferromagnetic transition are typical for LSMO films with reduced stroncium content [64]. Significant drop of the strip resitance seen just below PM-FM transition temperature of the manganite demonstrates shunting of the YBCO stripe resistance by the underlying conducting LSMO film. We point out existence of rather low resistance (<10^{-4} Ωcm^2) between conducting YBCO and LSMO layers (the barrier resistance) both in YBCO/LSMO and YBCO/STO/LSMO heterostructures. Similar low barrier resistance (10^{-4}-10^{-5} Ωcm^2) have been reported for the tunnelling in PVD grown YBCO/STO/LSMO structures [47,48].

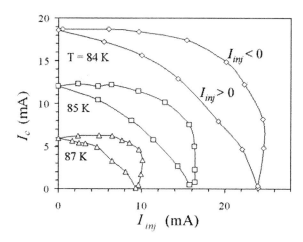

Fig. 35. Critical current as a function of injection current measured for the YBCO stripe of the YBCO/LSMO heterostructure at different temperatures.

Fig. 35 shows I$_c$ versus injection current I$_{inj}$ of a YBCO/LSMO heterostructure measured at several temperatures below T$_c$ for two polarities of I$_{inj}$ (>0 and <0). I$_c$ values were determined from V-I characteristics measured passing the current between contacts 1 and 3 (I$_{13}$ or I$_{tr}$) and measuring the voltage between the contacts 2 and 4 (V$_{24}$). The criterion of V$_{24}$=1 μV was used for I$_c$ calculation. Injection current was introduced through the contact 0. In all cases higher I$_c$ values were measured when directions of I$_{tr}$ and I$_{inj}$ in the superconducting layer were different. Rather similar dependencies were measured for the same heterostructure in liquid nitrogen (at T=77 K) applying different magnetic fields up to 0.5 T.

The expected effect of spin-polarized carrier injection on major superconductor parameters needs to be separated from a possible heating effect. Certainly, both I$_c$ and T$_c$ of a HTSC film may be reduced additionally by I$_{inj}$ due to a current-induced heating of higher resistance manganite layer [49,65]. Assuming that current density of injected carriers crossing the FM/HTSC (FM/I/HTSC) interface is uniform in the whole interface area (0.1 mm x 3.0 mm) one needs to take into account summation of currents and possible variation of I$_{inj}$ in the superconductor along the stripe edge [65]. The observed nonlinear decrease of I$_c$ with I$_{inj}$ in our heterostructures increasing both for I$_{inj}$>0 and <0 can not be explained by any current summation effect. Similar I$_c$-I$_{inj}$ dependencies were measured for the same heterostructure at 83 K in a self-field and at 77 K in an external magnetic field of 0.5 T i.e. under different cooling conditions (vapour and liquid nitrogen, respectively). Thus we avoid possible influence of current-induced heating at least for I$_{inj}$<20 mA. Slight increase of gain G = ΔI$_c$/I$_{inj}$ with temperature decreasing below T$_c$ is in consistence with earlier reports for similar HTSC/I/FM oxide heterostructures [47,49]. Relatively low G values (0.5-1.0) measured for our YBCO/LSMO structures may be understood assuming the presence of an intermediate layer with depressed FM properties at the YBCO/LSMO interface. We do believe that geometry of heterostructures and their preparation conditions may be optimised to achieve higher current gain in SPQID based on YBCO/LSMO heterostructures.

The sample geometry has significant effects on the measurement results in spin-injection devices [65]. The entire cross section of the superconductor must be perturbed in order to suppress I_c with a maximum gain. In the sample geometry where manganite layer is a bottom layer and YBCO is a top layer it is difficult to obtain uniform injection of polarized carriers into a large area of superconductor and to avoid the possible effect of Joule heating due to current flow along a high resistance manganite layer. Device structure in which manganite is the top layer allows formation of a top injection scheme. A low resistance contact to the LSMO layer allows current to enter the superconductor uniformly and the entire volume of the superconductor underneath the injector may be perturbed. However, such configuration with a bottom YBCO layer needs very smooth surface of YBCO layer what is difficult to obtain especially by MOCVD techniques. MOCVD grown YBCO layers exhibiting good superconducting parameters usually have Cu-rich precipitates on the surface which result in disturbed interfaces with overlaying layers.

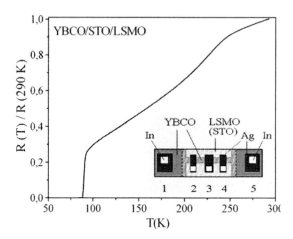

Fig. 36. Resistance vs. temperature of YBCO stripe measured for a LaAlO$_3$//YBCO/STO/LSMO structure (R= U_{24}/ I_{15})

Recently we tried to grow smooth YBCO films on LaAlO$_3$ substrates and to form spin-injection devices from YBCO ($\sim$150 nm)/STO($\sim$5 nm)/LSMO($\sim$130 nm) trilayers containing YBCO as a bottom layer. Deposition at 820 °C under optimized vapour phase composition allowed growing single phase epitaxial YBCO films with significantly improved surface quality. Roughness average (Ra) values measured by AFM were 1.5-2 nm for YBCO films (thickness $\sim$150 nm) and 1-1.4 nm for LSMO films (thickness $\sim$150 nm) grown on LaAlO$_3$. As expected, such a smooth YBCO films had reduced superconducting properties (T$_c$ varied between 85 and 87 K with ΔT$_c$=0.5-1 K and J$_c$ =0.5-1 MA/cm^2). Fabrication of the heterostructures consisted of several technological procedures. Ti coatings were firstly deposited onto masked LaAlO$_3$ substrates by means of DC magnetron sputtering. In following, a stripe-like superconducting YBCO film (0.5 mm wide and 5.0 mm long) was grown *in situ* at 820 °C onto uncoated parts of the substrate since the YBCO film deposited on Ti (TiO2)-coated substrate part was found to be highly resistive. Ultrathin STO barrier ($\sim$ 5 nm) and the LSMO top layer were deposited subsequently onto the YBCO film. Finally, Ag coatings were sputtered onto the top LSMO layer to fabricate low resistance contacts for electrical measurements (see scheme in the inset of Fig. 36).

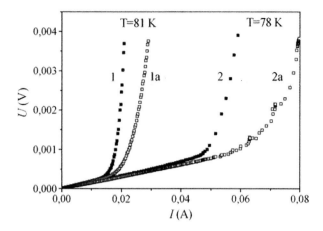

Fig. 37. Voltage vs. current for a LaAlO$_3$//YBCO/STO/LSMO structure at 81 K (1,1a) and 77 K (2,2a) by injecting spin polarized (1,2 – I$_{3,4}$, U$_{1,5}$) and nonpolarized (1a,2a – I$_{1,5}$, U$_{3,4}$) current.

Resistance anomaly seen in Fig. 36 at ~250 K is due to the characteristic paramagnetic-ferromagnetic transition of the overlaying manganite film while sharp resistance drop at ~87 K demonstrates superconducting transition of the underlaying YBCO fillm. The interface resistance between HTS and manganite layers (R$_i$ = U$_{35}$S/I$_{13}$, here S is the interface square) was found to be relatively low (~2 × 10^{-5} Ω cm^2 at 100 K).

Fig. 37 shows typical voltage versus current of the patterned superconducting YBCO film of the YBCO/STO/LSMO heterostructure. Nonzero resistance of the heterostructure (at T < Tc) seen in the figure may be explained assuming shunting of the YBCO film resistance by highly resistive (>10^5 Ω) films of polycrystalline YBCO and LSMO deposited onto Ti-coated parts of the heterostructure. In such a case, critical current of the YBCO stripe has been estimated from the observed onset of the nonlinearity in the measured U-I curves. Our estimations gave Jc values of about 10^5 A/cm2 at 77 K for the YBCO stripes. In most cases, the Ic values for spin-polarized carriers injected from the LSMO layer were lower compared to those corresponding to nonpolarized electrons. Relatively low current gain (G=ΔI$_{c\text{-sp}}$/I$_{c\text{-np}}$≤1.3) estimated in this work for the YBCO/STO/LSMO heterostructures may, probably, be understood assuming the presence of an intermediate layer with depressed FM properties at the YBCO/LSMO interface. In our opinion, the smoothness of the bottom YBCO layer (Ra=1.5-2 nm) is still insufficient and needs to be improved for obtaining high quality interfaces in heterostructures. Both technological processes and investigations of the structures are in progress in order to enhance the role of spin-polarized carrier injection on critical parameters of the YBCO films. On the other hand, possible current summation and heating effect would be considered particularly to eliminate doubts in measurements results.

3.3.5   Superlattices

In the case of superconductors, varying the thickness and the nature of the stacking components can artificially modulate the anisotropic nature of high HTS materials. The study of these artificially modulated structures can contribute to the understanding of the basic properties of high T$_c$ superconductors. The superconducting multilayer (superlattice) which has been studied in our work is (YBa$_2$Cu$_3$O$_{7-x}$ / PrBa$_2$Cu$_3$O$_{7-x}$)$_n$ ((YBCO/PBCO)$_n$) [66-69].

This structure is very convenient because of the direct material compatibility, which allows the switch from a superconductor to a metal / semiconductor / insulator just by changing one element without structural change.

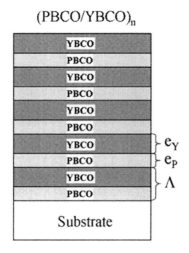

Fig. 38. Scheme of (PBCO/ YBCO)$_n$ superlattice (only 5 periods are shown). Λ is a superlattice period, e_Y and e_P are thicknesses of YBCO and PBCO layers, respectively.

Table 7. Deposition conditions used for superlattices and some properties of single YBCO and PBCO layers (~0.3 μm) grown under these conditions

Deposition conditions or film properties	YBa$_2$Cu$_3$O$_{7-x}$	PrBa$_2$Cu$_3$O$_{7-x}$
Precursors	Y(thd)$_3$(triglyme) Ba(thd)$_2$ (triglyme) Cu(thd)$_2$	Pr(thd)$_3$ Ba(thd)$_2$ (triglyme) Cu(thd)$_2$
Substrate	LaAlO$_3$ (001)	LaAlO$_3$ (001)
Substrate temperature (°C)	840	840
Evaporator temperature (°C)	250	250
Total flow rate (Ar+O$_2$) (l/h)	26	26
Total pressure (Torr)	5	5
Solution composition (Y(Pr) : Ba : Cu molar ratio)	1 : 2.2 : 2.3	1.3 : 2.2 : 2.0
Ba concentration (mol/l)	0.008	0.008
Solvent	monoglyme	monoglyme
Mass of injected microdose (mg)	1.5	1.1
Injection frequency (Hz)	2	2
Surface	smooth	smooth
Microstructure	Epitaxial c-axis textured layer	epitaxial c-axis textured layer
Additional phases	traces of Y$_2$O$_3$	traces of Y$_2$O$_3$
Calculated c-parameter (Å)	11.72	11.67
Resistivity at 100 K (Ω cm)	1.5×10^{-4}	0.5
Resistivity ratio ($\rho_{300 K}/\rho_{100 K}$)	3	
T$_c$ (ΔT$_c$) (K)	~ 92 (≤0.3)	
J$_c$ at 77 K (MA/cm^2)	~ 4 - 5	

A schematic view of such superlattice is presented in Fig. 38. The superlattice period Λ is the sum of the thicknesses of two different layers, each containing a number of stacked crystal cells. Main difficulties of constructing of such artificial structures are the need of very smooth surfaces (interfaces) and precise control of the thicknesses of layers in multilayer.

At first, deposition conditions were carefully optimized in order to epitaxially grow YBCO and PBCO layers with good electrical properties and smooth surface at the same time. Moreover, deposition conditions of two different materials had to be concerted in order to stack repeatedly (in-situ) two materials.

The optimized growth conditions for both materials are summarised in Table 7 where some properties of single YBCO and PBCO layers grown under these conditions are given as well. The dilution of the solution and the fow rate of the precursors (mass for each injection and injection frequency) were optimized to reduce the average growth rate to 0.5 - 1 Å per droplet. This low growth rate allows improving of the thickness control of each layer in the multilayer. For multilayers growth, YBCO and PBCO solutions were alternatively injected in the evaporation zone. A break time of 25 s was required for flushing out the vapors between YBCO and PBCO deposition. This break allows to avoid mixing of YBCO and PBCO vapours, leading to $Y_xPr_{1-x}Ba_2Cu_3O_{7-x}$ deposition. Deposition of YBCO and PBCO sequence was repeated 20 times (or only 10 times for periods superior to 300 Å).

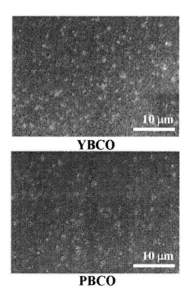

Fig. 39. SEM surface view of single YBCO and PBCO layers (~0.3 μm) on LaAlO$_3$

Single YBCO and PBCO layers (~ 0.3 μm) grown under these optimized conditions have a rather smooth surface (Fig. 39). These rather smooth YBCO layers showed excellent superconducting properties which were better than we obtained for smooth YBCO layers in the study of spin-injection devices. YBCO layers exhibit very sharp superconducting transition (≤0.3 K) at ~92 K and high critical current density (~ 4 - 5 MA/cm^2 at 77 K). These properties are comparable with the best results obtained by PVD methods. PBCO layers show a semi-conducting behavior, with a resistivity close to 0.5 Ω cm at 100 K.

Based on the preliminary experimental data on growth rates of single layers (~0,8 Å/pulse for YBCO and ~0,4 Å/pulse PBCO) a series of YBCO/PBCO superlattices were elaborated. The number of YBCO solution droplets injected for one sequence (N_Y) was varied from 40 to 400 while the number of PBCO solution droplets (N_P) was kept equal to 100. The period Λ can be described by the equation:

$$\Lambda = e_Y + e_P = N_Y v_Y + N_P v_P,$$

where e_Y and e_P are the YBCO and PBCO thickness, respectively, v_Y and v_P are the growth rates of each compound, expressed in Å per injection.

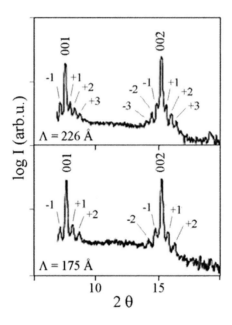

Fig. 40. $\theta/2\theta$ scans of two (YBCO/PBCO)$_n$ superlattices with different periodicity.

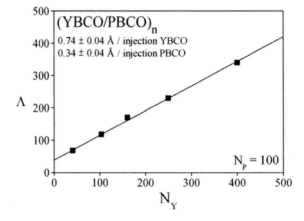

Fig. 41. Superlattice period (Λ, Å) versus YBCO injection number (N_Y). PBCO droplets number is constant ($N_P = 100$).

The only $(00l)$ sharp reflections found in XRD spectra and the pure φ-scans of (102) reflection demonstrate that all multilayers are grown epitaxially on $LaAlO_3$ (001) substrates with a cube-on-cube epitaxial relation with the substrate pseudo-cubic lattice. Bidirectional twinning was clearly observed in thicker multilayers and became not distinguished in the two thinnest multilayers. Fig. 40 shows X-ray diffraction patterns in logarithmic scale, close to (001) and (002) reflections, for two multilayers. The presence of clear satellite peaks indicate regular stacking of YBCO and PBCO cells in superlattice. The period of each superlattice was deduced from the satellite positions with the standard formula:

$$\Lambda = (1/d_i - 1/d_{i-1})^{-1},$$

where d_i and d_{i-1} are the positions of two adjacent peaks. A linear dependence (Fig. 41) is observed between the period Λ and the number of YBCO solution droplets injected during one sequence N_Y (with N_P fixed at 100 injections). This indicates a reproducibility of the thickness deposited at each injection and an easy control of the superlattice period. Calculated average growth rate per injection is: $v_Y = 0.74 \pm 0.04$ Å/inj. for YBCO and $v_P = 0.39 \pm 0.04$ Å/inj. for PBCO. From these values, the thicknesses of YBCO and PBCO layers in superlattices (e_Y and e_P) were calculated. These thicknesses and some other properties of deposited superlattices are presented in Table 8. The calculation indicates that this series consists of superlattices with 37 Å of PBCO (equivalent to 3 cells) alternated with thickness of YBCO layer varying from 30 to 296 Å (approx. from 3 to 28 cells).

Table 7. Characteristics of the (YBCO/PBCO)$_n$ supperlattices grown on $LaAlO_3$ substrates

Sample	1	2	3	4	5
N_P	100	100	100	100	100
N_Y	400	250	160	100	40
n	10	20	20	20	20
Λ (Å)	331 ± 20	226 ± 15	175 ± 10	119 ± 6	67 ± 2
$e_P \pm 3$ % (Å)	37	37	37	37	37
$e_Y \pm 3$ % (Å)	296	185	118	74	30
c-param ± 0.005 (Å)	11.685	11.680	11.675	11.670	11.665
FWHM, RC (007)	0.45	0.44	0.59	-	0.59
FWHM, φ-scan (102)	1.21	1.28	0.99	0.73	0.63
T_c (K)	90.2	89.8	84.4	83.3	40
ΔT_c (K)	0.4	0.3	0.8	1.6	5.0

FWHM – the ful width at half maximum of rocking curve (RC) or φ-scan, N – the number of YBCO or PBCO injections during one sequence, n - the number of periods in the multilayer, Λ - the superlattice period decuced from the satellite positions in XRD $\theta/2\theta$ scans, e – the YBCO or PBCO thickness deduced from the average growth rate, T_c – the critical temperature, ΔT_c – the full width of superconducting transition.

The critical temperature and the transition width of different superlattices are presented in Table 8. When the thickness of the YBCO layer decreases from 296 Å to 30 Å, T_c decreases from 90.2 K to 40 K. For a YBCO thickness equal to 296 Å (approx. 28 cells) or 185 Å (approx. 19 cells), superconducting properties of the supperlattices are equivalent to bulk's ones ($T_c \sim 90$ K). In the case of thinner YBCO layers (≤ 15 cells), T_c and ΔT_c of the multilayer are strongly affected.

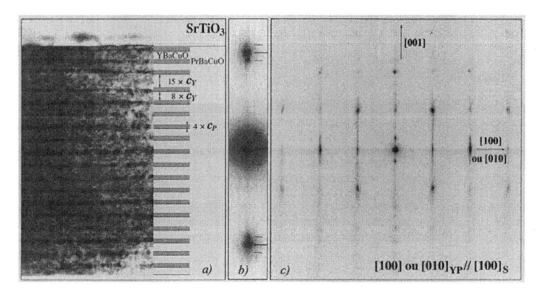

Fig. 42. TEM cros section view (a) and electron diffraction pattern (b,c) for a (YBCO/PBCO)$_n$ superlattice of the periodicity of about 140 Å grown on SrTiO$_3$ substrate. (b) is an enlarged view of the patern (c) along [001] direction.

Similar series of YBCO/PBCO superlattices has been grown on SrTiO$_3$ (001) substrates and results similar to those on LaAlO$_3$ were obtained. Fig. 42 shows a TEM image of the cross-section and a pattern of electron diffraction for a multilayer grown on SrTiO$_3$ substrate. These observations allow distinguish alternate periods of the stacking of two materials and to determine corresponding thicknesses of YBCO and PBCO layers. 18 periods can be seen in the multilayer and the first layer in contact with the substrat is PBCO. The stacking is regular and each period consists of 4 cells of PBCO and 8 cells of YBCO and the determined periodicity is $\sim$ 140 Å. The periodicity of this superlattice determined by X-ray diffraction is 152 ± 10 Å. However, the figure reveals a defect of the stacking of two materials: YBCO layer situated in the fourth period consist of 15 cells instead of 8. The deposition process is governed by computer, so the most probable reason may be the temporary perturbation of the functioning of injectors. Such perturbation is rare, but not excluded. The electron diffraction pattern (Fig. 42) shows satellite spots demonstrating the regular stack of cells in multilayer.

Physical properties of the grown (YBCO/PBCO)$_n$ superlattices were studied in details [70-72] in order to determine the complex physical properties generated by this artificial stacking of insulating and superconducting layers. PI-MOCVD technique has also been successfully applied for the growth of multilayers and superlattces of other functional materials: multilayers (Ta$_2$O$_5$/SiO$_2$)$_n$ [73], superlattices of (SrTiO$_3$/BaTiO$_3$)$_n$ and (La$_{1-x}$Sr$_x$MnO$_3$/SrTiO$_3$)$_n$ [69,74-79].

3.4 YBCO Coated Conductors

3.4.1 Introduction

Nowadays, the superconducting technology (NMR, motors, SMES, Fault Current Limiters FCL) are at the demonstration state or remain as niche market. These systems are based on low temperature superconductors or on Bi-based high temperature superconductors. In recent years, research efforts have been made to develop new technologies to elaborate tape-shaped 'high temperature superconductors (HTS) coated conductors' essentially based on the $YBa_2Cu_3O_7$ (YBCO) compound. The main interest of YBCO is its strong pinning properties in high magnetic fields at 77 K. It is not easy to fabricate YBCO based conductor cables with high critical-current density (J_c), because the grain-boundaries misorientation angle needs to be lower than few degrees due to the intergranular weak links of this material. Whereas, epitaxially grown films show high J_c, typically $> 10^6$ A/cm^2 when the grain-boundaries misorientation is inferior to 5 degrees.

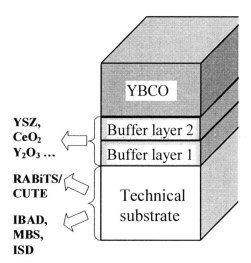

Fig. 43: Typical architecture of a tape-shaped coated conductor on a metallic substrate

Thus YBCO conductors have been developed as tape-shaped coated conductors on different metallic substrates by using deposition techniques of the superconducting material. It is generally necessary to deposit one or several buffer layers, either to improve the matching between the substrate and YBCO, or as chemical barrier (Fig. 43).

There are several approaches to elaborate YBCO coated conductors:

- Deposition of YBCO layers on polycrystalline metal substrates with artificially textured buffer layer deposited by Ion Beam Assisted Deposition (IBAD) [80,81], Modified Bias Sputtering (MBS) [82] or Inclined Substrate Deposition (ISD) [83].
- Deposition of YBCO films on textured metal substrate like Cube Texture (CUTE) or Rolling Assisted Biaxially Textured Substrates (RABiTS) [84] with or without buffer layer. One difficulty with RABiTS nickel-based coated conductors is the formation of polycrystalline or non-epitaxial NiO on the surface of the nickel tape during the

deposition of oxide buffer layers. The controlled oxidation of nickel to grow biaxially textured NiO is named Surface Oxidation Epitaxy (SOE) [85,86].

In the present section of the chapter, we detail the chemical engineering of Metal Organic Chemical Vapor Deposition (MOCVD) used to form the complex architecture of coated conductors: from buffer layers (CeO_2, YSZ, Y_2O_3...) to the superconducting phases (Y-123), and even metallic shunt layer. For people interested in coated conductors, more information is available in "Coated Conductors and HTS Materials by chemical deposition processes" [87].

3.4.2   MOCVD Process in Reel-to-Reel Mode

As described previously in this chapter, high quality YBCO films have been deposited, in static mode, using the PI-MOCVD systems on single crystals. Moreover, promising results were obtained on metallic substrates [88,89]. These systems have also been used to deposit high quality buffer layers [90,33,31,91]. In order to deposit HTS thin or thick films on moving tapes a special reel-to-reel deposition system based on single solution source, has been developed at LMGP in the context of the European READY project [92]. In a general way, this reactor (Fig. 44) consist of a process chamber (1) containing all the necessary elements for the MOCVD process: an evaporator with incorporated injectors and heated vapour transport routes, a resistive furnace (2), a showerhead for precursors vapours distribution (3), a pumping system (4), a pressure control system, a gas panel, and a reel-to-reel system (5). The whole system is under vacuum. The geometry and deposition parameters (essentially gas distribution) have to be optimised to improve the deposition yield in the reaction zone. The temperature homogeneity in the deposition zone has to be better than ± 3 °C. Using lateral furnace allows to improve the thermal profile in the deposition zone, and to perform a preheating of the substrates.

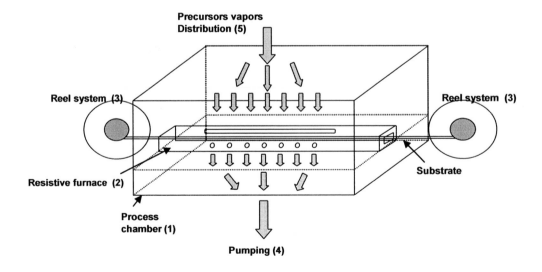

Fig. 44: Schematic representation of a MOCVD reactor for coated conductors

This prototype consists of only one deposition chamber, and buffer layers and YBCO films depositions are performed sequentially. A reel and de-reel system is provided to transfer tapes in two ways. After YBCO deposition, the lateral furnace can be also used as post deposition annealing furnace to perform the oxygen loading of the films. The travelling velocity of the tape can vary from 0.5m/h to 20m/h, depending on the deposition rate and on the film thickness. When depositing on small samples, a stainless steel or a nickel alloy (5%W) tape is used as substrate holder. The samples are glued or mechanically fixed to this holder, and the deposition takes place in a reel-to-reel mode.

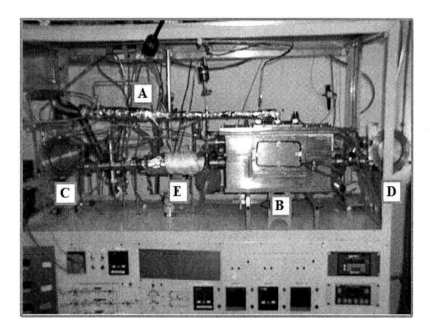

Fig. 45. Reel-to-reel PI-MOCVD deposition system at LMGP. Description of the labelled elements: A: vapour source, B: deposition chamber, C: take-up reel, D: play-out reel, E: annealing chamber.

The reactor developed at LMGP is presented in Fig. 45. The experience obtained in static substrates deposition has been transferred to this continuous system. Buffer layers and superconducting films have been successfully deposited in this reel-to-reel system. CeO$_2$ and YBCO layers were deposited on 2m-long tape in a continuous mode. The average value of the in-plane texture from φ-scan measurements was FWHM = $9.23°\pm0.48°$ for CeO$_2$ and FWHM= $8.26°\pm0.45°$ for YBCO on IBAD substrates. YSZ on NiO was also deposited on 2m-long tapes. An important feature of this experiment is that the in-plane texture of the YSZ on NiO layers improves in comparison with the static mode. This fact is probably due to shorter remaining times of the substrate in the high temperature zone, essentially before deposition.

3.4.3   Coated Conductors Architecture by PI-MOCVD

Several architectures on metallic substrate have been tested using PI-MOCVD. Nevertheless, a key factor for the substrate choice in PI-MOCVD process is the use of oxidizing atmosphere at high temperature.

Table 9. Typical deposition parameters for some buffer layers and YBCO by PI-MOCVD

Parameters	Y_2O_3	YSZ	CeO_2	MgO	YBCO
Precursor	$Y(thd)_3$	$Y(thd)_3$, $Zr(thd)_4$	$Ce(thd)_4$	$Mg(thd)_2$	$Y(thd)_3$, $Cu(thd)_2$ $Ba(thd)_2$
Solvent	monoglyme	monoglyme	heptane	monoglyme	monoglyme
Injection frequency (Hz)	2-3	3	4	1-2	1
Injection time (ms)	1-2	2	1	1	4
Evaporation temperature °C)	250	300	230	280	300
Deposition temperature (°C)	600-800	600-700	800-850	700-800	775-825
Deposition pressure (Torr)	5	5	5	5	5
% O_2 in Ar+O_2	40	42	42	30	41

The coated conductor architecture can be based on several oxide buffer layers (Y_2O_3, YSZ, Gd_2O_3, GdSZ, MgO, $SrTiO_3$ and CeO_2). All theses oxides can be synthesised by PI-MOCVD and the precursors for these materials are commercially available. The deposition conditions have been optimised for these materials from the HTS application point of view (biaxial texture, dense film, optimal thickness), and as a function of substrates. Typical deposition parameters for the usual buffer layers and for superconducting layers are summarized in Table 9. The deposition rates depend on the deposition parameters and typically vary from 1 μm/h to 6μm/h.

Silver substrate

a) Ag (200)
before deposition

b) YBCO (103)

c) Ag (200)
after deposition

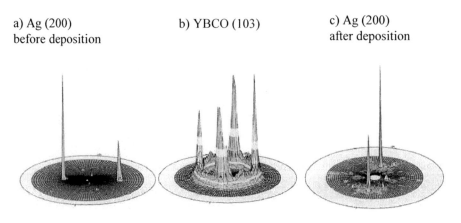

Fig. 46: Pole figures of reflections corresponding to: a) silver substrate before deposition, b) YBCO layer on this substrate, and c) silver substrate after deposition.

In this case, no buffer layer is needed because the direct deposition on silver {110}<011> textured substrates is possible thanks to the chemical stability of silver with oxygen. YBCO films were biaxially textured (Fig. 46), but the in-plane orientation was not complete. The J$_c$ value obtained for this sample was ~0.2 MA/cm^2 [93].

Ni substrate

A biaxially textured tape can easily be obtained from nickel or nickel alloy with {100}<001> orientation at high processing speed. The fabrication of biaxially textured nickel tape takes place by consecutive rolling of polycrystalline randomly oriented high purity (>99.99%) bars to total deformation greater than 90%. Subsequent annealing of the substrate is then performed in a controlled atmosphere. Tapes prepared in this way present single cube texture, with FWHM values of 6° from Ni (111) $\tilde{\varphi}$-scan.

The direct deposition of YBCO on nickel-based RABiTs substrates is not possible by MOCVD due to the poisoning effect leading to non superconducting Y$_2$Ba(Cu,Ni)O$_X$ phase; instead, buffered substrates are more suitable. Buffer layers architectures were prepared combining PI-MOCVD and other techniques.

a) Surface oxidation epitaxy (SOE), Ni/NiO.

In SOE processing, a cube-textured nickel tape is inserted into an electric furnace heated to 1000-1300°C for metal oxidation. Nickel oxide can epitaxially grow on the nickel surface if the oxidation temperature and the pressure are optimized. The sharpness of the cube texture of these NiO layers is strongly dependent on the substrate texture and on the grain size of the substrate material, as well as the processing conditions during the oxidation [94]. Ni/NiO substrates prepared by SOE at IFW as described by D. Selbmann [94,95] were used. The nickel-based tapes were microalloys of 0.1%at Mo or W. YBCO deposited on biaxially textured NiO lead to superconducting films, but the quality is limited by the nickel diffusion in YBCO. Several buffer layers architectures have been studied to optimise the network matching and to solve the interdiffusion phenomena.

MgO/YBCO films on Ni/NiO. MgO(400nm)/YBCO(700nm) layers [88] were grown by PI-MOCVD on the top of these substrates. The buffer layers were deposited at a growth rate of 1.5 µm/h, and the YBCO films at 5 µm/h. A T$_c$(onset) of 87K and a critical current density J$_c$=10^4 A/cm^2 at 77 K were measured by AC susceptibility. This low value is due to a bad texture of the YBCO: FWHM φ-scan (103) =22.8°.

YSZ/YBCO films on Ni/NiO. A 200nm-thick (200)-textured YSZ film was deposited to serve as a chemical and mechanical diffusion barrier [96]. The superconducting transition for the YSZ(200nm)/YBCO(750nm) structures was rather sharp (Tc~88-89K) but the J$_c$ values were low (J$_c$~10^4 A/cm^2 at 77 K), due to the lattice mismatch between YSZ and YBCO (~6 %), and because of the Zr diffusion in the YBCO film. On the other hand, TEM observations showed that long dislocations on Ni/NiO were stopped at the NiO-YSZ interface and did not propagate across the boundaries. At the NiO-YSZ interface a large 40 nm thick, well crystallised layer nucleated beneath the interface, whereas the dislocations pile-up stopped at the interface. In fact, TEM observation showed that though the YBCO-YSZ interface remained sharp the interface look irregular, but no sign of inter-diffusion was detected. YBCO on YSZ was polycrystalline and the crystals were separated by almost vertical boundaries. Bright lattice fringes (~14 Å) were detected corresponding to YBCO (001) orientation. The planes are not perfectly aligned across the boundaries, with a misorientation as much as 10° (comparable with XRD in-plane texture of 9.3°).

YSZ/CeO$_2$/YBCO films on Ni/NiO. We added a 150 nm CeO$_2$ film which capped the grooves, and diminished the lattice mismatch with the YBCO, giving denser and smoother YBCO coated Ni tapes. Small grains size of the CeO$_2$ films lead to a flatter surface (rms roughness of the grain 12 nm and grooving from 31 nm to 45 nm deep) with CeO$_2$ (111) rare misoriented spots. The orientation of the grains layer after layer has been followed by BKD (Backscattered Kikuchi Diffraction) showing a final mean grain misorientation of 7°. J$_c$ values of such architecture were 0.45 MA/cm^2 at 77 K.

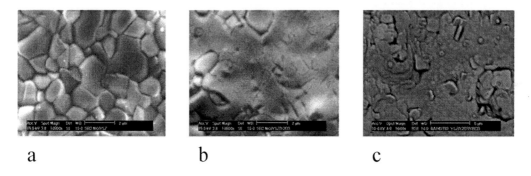

a   b   c

Fig. 47: Flattening effect of the Y$_2$O$_3$ buffer layer. Surface morphology of (a) Ni/NiO//YSZ, (b) Ni/NiO//YSZ/Y$_2$O$_3$, and (c) Ni/NiO//YSZ/Y$_2$O$_3$/YBCO as observed by SEM

YSZ/Y$_2$O$_3$/YBCO films on Ni/NiO. Adding a second buffer layer of Y$_2$O$_3$ leads to J$_c$ values of 0.5 MA/cm^2 at 77 K. The beneficial effect of the Y$_2$O$_3$ layer, in addition to the diffusion barrier role, was also the surface flattening effect [97]. This effect is shown if Fig. 47. NiO thermally grown on Ni substrates presents some grooves, which are transferred to the subsequent layers. The YSZ layers reproduce the morphology of the NiO surface (Fig. 47a). The second buffer layer fills the grooves and improves the surface roughness (Fig. 47b)

b) PLD CeO$_2$/YSZ/CeO$_2$/Ni by Cryoelectra

YBCO layers were also deposited on the top of Ni substrates ex-situ buffered with CeO$_2$/YSZ/CeO$_2$ by PLD, and provided by Cryoelectra GmbH [97]. J$_c$ values close to 1 MA/cm^2 were obtained on these samples. When depositing the CeO$_2$(150nm)/YSZ(200nm)/CeO$_2$(150nm)/YBCO structure by PI-MOCVD on Ni/NiO substrate, J$_c$ values of 9×10^5 A/cm^2 at 77 K were obtained.

c) PLD buffered NiW by American Superconductor

YBCO films were deposited on buffered NiW tapes supplied by American Superconductor. The whole structure consisted of NiW/Ni/Y$_2$O$_3$/YSZ/CeO$_2$//YBCO [98]. Early experiments gave T$_c$ (onset) = 92.5 K with a sharp transition (~0.5 K), and J$_c$ values of 0.8 MA/cm^2 at 77 K. The in-plane texture of the YBCO layer has a FWHM = 6.1°. CeO$_2$ and YBCO films were characterized by BKD (Fig. 48). The image quality was measured for each individual Kikuchi pattern during the indexation. On Fig. 48b, black spots denote a bad IQ (Quality Index). Bad IQ implies that the pattern recognition software did not obtain a solution. Thus defects (cracks, pores, amorphous grains) will trigger a larger scatter of the electrons and will clearly appear on the diffraction pattern. Furthermore, we find that almost all the grains indexed are (001) oriented and no (110) orientation of the grain was detected.

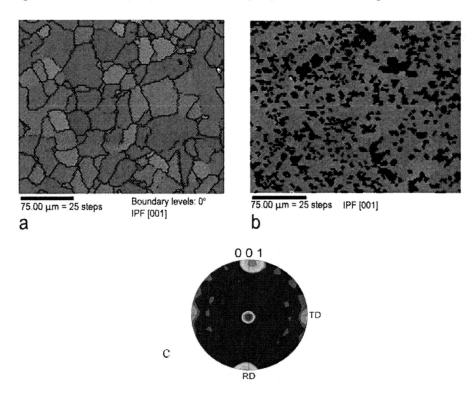

Fig. 48. BKD of the CeO$_2$ buffered tapes (NiW/Ni/Y$_2$O$_3$/YSZ/CeO$_2$) and YBCO films: a) Inverse pole figure of CeO$_2$, b) Inverse pole figure of YBCO, c) Pole figure of YBCO.

YSZ-IBAD substrates

An alternative substrate is buffered substrate prepared by IBAD (Ion Beam Assisted Deposition usually based on YSZ on Stainless Steel, YSZ/SS) [99].

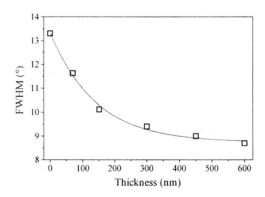

Fig. 49. Influence of thickness on the in-plane texture of the CeO$_2$ films deposited on YSZ/SS IBAD substrates. FWHM values were obtained from the CeO$_2$(111) reflection in φ-scan mode.

CeO$_2$ and YSZ films were deposited by PI-MOCVD, in continuous mode, on the top of YSZ/SS IBAD substrates. The in-plane texture versus thickness of the CeO$_2$ layer follows an exponential decay [100]. This result was previously reported by J. Wiesmann [101], when increasing thickness of YSZ layer by IBAD: a saturation of the in-plane texture was observed at a FWHM = 7° for 5.5 µm-thick films. In our study, we started working with a YSZ(800nm)/SS structure prepared by IBAD. The in-plane texture of the initial YSZ layer showed a FWHM = 13.5° value from the YSZ(111) reflection in φ-scan mode. The saturation in texture was found at a FWHM=8.7° for a 600nm-thick CeO$_2$ film (Fig. 49), and at a FWHM=10° for a 300nm-thick YSZ film. In addition to these lower saturation thickness values, the growth rate of the films by MOCVD was around 10 µm/h, as compared to 0.3-0.7 µm/h by IBAD. On the other hand, AFM measurements of the surface roughness showed lower values for CeO$_2$ films than for YSZ; denser YBCO films were consequently obtained on these samples.

Table 10: J$_c$ and T$_c$ values of samples grown in reel-to-reel mode. Bold characters indicate deposition by PI-MOCVD

Architecture	J$_c$ (A/cm^2)	T$_c$ onset (K)	In-plane texture of MOCVD layers (FWHM in°)
Ag//**YBCO**	2×10^5	92	
Ni/NiO//**MgO/YBCO**	1×10^4	87	8.4/22.8
Ni/NiO//**YSZ/CeO$_2$/YBCO**	4.5×10^5	90	10.5/8.2/7.4
Ni/NiO//**YSZ/Y$_2$O$_3$/YBCO**	5×10^5	91	10.6/8.5/7.5
Ni/NiO//**CeO$_2$/Y$_2$O$_3$/CeO$_2$/YBCO**	9×10^5	89.8	10.7/11/10.2
Ni/CeO$_2$/Y$_2$O$_3$/CeO$_2$//**YBCO**	1×10^6	90	6.3
NiW/Ni/Y$_2$O$_3$/YSZ/CeO$_2$//**YBCO**	8×10^5	92.5	6
SS/YSZ//**YSZ/CeO$_2$/YBCO**	8×10^5	90	11/9.2/8.2
SS/YSZ//**CeO2/YBCO**	1.3×10^6	91	8.5/8.1

Firstly, the optimal architecture on YSZ IBAD substrates depends on the epitaxial quality of the buffer layer. For IBAD substrate with in-plane texture of the initial YSZ layer higher than 10°, the PI-MOCVD architecture is YSZ(200nm)/CeO$_2$(150nm)/YBCO. For better quality substrates, only a CeO$_2$ buffer layer is needed to enhance the matching and the chemical stability with YBCO.

For low quality substrate, so two-buffer layered architecture, the structures deposited lead to critical current densities of 0.8 MA/cm^2 at 77K with a T$_c$= 90K. The tape velocity used for the deposition of the buffer layer ranged between 3-4 m/h, and for the superconducting layer between 1-2 m/h. On the other hand, the deposition of CeO$_2$(300nm)/YBCO(750nm) structures on the top of YSZ/SS substrates of good quality leads to J$_c$ values of 1.3 MA/cm^2 at 77 K.

Table 10 summarizes the J$_c$ and T$_c$ values obtained for the different coated conductor architectures using PI-MOCVD.

3.4.4   Transport Measurements on PI-MOCVD Coated Conductors: Silver Metallization by PI-MOCVD

For cables fabrication, the main performance of coated conductors is the current transported. From this point of view, measurements of I_c are crucial to validate the architectures and the process. Nevertheless, the use of HTS coated conductors in high current applications requires a shunt layer to avoid hot spots, which would provoke local destruction, as well as to thermally stabilize the superconductor during the dissipative state. An additional aspect is to facilitate the current contacts at both tapes ends by soldering power leads. Usually, silver or gold are chosen as contact metal, and they are deposited by Physical Vapour Deposition techniques as e-beam or thermal evaporation. We have developed a MOCVD process for silver deposition base on a silver carboxylate precursor [102,103]. First, we optimized the Ag deposition conditions on YBCO films. Best deposition conditions were then tested in continuous mode into a reel-to-reel MOCVD reactor.

Nevertheless, silver layers should be compatible with the metallization requirements; the most important one is a low contact resistivity between the shunt and the HTS film. The contact resistance directly influences the power dissipation, essentially in the current contacts. Usokin et al [104] have demonstrated that the threshold of tape damage can be estimated to ~1 W/cm^2, which mean that for reasonable contact lengths of 30 mm, the contact resistivity has to be lower than $10^{-5} \Omega$ cm^2. Transported current is proportional to the cross section of the HTS coating. Another way to present the coated conductor performance is using the figure of merit of critical current divided by effective tape width (I_c/w), because the width is the primary determinant of the cross section of the HTS coating. Taking into account this parameter, the best values obtained until now are 391 A/cm-width for a 20 cm-length sample, i.e. I_c= 137A for 4 mm-wide, and 223 A/cm-width for a 10 m-length sample; these values are equivalent to a J_c of 2 MA/cm^2. These samples were prepared by PLD on IBAD substrates [105].

Table 11: Transport current measured on PI-MOCVD coated conductors

Substrate; width	Buffer layer	Superconductor	I_c (A) at 77K	I_c/w (A/cm)
NiMo/NiO ; 10mm	-	YBCO (0,75 µm)	0.4	0.4
NiMo/NiO; 10mm	YSZ	YBCO (0,75 µm)	2	2
NiW-AmSup; 5mm	-	YBCO (0,6µm)	5.5	11
IBAD-YSZ; 3,5mm	-	YBCO (0,6 µm)	2.2	5.5
IBAD-YSZ; 10mm	Y_2O_3	YBCO(0,75 µm)	10	10
IBAD-YSZ; 4mm	CeO_2	YBCO (1,7 µm)	>28	>70
IBAD-YSZ; 10mm	CeO_2	YBCO (1,7 µm)	>51	>51
IBAD-YSZ; 4 mm	CeO_2	YBCO (1µm)	>68	>170
LAO; 5mm	-	YBCO (0,4 µm)	27.6	55.2
LAO; 5mm	-	YBCO (0,6 µm)	77	154
LAO; 5mm	-	YBCO (1,7 µm)	33	66
LAO; 5mm	-	YBCO (2,3 µm)	14	28

We measured the transport current on different architectures prepared on short (between 4 and 20cm) samples. Silver was deposited on the top of the coated conductor architectures and the transport properties in self-field were measured at 77 K. Main results are summarised in Table 11. It is important to notice than values preceded by ">" means that the superconductor transition for critical current was not reached because the contact resistance was too high and contact transited by temperature (Joule effect). Depositions on single crystal of $LaAlO_3$ are also included for comparison. Best results were obtained for IBAD substrates as found in susceptibility measurements, equivalent to 1.7 MA/cm^2. An important issue is the effect of YBCO thickness on transport results: current does not increase with thickness because a degradation of the epitaxial quality of the YBCO layer appears for thickness higher than 1 µm.

The results shown here are not intrinsic to buffer layer architecture or superconducting layer, but they are also depending on buffered substrates quality and on deposition run. Coated conductors are in optimisation stage and results still improve with experience.

Few groups in the world use MOCVD for coated conductors fabrication, and most of them use the MOCVD technique only for YBCO deposition, buffer layers being processed by physical techniques.

In Europe, MOCVD is used until now at laboratory scale; however, Nexans is involved as partner with activity in several projects on this topic.

In USA, Intermagnetic General Corporation (IGC)-SuperPower is the only industrial using MOCVD process. They have reported up to now the longest length for YBCO films deposited by MOCVD. They have developed a pilot system to deposit YBCO [106]. In this system, the usual linear tape speed is 10m/h and they get a deposition rate of 120 Å/s (tape speed up to 30 m/h can be used, but with a consequently decrease in the deposition rate). They reported during the ASC 2004 critical currents of 116 A over 1.86-meter length 4 mm-wide IBAD tape (in fact IBAD-MgO/CeO2-PLD), so 290 A/cm-wide, and for longer tapes, currents ranging between 65 A and 193 A for a 25 m-length tape [107]. Last communication by Press Release (January 2005) shows I_c ranging between 95 and 160 A for a 97 m-length tape. Mean I_c value is 150 A/cm-width.

The University of Houston is working in photo activated MOCVD to increase the deposition rate. They also presented in ASC2004 a Reel-to-reel MOCVD system for long length tapes.

In Asia, several countries work on MOCVD. Most of the activities are concentrated in Japan. ISTEC (International Supeconductivity Technology Center) (Superconductivity Research Laboratory-Fujikura-Furkawa Electric Co -Showa Electric) works on MOCVD as an innovative process as compared to PLD. Chubu Electric Power Co. reported the successful deposition of 100 meters length of 0.4 µm-thick YBCO on rolled non-textured Ag tape using a six-stage MOCVD system at a travelling speed of 10 m/hr. Due to the non-textured substrate, superconductor characteristics were rather low $6.1 \cdot 10^4 A/cm^2$ (I_c 2.43 A) [108]. Last results have been recently published [109]. A tape of PLD-CeO_2(1µm)/IBAD-$Gd_2Zr_2O_7$(1µm)/Hastelloy(0.1mm) has been recovered with 1µm-thick YBCO by MOCVD. Ic of the tape which was prepared at a speed of 5m/h x 2 passes was 101 A/cm. These results show that high speed depositions are compatible with good superconductor quality and that MOCVD can be implemented as a multiple-stage. KAERI (Korea Atomic Energy Research Institute) is developing a reel-to-reel system based on previous results on static reactors [110]. They have deposited several buffer layers on RABiTs.

4 Industrial Developments of Pulsed Injection MOCVD Technique

After the invention of PI-MOCVD principles in 1992-1993 years, the patent was licensed in 1999 to the French company JIPELEC, and later to the Germany company Aixtron AG. These two companies worked on several industrial developments of PI-MOCVD technique.

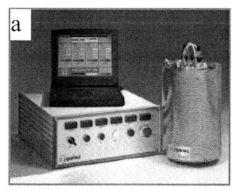

Fig. 50. Industrial liquid precursor delivery and vaporization system for MOCVD, elaborated at Jipelec (France): a) INJECT LAB, b) INJECT PRO, specifically developed for low vapor pressure precursors. By courtesy of Mr Decams and Mr Guillon, Jipelec, France.

Fig. 51. PI-MOCVD systems elaborated by Jipelec in the frame of European FP5 projects: a) reel-to-reel system for the preparation of long length YBCO coated conductors, b) static-mode reactor for the deposition of oxide layers including HTSC on large area substrates up to 4 inches in diameter. By courtesy Mr Decams and Mr Guillon of Jipelec, France

In a first stage, Jipelec conceived (Fig. 50) and launched a serial production of liquid precursor delivery and vapour generation systems based on the new pulsed injection principle. These sources contain single or multiple injectors and can be used for MOCVD deposition of various mono and multicomponent oxide layers and multilayers [111]. Then, based on these sources and focused on HTS, Jipelec developed a PI-MOCVD reel-to-reel system for YBCO coated conductors (Fig. 51a) and a static-mode system for the depositions of oxide layers including YBCO on large area substrates up to 4 inches (Fig. 51b).

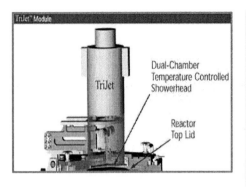

Fig. 52. Aixtron Trijet vapour delivery system installed in Tricent MOCVD cluster for oxide films. By courtesy of Aixtron, Germany.

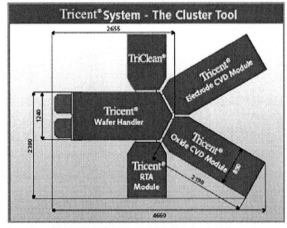
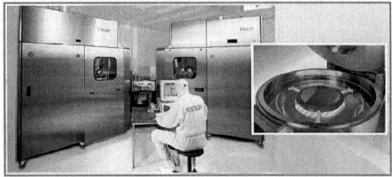

Fig. 53. Aixtron Tricent® system – the cluster tool for the deposition of high-k, ferroelectric and electrode films on up to 300 mm wafers. HTSC oxides, i.e. YBCO can be deposited in this system as well. By courtesy of Aixtron, Germany.

Aixtron fabricates several MOCVD systems for oxides based on the pulsed liquid injection principle [112]. The vapour delivery system (Fig. 52) developed by Aixtron is called TriJet™. This source is currently installed in Aixtron Tricent® MOCVD cluster for the deposition of high-k, ferroelectric and electrode films on wafers up to 300 mm in diameter (Fig. 53). HTSC oxides, i.e. YBCO can be deposited in this system as well.

Tricent® system is flexible in wafer sizes from 6", 8" to 300 mm. The use of TriJet™ ensures contactless precursor evaporation and results in long-term stability, high precursor efficiency, and high deposition rates. The MOCVD cluster also contains a dual-Chamber temperature-controlled showerhead featuring which separates inlets of precursors and oxide agents for homogeneous and repeatable mixing of reaction species. These issues result in highly reproducible deposition of homogeneous films. Wafer rotation provides best product uniformity. Cluster Tool architecture brooks automation wafer handling platform (for wafers up to 300mm) integrating up to four Tricent® MOCVD modules. The material applications claimed by Aixtron for this technology are high-k, ferroelectrics, electrodes, barrier layers, sensor, HTS.

Fig.54. Reactor module of the Aixtron AIX200FE system for multicomponent oxide MOCVD. By courtesy of Aixtron, Germany.

Aixtron is also producing a research reactor AIX200FE for multicomponent oxide MOCVD deposition, including HTSC materials. The reactor is equipped with the same pulsed liquid precursor delivery system, TriJet™. Sample size for this reactor is up to 2" x 2". The AIX200FE is provided with IR lamp heating and Gas Foil Rotation® to ensure high growth and temperature uniformity. The reactor consists of electronic computer control part, gas mixing system and reactor module (Fig. 54).

5 Conclusion

Pulsed Injection MOCVD technique is a very versatile proven technology for the deposition of High-T$_c$ superconductors and related multilayered superconducting structures. The main features of this technique are "digital" control of precursors vapor generation and film growth as well as high reproducibility of precursors feeding rate, MOCVD process and film properties. High quality YBCO films can be reproducibly obtained on small and large area single crystal substrates. Film quality is comparable to this reached in the best samples obtained by PVD techniques, while the reproducibility is higher. The PI-MOCVD technique also has showed its performance for YBCO coated conductor deposition. Although the properties of prepared tape-like coated conductors are, for instance, slightly lower compared to the best results reported, they are in optimisation stage and results still improve with experience. In general, PI-MOCVD technique becomes a powerful tool for the preparation of various oxide layers and sophisticated artificial multilayered heterostructures including superlattices.

References

[1]   Leskela, M.; Molsa, H.; Niinisto, L. *Supercond. Sci. Technol.* 1993, 6, 627-656 (review).

[2]   Watson, I.M. *Chem. Vap. Deposition* 1997, 3, 9-26 (review).

[3]   Tiitta, M.; Niinisto, L. *Chem. Vap. Deposition* 1997, 3, 167-182 (review).

[4]   Otway, D..J.; Obi, B.; Rees W.S. *J. Alloys Compounds* 1997, 251, 254-263.

[5]   Marks, T.J.; Belot, J.A.; Reedy, C.J.; McNeely, R.J.; Studebaker, D.B.; Neumayer, D.A.; Stern, C.L. *J. Alloys Compounds* 1997, 251, 243-248.

[6]   Tasaki, Y.; Yoshizawa, S.; Koyama, K.; Fujino, Y. *IEEE Trans. Appl. Supecond.* 1999, 9, 2367-2370.

[7]   Hubert-Pfalzgraf, L.G.; Guillon, H. Appl. *Organometal. Chem.* 1998, 12, 221-236.

[8]   Guillon, H.; Daniele, S.; Hubert-Pfalzgraf, L.G.; Bavoux, C. *Inorg. Chim. Acta* 2000, 304, 99-107.

[9]   Lackey, W.J. Carter, W.B.; Hanigofsky, J.A.; Hill, D.N.; Barefield, E.K.; Neumeier, G.; O'Brien D.F.; Shapiro, M.J.; Thompson, J.R.; Green, A.J.; Moss, T.S.; Jake, R.A., Efferson, K.R. *Appl. Phys. Lett.* 1990, 56, 1175-1177.

[10]   Molodyk, A.A. *J. Alloys Compounds* 1997, 251, 303-307.

[11]   Hiskes, R.; DiCarolis, S.A.; Jacowitz, R.D.; Lu, Z.; Feigelson, R.S.; Route, R.K; Young, J.L. *J. Crystal Growth* 1993, 128, 781-787.

[12]   Jergel, M. *Supercond. Sci. Technol.* 1995, 8, 67-78 (review).

[13]   Matsuno, S.; Uchikawa, F.; Utsunomiya, S.; Nakabayashi, S. *Appl. Phys. Lett.* 1992, 60, 2427-2429.

[14]   Weiss, F.; Frohlich, K.; Haase, R.; Labeau, M.; Selbmann, D.; Senateur, J.P.; Thomas, O. *J. Phys. IV* 1993, 3, C3, 321-328.

[15]   Gorbenko, O.Y.; Fuflygin, V.N., Erokhin, Y.Y.; Graboy, I.E.; Kaul, A.R.; Tretyakov, Y.D.; Wahl, G.; Klippe, L. *J. Mater. Chem.* 1994, 4, 1585-1589.

[16]   Matsuno, S.; Unemura, T.; Uchikawa, F.; Ikeda, B. *Jpn. J. Appl. Phys.* 1995, 34, 2293-2299.

[17]   Studebaker, D.B.; Zhang, J.; Marks, T.J.; Wang, Y.Y.; Dravid, V.P.; Schindler, J.L.; Kannewurf C.R. *Appl. Phys. Lett.* 1998, 72, 1253-1255.

[18]   Senateur, J.P.; Madar, R.; Thomas, O.; Weiss, F.; Abrutis, A. *Patent* No **93**/08838, PCT FR94/00858 (Europe, USA).

[19]   Senateur, J.P.; Abrutis, A.; Felten, F.; Weiss, F.; Thomas, O.; Madar, R. In Advances in Inorganic Films and Coatings; Vincenzini, P.; Ed.; *Techna srl.: IT,* 1995, pp 161-166.

[20]   Abrutis, A.; Senateur, J.P.; Weiss, F.; Kubilius, V.; Bigelytė, V.; Šaltyė, Z.; Vengalis, B.; Jukna, A. *Supercond. Sci. Technol.* 1997, 10, 959-965.

[21]   Abrutis, A.; .Senateur, J.P.; Weiss, F.; Bigelyte, V.; Teiserskis, A.; Kubilius, V.; Galindo, V.; Balevicius, S. *J. Cryst. Growth* 1998, 191, 79-83.

[22] Abrutis, A.; Plausinaitiene, V.; Teiserskis, A.; Saltyte, Z.; Kubilius, V.; Bartasyte A.; Senateur, J.P. *J. Phys IV,* 2001, 11, Pr11, 215-219.

[23] Robles, J.J.; Bartasyte, A.; Ng, H.P.; Abrutis, A.; Weiss, F. Pysica C, 2003, 400, 36-42

[24] Jeschke, U; Schneider, R; Ulmer, G; Linker, G. *Phusica C,* 1995, 243, 243-251.

[25] Villard ,C.; Koren, G.; Cohen, D.; Polturak, E.; Thrane, B.; Chateignier, D. *Phys. Rev. Lett.* 1996, 77, 3913-3916.

[26] Schweitzer, D.; Bollmeier, T.; Stritzker, B.; Rauschenbach, B. *Thin Solid Films* 1996, 280, 147-151.

[27] Scherer, T.; Marienhoff, P.; Herwig, R.; Neuhaus, M.; Jutzi, W. *Physica C* 1992, 197, 79-83.

[28] Koren, G.; Polturak, E.; Levy, N.; Deutscher, G.; Zakharov, N.D. *Appl. Phys. Lett.* 1998, 73, 3763-3765.

[29] Abrutis, A.; Kubilius, V.; Teišerskis, A.; Bigelytė, V.; Galindo, V.; Weiss, F.; Senateur, J.P.; Vengalis, B.; Jukna, A.; Butkutė, R.. *Thin Solid Films* 1997, 311, 251-258.

[30] Abrutis, A.; Kubilius, V.; Bigelytė, V.; Teišerskis, A.; Šaltytė, Z.; Senateur, J.P.; Weiss F. *Mater. Lett.* 1997, 31, 201-207.

[31] Abrutis, A.; Plausinaitiene, V.; Teiserskis, A.; Kubilius, V.; Senateur, J.P.. Weiss, F. *Chem. Vap. Deposition,* 1999, 5, 171-177.

[32] Abrutis, A. *J.Phys.IV* 1999, 9, Pr8, 683-687.

[33] Abrutis, A.; Plausinaitiene, V.; Teiserskis, A.; Kubilius, V.; Saltyte, Z.; Senateur, J.P.; Dapkus, L. *J. Phys. IV,* 1999, 9, Pr9, 689-695.

[34] Chern, C.S.; Martens, J.S.; Li, J.Q.; Gallois, B.M.; Lu, P.; Kear, B.H. *Supercond. Sci. Technol.* 1993, 6, 460-463.

[35] Lu, Z.; Truman, J.K.; Johansson, M.E.; Zhang, D.; Shih, C.F.; Liang, G.C. *Appl. Phys. Lett.* 1995, 67, 712-714.

[36] Schulte, B.; Richards. B.C.; Cook, S.L. J. Alloys Compounds 1997, 251, 360-365.

[37] Sekizawa, S.; Naruse, H.; Sudoh, K.; Yoshida, Y.; Takai, Y. *Physica C* 2003, 392, 1270-1275.

[38] Endo, K.; Badica, P.; Itoh, J. *Physica C* 2003, 386, 323-326.

[39] Abrutis, A.; Teiserskis, A.; Kubilius, V.; Saltyte, Z.; Senateur, J.P.; Weiss, F.; Dubourdieu, C. Inst. *Phys. Conf. Ser.* 1999, 167, 223-226.

[40] Abrutis, A.; Plausinaitiene, V.; Teiserskis, A.; Saltyte, Z.; Kubilius, V.; Vengalis, B.; Dapkus, L.; Senateur, J.P. *J. Phys IV* 2001, 11, Pr11, 221-225.

[41] Abrutis, A.; Plausinaitiene, V.; Pasko, S.; Teiserskis, A.; Kubilius, V.; Saltyte, Z. ; Sénateur, J.P. *J. Phys. IV* 2001, 11, Pr3, 1169-1173.

[42] Lindner, J.; Weiss, F.; Senateur, J.P.; Abrutis, A. *Integrated Ferroelectrics* 2000, 30, 301-308.

[43] Weiss, F.; Senateur, J.P.; Lindner, J.; Galindo, V.; Dubourdieu, C.; Abrutis, A. *J.Phys.IV* 1999, 9, Pr8, 283-293.

[44] Lu, Y.; Li, X.W.; Gong, G.Q.; Xiao, G.; Gupta, A.; Lecoeur. P.; Sun, J.Z.; Wang, Y.Y.; Dravid. V.P. *Phys. Rev. B* 1996, 54, 8357-8360.

[45] Viret, M.; Drouet, M.; Nassar, J.; Contour, J.P.; Fermon, C.; Fert, A. *Europhys. Lett.* 1997, 39, 545-549.

[46] Mathur, N.D.; Burnell, G.; Isaac, S.P.; Jackson, T.J.; Teo, B.S.; MacManus Driscoll, J.L.; Cohen, L.F.; Evetts, J.E.; Blamire, M.G. *Nature* 1997, 387, 266-268.

[47]   Dong,Z.W.; Ramesh, R.; Venkatesan, T.; Johnson, M.; Chen, Z.Y.; Pai, S.P.;
       Talyansky, V.; Sharma, R.P.; Shreekala, R.; Lobb, C.J.; Greene, R.L. *Appl. Phys. Lett.*
       1997, 71, 1718-1720.

[48]   Vasko, V.A.; Larkin, V.A.; Kraus, P.A.; Nikolaev, K.R.; Grupp, D.E.; Nordman, C.A.;
       Goldman, A.M. *Phys. Rev. Lett.* 1997, 78, 1134-1137.

[49]   Yeh, N.C.; Vasquez, R.P.; Fu, C.C.; Samoilov, A.V.; Li, Y.; Vakili, K. *Phys. Rev. B*
       1999, 60, 10522-10526.

[50]   Koller, D; Osofsky, M.S.; Chrisey, D.B.; Horwitz, J.S.; Soulen, R.J.; Stroud, R.M.;
       Eddy, C.R.; Kim, J.; Auyeung, R.C.Y.; Byers, J.M.; Woodfield, B.F.; Daly, G.M.;
       Clinton, T.W.; Johnson, M. *J. Appl. Phys.* 1998, 83, 6774-6776.

[51]   Stroud, R.M.; Kim, J.; Eddy, C.R.; Chrisey, D.B.; Horwitz, J.S.; Koller, D.; Osofsky,
       M.S.; Soulen, R.J.; Auyeung, R.C.Y. *J. Appl. Phys.* 83, 7189-7191.

[52]   Fu, C.C.; Huang, Z.; Yeh, N.C. *Phys. Rev. B* 2002, 65, Art. No. 224516.

[53]   Okuda, N.; Matsuzaki, T.; Shinozaki, K.; Mizutani, N.; Funakubo, H. *Trans. Mater.
       Res. Soc. Jap.* 1999, 24, 51-54.

[54]   Okuda, N.; Saito, K., Funakubo, H. *Jpn. J. Appl. Phys.* 2000, 39, 572-576.

[55]   Yang, H.C.; Lin, J.I.; Horng, H.E. *Chin. J. Appl. Phys.* 1998, 36, 388-393.

[56]   Itoh, M.; Sudoh, K.; Ichino, Y.; Yoshida, Y.; Takai, Y. *Physica C* 2003, 392, 1265-
       1269.

[57]   *LANL Annual Peer Review*, August 2001.

[58]   Plausinaitienė, V. ; Abrutis, A.; Vengalis, B.; Butkute, R.; Senateur, J.P.; Saltyte, Z.;
       Kubilius, V. *Physica C* 2001, 351, 13-16.

[59]   Vengalis, B.; Plausinaitiene, V.; Abrutis, A.; Saltyte, Z.; Butkute, R.; Petrauskas, V.;
       Maria, J.; Bonfait, G. *J. Phys*, 2001, 11, Pr11, 53-57.

[60]   Vengalis, B.; Plausinaitiene, V.; Abrutis, A.; Saltyte, Z. *Acta Physica Polonica A* 2005,
       107, 286-289.

[61]   Sarkar, S.; Raychaudhuri, P.; Mal, P.K.; Bhangale, A.R.; Pinto, R. *J. Appl. Phys.* 2001,
       89, 7502-7504.

[62]   Sawa, A.; Kashiwaya, S.; Obara, H.; Yamasaki, H.; Koyanagi, M.; Yoshida, N.;
       Tanaka, Y. *Physica C* 2000, 339, 287-297.

[63]   Chen, Z.Y.; Biswas, A.; Zutic, I.; Wu, T.; Ogale, S.B.; Greene, R.L.; Venkatesan, T.
       Phys. Rev. B 2001, 63, Art. No. 212508.

[64]   Abrutis, A.; Plausinaitiene, V.; Kubilius, V.; Teiserskis, A.; Saltyte, Z.; Butkute, R.;
       Senateur, J.P. *Thin Solid Films* 2002, 413, 32-40.

[65]   Gim, Y.; Kleinsasser, A.W.; Barner, J.B. *J. Appl. Phys.* 2001, 90, 4063-4077.

[66]   Galindo, V.; Senateur, J.P.; Weiss, F.; Abrutis, A. In Book High-Temperature
       Superconductors and Novel Inorganic Materials; Van Tendeloo, G.; Ed.;. Kluwer
       Academic Publishers: HO, 1999, pp 85-90.

[67]   Weiss, F.; Senateur, J.P.; Lindner, J.; Galindo, V.; Dubourdieu, C.; Abrutis, A.
       J.Phys.IV 1999, 9, Pr8, 283-293.

[68]   Galindo, V.; Senateur, J.P.; Abrutis, A.; Teiserskis, A.; Weiss, F. *J. Crystal Growth*
       2000, 208, 357-364.

[69]   Weiss, F.; Lindner, J.; Senateur, J.P.; Dubourdieu, C.; Galindo, V.; Audier, M.;
       Abrutis, A.; Rosina, M.; Frohlich, K.; Haessler, W.; Oswald, S.; Figueras, A.; Santiso,
       J. Surf. Coat. Technol. 2000, 133, 191-197.

[70] Galindo, V.; Senateur, J.P.; Alves, E.; da Silva, R.C.; Silva, J.A.; Cruz, M.M.; Godinho, M.; Casaca, A.; Bonfait, G. *J. Low Temper. Phys.* 1999, 117, 657-661.

[71] Casaca, A.; Bonfait, G.; Galindo, V.; Senateur, J.P.; Feinberg, D. *J. Low Temper. Phys.* 1999, 117, 1459-1463.

[72] Casaca, A.; Bonfait, G.; Galindo, V.; Senateur, J.P.; Feinberg, D. *Physica B* 2000, 284, 601-603.

[73] Felten, F.; Senateur, J.P.; Labeau, M.; Yu-Zhang, K.; Abrutis. A. *Thin Solid Films* 1997, 296, 79-81.

[74] Senateur, J.P.; Dubourdieu, C.; Weiss, F.; Rosina, M.; Abrutis, A. *Adv. Mater. Opt. Electron.* 2000, 10, 155-161.

[75] Lindner, J.; Weiss, F.; Senateur, J.P.; Haessler, W.; Koebernik, G.; Figueras, A.; Santiso, J. *J. Phys. IV* 1999, 9, P8, 411-418.

[76] Lindner, J.; Weiss, F.; Senateur, J.P.; Haessler, W.; Kobernik, G.; Oswald, S.; Figueras, A.; Santiso, *J. Integrated Ferroelectrics* 2000, 30, 53-59.

[77] Oswald, S.; Hassler, W.; Reiche, R.; Lindner, J.; Weiss, F. *Microchim. Acta* 2000, 133, 303-306.

[78] Dooryhee, E.; Hodeau, J.L.; Nemoz, M.; Rodriguez, J.A.; Dubourdieu, C.; Pantou, R.; Rosina, M.; Weiss, F.; Senateur, J.P.; Audier, M.; Roussel, H.; Lindner, J. *J. Phys. IV* 2001, 11, Pr11, 267-272.

[79] Dubourdieu, C.; Senateur, J.P.; Weiss, F. In Crystal Growth in Thin Solid Films: Control of Epitaxy, Guilloux-Viry, M.; Perrin, A.; Ed.; ISBN 81-7736-095-7; *Research Signpost*: Kerala, IN, 2002; pp 169-206.

[80] Iijima, Y.; Tanabe, N.; Ikeno, Y.; Kohno, O. *Physica C* 1991, 185, 1959-1960.

[81] Iijima, Y.; Onabe, K.; Futaki, N.; Tanabe, N.; Sadakata, N.; Kohno, O.; Ikeno, Y. N., *J.Appl.Phys.* 1993, 74, 1905-1911.

[82] Fukutomi, M. ; Aoki, S.; Komari, K.; Chatterjee, R.; Maeda, H. *Physica C* 1994, 219, 333-339.

[83] Bauer, M.; Semerad, R.; Kinder, H. IEEE Trans. Appl. Supercond. 1999, 9, 1502-1505.

[84] Goyal, A. ; Norton, D.P.; Kroeger, D.M.; Christen, D.K.; Paranthaman, M.; Specht. E.D.; Budai, J.D.; He, Q. ; Saffian, S.; List, F.A.; Lee, D.F.; Hatfield, E.; Martin, P.M.; Klabunde, C.E.; Mathis, J.; Park, C. *J. Mater. Res.* 1997, 12, 2924-2940.

[85] Matsumoto, K.; Niiori, Y.; Hirabayashi, I.; Koshizuka, N.; Watanabe, T.; Tanaka, Y.; Ikeda, M. In Advances in Superconductivity. Proceedings of the 10th International Symposium; Springer-Verlag: Tokyo, JP, 1998, Vol 2, pp 611-614.

[86] Matsumoto, K.; Kim, S. B.; Wen, J. G.; Hirabayashi, I.; Tanaka, S.; Uno, N.; Ikeda, M.; IEEE Trans. *Appl. Supercond.* 1999, 9, 1539-1542.

[87] Beauquis, S.; Jiménez, C.; Weiss, F. In High temperature superconductivity I: Materials; Narkilar, A.V.; Ed.; ISBN 3-540-40631; Springer-Verlag: Berlin Heidelberg New York, 2004, pp 115-167.

[88] Jimenez, C.; Weiss, F.; Senateur, J.P.; Abrutis, A.; Krellmann, M.; Selbmann, D.; Eickemeyer, J.; Stadel, O.; Wahl, G. *IEEE Trans. App Supercond.*, 2001, 11, 2905 -2908.

[89] Schmatz, U.; Weiss, F.; Klippe, L.; Stadel, O.; Wahl, G.; Krellmann, M.; Selbmann, D.; Hubert-Pfalzgraf, L.; Guillon, H.; Peña, J.; Vallet-Regi, M. *Electrochem. Soc. Proced.* 1997, 97-25, 1005-1011.

[90]   Krellmann, M.; Selbmann, D.; Schmatz, U.; Weiss, F., *J. Alloys Compounds* 1997, 251, 307-309.

[91]   Meffre, W. PhD Thesis INPG, 1999, 85.

[92]   Brite Euram READY Project, BRPR-CT98-076

[93]   Teišerskis, A.; Abrutis, A.; Saltyte, Z.; Jiménez, C.; Weiss, F.; Sénateur, J.P.; Suo, H.L.; Genoud, J.Y.; Flükinger, R. In Thin films deposition of oxide multilayers. Industrial-scale processing, Vengalis, B.; Abrutis, A; Ed.; ISBN 9986-19-394-X; Vilnius University Press: Vilnius, LT, 2000; pp 57-60.

[94]   Selbmann, D.; Eickmeyer, J.; Wendrock ,H.; Jimenez, C.; Donet, S.; Weiss, F.; Miller, U.; Stadel, O. *J. Phys. IV* 2001, 11, Pr11, 239-245.

[95]   Eickemeyer, J.; Selbmann, D.; Opitz, R.; Wendrock, H.; Maher, E.; Miller, U.; Prusseit, W. *Physica C* 2002, 372-376, 814-817.

[96]   Donet, S.; Weiss, F.; Sénateur, J.P.; Chaudouet, P.; Abrutis, A.; Teišerskis, A.; Saltyte, Z.; Selbmann, D.; Eickemeyer, J.; Stadel, O.; Wahl, G.; Jimenez, C.; Miller. U. *Physica C* 2002, 372-376, 652-655.

[97]   Donet, S.; Weiss, F.; Senateur, J.P.; Chaudouët, P.; Abrutis, A.; Teišerskis, A.; Saltyte, Z.; Selbmann, D.; Eickemeyer, J.; Stadel, O.; Wahl, G.; Jimenez, C.; Miller, U. *J. Phys. IV* 2001, 11, Pr11, 319-323.

[98]   Donet, S.; Weiss, F.; Chaudouet, P.; Selbmann, D.; Prusseit, W.; Bruzek, C.E.; Saugrain, J.M. In ChemicaL Vapor Deposition and EUROCVD 14 Proceedings; Allendorf, M.D.; Maury, F.; Teyssandier, F.; Ed.; ISBN 1-56677-378-4; *The Electrochemical Soc. Inc:* Penington, US, 2003, Vol. 2003-8, pp 1522-1527.

[99]   Dzick, J.; Hoffmann, J.; Sievers, S.; Lautshcor, L.O.; Freyhardt, H.C. *Physica C* 2002, 372-276, 723-728.

[100] Donet, S.; Weiss, F.; Chaudouët, P.; Beauquis, S.; Abrutis, A.; Freyhardt , H.C.; Usokin, A.; Selbmann, D.; Eickemeyer, J.; Jimenez, C.; Bruzek, C.E.; Saugrain, J.M. *IEEE Trans. Appl. Supercond.* 2003, 13, 2524-2527.

[101] Wiesmann, J.; Dzick, J.; Hoffman, J.; Heinemann, K.; Freyhardt, H.C. *J. Mat. Res.* 1998, 13, 3149-3152.

[102] Abourida, M.; Guillon, H.; Jiménez, C.; Decams, J.M.; Weiss, F.; Valet, O.; Doppelt, P. In ChemicaL Vapor Deposition and EUROCVD 14 Proceedings; Allendorf, M.D.; Maury, F.; Teyssandier, F.; Ed.; ISBN 1-56677-378-4; *The Electrochemical Soc. Inc:* Penington, US, 2003, Vol. 2003-8, pp 938-945.

[103] Samoilenkov, S.; Stefan, M.; Wahl, G.; Paramonov, S.; Kuzmina, N.; Kaul, A. *Chem. Vap. Deposition* 2002, 2, 74-78.

[104] Usoskin, A.; Issaev, A.; Freyhardt, H.C.; Leghissa, M.; Oomen, M.P.,; Neumueller, H.-W. *Physica C* 2002, 372-376, 857-862.

[105] Usoskin, A.; Isaev, A.; Knoke, J.; Freyhardt, H.C. Proceedings of the 6th EUCAS 2003, Sorrento, IoP, *Conf. Serie* Number **181**, 1737- 1740.

[106] Motowidlo, L.R.; Selvamanickam, V.; Galinski, G.; Vo, N.; Haldar, P.; Sokolowski, R.S. *Physica C* 2000, 333, 44-50.

[107] Selvamanickan V.; Knoll A.; Xie Y.; Li Y.; Chen Y.; Reeves J.; Xiong X.; Qiao Y.; Slagaj T.; Lenseth K.; Hazelton D.; Reis C.; Yumura H.; Weber C. *IEEE Trans. Appl. Supercond.* 2005, 15, 2596-2599.

[108] Onabe K.; DoiT.; Kashima N.; Nagaya S.; Saitoh, T.; *Physica C* 2003, 392-296, 863-866.

[109] Kashima N.; Niwa T.; Mori M.; Nagaya S.; Muroga T.; Miyata S.; Watanabe T.; Yamada Y.; Izumi T.; Shiohara Y. *IEEE Trans. Appl. Supercond.* 2005, 15, 2763-2766.

[110] Kim, C.J.; Kim, H.J.; Sun, J.W.; Ji, B.K; Kim, H.S.; Jo, J.; Jun, B.H.; Jung, C.H.; Park, S.D.; Park, H.W.; Hong, G.W. *Physica C* 2003, 386, 327-332.

[111] http://www.jipelec.com/company/index.htm

[112] http://www.aixtron.com

In: Recent Developments in Superconductivity Research   ISBN 978-1-60021-462-2
Editor: Barry P. Martins pp. 201-243   © 2007 Nova Science Publishers, Inc.

Chapter 7

VORTEX STATE MICROWAVE RESPONSE IN SUPERCONDUCTING CUPRATES AND MGB$_2$

E. Silva[1,*] *N. Pompeo*[2] *S. Sarti*[3,†] *C. Amabile*[4]
[1]Dipartimento di Fisica "E. Amaldi" and Unità CNISM,
Università Roma Tre, Via della Vasca Navale 84, 00146 Roma, Italy
[2]Dipartimento di Fisica "E.Amaldi" and Unità CNISM,
Università Roma Tre, Via della Vasca Navale 84, 00146 Roma, Italy
[3]Dipartimento di Fisica and Unità CNISM,
Università "La Sapienza", P.le Aldo Moro 2, 00185 Roma, Italy
[4]Dipartimento di Fisica,
Università "La Sapienza", P.le Aldo Moro 2, 00185 Roma, Italy

Abstract

We investigate the physics of the microwave response in $YBa_2Cu_3O_{7-\delta}$, $SmBa_2Cu_3O_{7-\delta}$ and MgB_2 in the vortex state. We first recall the theoretical basics of vortex-state microwave response in the London limit. We then present a wide set of measurements of the field, temperature, and frequency dependences of the vortex state microwave complex resistivity in superconducting thin films, measured by a resonant cavity and by swept-frequency Corbino disk. The combination of these techniques allows for a comprehensive description of the microwave response in the vortex state in these innovative superconductors. In all materials investigated we show that flux motion alone cannot take into account all the observed experimental features, neither in the frequency nor in the field dependence. The discrepancy can be resolved by considering the (usually neglected) contribution of quasiparticles to the response in the vortex state. The peculiar, albeit different, physics of the superconducting materials here considered, namely two-band superconductivity in MgB_2 and superconducting gap with lines of nodes in cuprates, give rise to a substantially increased contribution of quasiparticles to the field-dependent microwave response. With careful combined analysis of the data it is possible to extract or infer many interesting quantities related to the vortex state, such as the temperature-dependent characteristic vortex frequency and vortex viscosity, the field dependence of the quasiparticle density, the temperature dependence of the σ-band superfluid density in MgB_2

---

*E-mail address: silva@fis.uniroma3.it
†E-mail address: stefano.sarti@roma1.infn.it

Keywords: superconductivity, surface impedance, microwaves, vortex motion, cuprates, $YBa_2Cu_3O_{7-\delta}$, $SmBa_2Cu_3O_{7-\delta}$, MgB_2, vortex viscosity.

1 Introduction

One of the most versatile experimental methods for the investigation of the physics of super-conductors is the measurement of the complex response to an alternating electromagnetic (e.m.) field in the radiofrequency (*rf*) and microwave ranges. The resulting data have been of great help in the understanding of the physics of conventional superconductors. Even confining the treatment to the linear response, in conventional superconductors microwave or *rf* measurements allowed, e.g., for the determination of the existence [1] and temperature dependence [2] of the superconducting gap, for the settling of the kind of dynamical fluctuations [3, 4, 5], for the determination of the penetration depth [6, 7, 8], for the determination of the thermodynamical critical field and of the third critical field for surface superconductivity [9], and for the determination of the upper critical field [10].

In type-II superconductors the investigation was extended to the mixed state: when the magnetic field $H_{c1} < H < H_{c2}$, with H_{c1} and H_{c2} the temperature dependent lower and upper critical field, respectively, the magnetic flux penetrates the superconductor as quantized flux lines, each carrying one flux quantum $\Phi_0 = 2.07 \times 10^{-15}$ T·m². Such flux lines, in presence of an electric current density **J**, are subjected to a Lorentz force (per unit length) $\mathbf{F_L} = \mathbf{J} \times \mathbf{\Phi_0}$ (where $\mathbf{\Phi_0} \parallel \mathbf{B}$). Moving vortices dissipate energy, so that an ideal type-II superconductor always has a finite resistivity in the vortex state. Dissipation can be viewed either as due to the continuous conversion of Cooper pairs into quasiparticles at the (moving) vortex boundary [11], or to Joule heating inside of the vortex core [12, 13]. In any case, the dissipation is expected to depend mainly on the fundamental properties of the superconductors, such as the quasiparticle density of states and relaxation time in the vortex core, and thus it is expected to be very similar in different samples of the same material.

A dissipationless regime can be achieved in dc by pinning vortices to defects [14, 15, 16]. When the force acting on vortices is alternating, pinning determines an additional imaginary (out-of-phase) response. As opposed to energy dissipation, the particular pinning mechanism and its efficacy are expected to be strongly sample dependent.

To take into account the energy dissipation and the effects of pinning a simple way is to build up an equation of motion for a single vortex. In a simple model, independently pinned vortices are considered and the equation of motion for the displacement **u** from the equilibrium position is written in the elastic approximation [1]: $\eta\dot{\mathbf{u}} + \kappa_p\mathbf{u} = \mathbf{F_L}$, where η is the so-called *vortex viscosity* (per vortex unit length) and takes into account the energy dissipation, and κ_p is the *pinning constant* (per vortex unit length) and takes into account the pinning of vortices. This simple equation of motion gives rise to a complex vortex motion resistivity due to an alternating current density $\mathbf{J}e^{i\omega t}$, with $\nu = \omega/2\pi$ the measuring frequency, that can be written as [22]:

$$\rho_{vGR} = \frac{\Phi_0 B}{\eta}\frac{1}{1 - i\frac{\kappa_p}{\eta\omega}} = \rho_{ff}\frac{1}{1 - i\frac{\omega_p}{\omega}} \tag{1}$$

---

[1]The vortex mass is usually neglected [17, 18], even if this is still a debated topic [19, 20, 21].

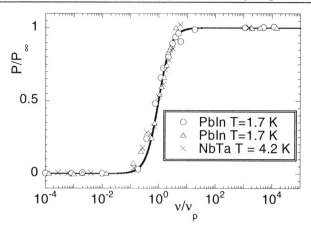

Figure 1: Normalized dissipated power in type-II superconductors at $H = \frac{1}{2}H_{c2}$ as a function of the measuring frequency normalized to ν_p (replotted from [22]) and fit by Eq.(1). $\nu_p = 3.9, 5.1, 15$ MHz [22].

where the last equality defines the so-called *flux-flow resistivity*, ρ_{ff}, and the *depinning frequency* (sometimes called "pinning frequency") $\nu_p = \omega_p/2\pi$. Referring to the physical meaning of the vortex parameters, η is expected to be a very similar, if not universal, function of the external parameters (e.g., the temperature) in different samples of the same material: it is a quantity related to the fundamentals of the physics of type-II superconductors. In particular, it was demonstrated by Bardeen and Stephen (BS) [13] that, in dirty superconductors, $\eta = \Phi_0 B_{c2}/\rho_n$, where ρ_n is the normal state dc resistivity and B_{c2} is the upper critical field.

In deep contrast, one expects largely different magnitudes, temperature and magnetic field dependence for κ_p (or ω_p) in different samples even of the same material, being the pinning constant related to extrinsic properties. Its study can, on one side, shed light on the characteristic features of the vortex matter, and on the other side can be of essential importance for the applications of superconductors [14, 15, 16].

The applicability of Eq.(1) was first brought to the attention of the scientific community with the seminal paper by Gittleman and Rosenblum [22]: as shown in Fig.1, the data for the dissipated power ($\propto \mathrm{Re}[\rho_{vGR}]$) in thin superconducting films followed very closely Eq.(1) over several orders of magnitude for the measuring frequency. The success of this extremely simple model determined its assumption for the interpretation of the data taken in high-T_c superconductors (HTCS). Many studies were carried out in HTCS, aimed at the determination of the vortex viscosity and pinning constant through measurements of the radiofrequency, microwave or millimeter-wave response [23, 24, 25, 26, 27, 28, 29, 30, 31, 32, 33, 34, 35, 36, 37, 38, 39, 40, 41, 42, 43, 44]. In particular, in the most studied compound YBa$_2$Cu$_3$O$_{7-\delta}$ (YBCO), it was found that the depinning frequency was located in the GHz range, thus making microwave techniques the ideal candidate method for the experimental investigation of the vortex motion. However, even after many years from the discovery of YBCO, the values given for vortex parameters as determined by microwave measurements (see [35] for a clear overview) presented a very puzzling framework. As reported in Figs.2

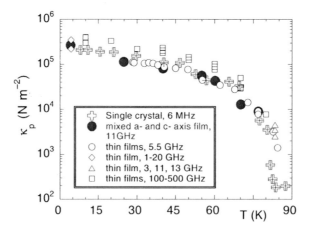

Figure 2: Pinning constant κ_p for many different YBCO samples: [23], crosses; [28], red full dots; [30], open circles; [31], diamonds; [33], open squares; [34], triangles. Data replotted from original papers. A remarkable collapse of the data on a single curve is evident.

and 3, with data replotted from original papers [23, 27, 28, 30, 31, 32, 33, 34] cited in [35], it was shown that while data for κ_p taken on different samples (c-axis and mixed a- and c-axis films from various sources, single crystals) lie all together on the same curve, the vortex viscosity presented a wide scattering of the data. As discussed above, due to the meaning of the vortex parameters this finding is not easily explained: one would rather expect, for a given material, a universal behavior for η and a sample-dependent k_p.

Similar oddities were systematically evident also in different frequency ranges in various HTCS. Sub-THz measurements of the vortex-state microwave resistivity of thin YBCO films [33] showed that vortex parameters obtained from conventional, vortex-motion-driven response, were in strong contrast with calculations of the same parameters from microscopic theories. Again, the pinning constant was found nearly sample-independent. The apparent vortex viscosity would differ from the microscopic calculation by more than an order of magnitude. An alternative analysis of the data [45] suggested that it was not possible to ignore, in the interpretation of the data, the effect of field-induced pair breaking. Accordingly to that picture, data of the imaginary conductivity in the same frequency range in $Bi_2Sr_2CaCu_2O_{8+x}$ (BSCCO) films in the vortex state [46] could be accounted for even without resorting to any vortex motion model, being the field-induced superfluid density suppression sufficient for a quantitative description.

Summarizing, it was clear that Eq.(1) did not provide a comprehensive explanation of the vortex state microwave response in HTCS. Even if much more complicated (and probably more realistic) vortex models could be invoked [47, 48], additional mechanisms had to be investigated.

Granularity has been sometimes indicated as a possible dominant source for the losses in the microwave response in superconducting films. Manifestations of granularity include weak-links dephasing [49, 50, 51, 52, 53, 54, 55], Josephson fluxon (JF) dynamics [56] and, as recently studied, Abrikosov-Josephson fluxon (AJF) dynamics [57, 58]. Weak-links

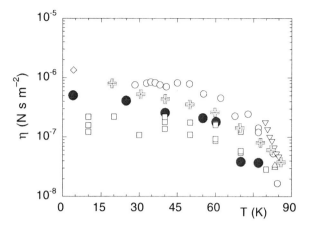

Figure 3: Vortex viscosity η for many different YBCO samples: [27], down triangles; [28], red full dots; [30], open circles; [31], diamonds;[32], crosses; [33], open squares; [34], triangles. Data replotted from original papers. A remarkable spread of the data is evident.

dephasing is charaterized by a very sharp increase of the dissipation at dc fields of order of, or less than, 20 mT, accompanied by a strong (and sometimes exceptionally strong) hysteresis with increasing or decreasing field [49, 50, 51, 52]. However, those effects are relevant in large-angle grain boundaries (or very weak links), such as those found in pellets and granular samples, and are not observed in good thin films. Josephson fluxon dynamics has been studied essentially in relation to nonlinear effects [59, 60], due to the short JF nucleation time. One of the characteristic features of AJF should be a rather pronounced sample dependence (defect-driven) of the microwave response, as in fact reported in Tl-2212 films [61]. This fact contrasts with the pseudo-universal behavior of the estimated κ_p, even if contributions from AJF cannot in general be excluded *a priori*.

Another intrinsic, unavoidable source for the measured microwave resistivity is obviously the conductivity due to charge carriers, that we write in terms of the so-called "two-fluid model" [11]: charge carriers are thought as given by a superconducting fraction x_s and a "normal" fraction (more appropriately, due to quasiparticle excitations) x_n. When the measuring frequency is much smaller than the quasiparticle relaxation rate this leads to the well-known dependence of the charge carriers conductivity at nonzero frequency:

$$\sigma = \sigma_1 - i\sigma_2 = \frac{ne^2}{m\omega}\left(\omega\tau_{qp}x_n - ix_s\right) = \frac{1}{\mu_0\omega}\left(\frac{2}{\delta_{nf}^2} - i\frac{1}{\lambda^2}\right) \tag{2}$$

where n is the charge carrier density, m is the carrier effective mass, τ_{qp} is the quasiparticle scattering time, and the last equality defines the temperature, field and frequency dependent normal fluid skin depth δ_{nf} and the temperature and field dependent London penetration depth λ. How this charge carriers conductivity combines with the vortex motion resistivity is discussed in Section 2.

It should be mentioned that measurements of σ_1 and σ_2 in zero applied field have been demonstrated to be of crucial importance in the determination of the peculiarities of HTCS. As a partial list of examples, microwave measurements of the low-temperature variation of the superfluid fractional density, expressed as $x_s(T) = \left[\frac{\lambda(0)}{\lambda(T)}\right]^2$, have shown a clear nonactivated behavior in YBCO [62, 63], thus giving strong evidence for an anisotropic superconducting gap, with lines of nodes. In particular, a linear decrease with temperature was found in clean crystals [62, 63], demonstrating the existence of zero-energy excitations at low temperatures. Moreover, it was shown by microwave spectroscopy of the real part of the response in YBCO [64] that the quasiparticle relaxation rate $1/\tau_{qp}$ decreased, with respect to its value above T_c, by up to two orders of magnitudes. Short relaxation rates below T_c were confirmed in YBCO films by millimeter-wave interferometry [65], in YBCO crystals by surface resistance measurements [66], and in YBCO films by far-infrared measurements [67]. Crystals consistently presented $1/\tau_{qp}$ at low temperatures up to two order of magnitudes smaller than in the normal state [64, 66]. Similar drops were found in some films [65], while other films presented a difference in $1/\tau_{qp}$ of only a factor of two [67].

All the above mentioned mechanisms present novel features in the recently discovered [68] metallic superconductor MgB_2. The well-established, albeit sensitive to interband scattering, two-gap nature of this compound [69, 70, 71] is expected to strongly affect the intrinsic properties, and specifically to the present study, the vortex viscosity and the superfluid density. Microwave measurements [72, 73, 74] in zero dc magnetic field showed peculiar temperature dependences of the surface impedance, due to the existence of the double gap. Measurements of the magnetic-field-dependent microwave surface impedance [75, 76] at a single frequency presented very puzzling data for the vortex parameters, in particular strong field dependence of the vortex viscosity (or, using the Bardeen-Stephen model, for the upper critical field itself). These findings have their counterpart in other experimental and theoretical results: the London penetration depth was found theoretically to strongly depend on impurity level [77], thus giving an explanation for the large range of reported values. μSR spectroscopy data required for their satisfactory explanation that the application of a magnetic field induces a strong increase in the penetration depth at low fields [78], or at least that two different characteristic lengths [79] were involved. Moreover, it was theoretically shown [80] that the structure of vortices is very different in two-gap, with respect to single-gap, superconductors. To our knowledge, there are no exhaustive theories at present for what concerns the dynamics of vortices in multigap superconductors. There are however several experimental indications that a sufficiently strong magnetic field (of order 1 T) can quench the two-gap nature of MgB_2: scanning tunnel [81] and point-contact [82] spectroscopies showed that the superconductivity coming from the π band is strongly suppressed with the application of an external field ~ 0.5 T. Neutron spectroscopy [83] brought evidence for a transformation of the vortex lattice at a similar field.

From all the above considerations it appears that, for different reasons, the microwave vortex state properties in YBCO (and, in general, in rare earth substituted materials RE-BCO) and MgB_2 are not yet unanimously understood. Aim of this study is to present microwave measurements at different frequencies and in wide magnetic field ranges, in order to identify the main fundamental mechanisms acting on the complex response. A related interest is to determine whether simple vortex motion models can be safely applied, possibly confining the complexity in some lumped parameter, or different approaches are needed.

This Chapter is organized as follows: Sec.2 presents an overview of the mean-field theory for vortex motion and two-fluid complex response, with some extension due to the peculiar electronic structure of the superconductors under study. Sec.3 presents the main properties of the samples under study, the electromagnetic response in thin films, and some detail on the resonant cavity and the Corbino disk employed for the measurements. Sec.4 presents and discusses the results in YBCO, MgB$_2$ and SmBCO. Conclusions are drawn in Sec.5.

2   Theoretical Background

Considering an electromagnetic field incident on a flat interface between a generic medium and a bulk, thick (with respect to the penetration depth and skin depth) (super)conductor, the response to the field is given by the surface impedance Z_s [84]. Z_s equals the ratio between the tangential components of the alternating electric and magnetic fields, $Z_s = \frac{E_{\parallel}}{H_{\parallel}}$. This expression can be easily cast in the form $Z_s = i\omega\mu_0\widetilde{\lambda}$, being $\widetilde{\lambda}$ an appropriate complex screening length. In the normal state, the screening length $\widetilde{\lambda} \to \frac{\delta_n}{\sqrt{2i}}$, where $\delta_n^2 = 2\rho_n/\mu_0\omega$ is the skin depth and ρ_n is the normal state resistivity. Deep in the superconducting state ($T \to 0$), the screening length is the London penetration depth λ. The complex response can be also described by a complex resistivity $\widetilde{\rho}$, related to the complex screening length $\widetilde{\lambda}$ through the relation $\widetilde{\rho} = i\omega\mu_0\widetilde{\lambda}^2$ (or equivalently to the surface impedance Z_s through the relation $Z_s = \sqrt{i\omega\mu_0\widetilde{\rho}}$). Thus, the response can be expressed formally either by the surface impedance, the complex resistivity, the complex conductivity, or the complex screening length.

Many models have been developed and used for the frequency response in the vortex state, with various degrees of complexity [39, 85, 86, 87, 88, 89, 90, 91]. A very general expression for the surface impedance of a semi-infinite superconductor in the mixed state has been calculated [88], and later extended [92, 93, 94], by Coffey and Clem (CC) within the limit of validity of the two-fluid model (that is, in the local response limit). The result was expressed in terms of the combination of three complex screening lengths, related to the different contributions to the overall e.m. response: the superfluid response, given by the temperature and field dependent London penetration depth $\lambda(T, B)$; the normal fluid skin depth, given by $\delta_{nf}(T, B, \omega) = [2/\mu_0\omega\sigma_{nf}(T, B)]^{1/2}$ with σ_{nf} the quasiparticle conductivity; and the vortex response, given by the complex vortex penetration depth $\widetilde{\delta}_v(T, B, \omega) = [2\rho_v(T, B, \omega)/\mu_0\omega]^{1/2}$. The resulting expression for the surface impedance reads:

$$Z_s(T, B) = i\omega\mu_0\widetilde{\lambda} = i\omega\mu_0 \left(\frac{\lambda^2(T, B) - (i/2)\widetilde{\delta}_v^2(B, T, \omega)}{1 + 2i\lambda^2(T, B)/\delta_{nf}^2(B, T, \omega)} \right)^{1/2} \tag{3}$$

In terms of the complex microwave resistivity of the superconductor one writes $\widetilde{\rho} =$

$\rho_1 + i\rho_2 = i\omega\mu_0\tilde{\lambda}^2$ and, after some algebra, one gets:

$$\rho_1 = \frac{1}{1 + 4(\lambda/\delta_{nf})^4} \left[r_1(B,T,\omega) + 2\left(\frac{\lambda}{\delta_{nf}}\right)^2 r_2(B,T,\omega) \right]$$

$$\rho_2 = \frac{1}{1 + 4(\lambda/\delta_{nf})^4} \left[r_2(B,T,\omega) - 2\left(\frac{\lambda}{\delta_{nf}}\right)^2 r_1(B,T,\omega) \right]$$

(4)

where $r_1 = \mathrm{Re}[\rho_v]$, $r_2 = \mathrm{Im}[\rho_v] + 2\rho_{nf}\left(\frac{\lambda}{\delta_{nf}}\right)^2$ and $\rho_{nf} = 1/\sigma_{nf}$.

Eqs.(4) contain in a selfconsistent way both the quasiparticle contribution (through λ and δ_{nf}) and the motion of vortices (through ρ_v). The model is founded on the interaction between charge carriers and a system of magnetic vortices moving under the influence of rf currents and pinning phenomena. Charge carriers are thought as bearing superconducting currents, represented by a superfluid characterized by the London penetration depth λ which gives rise to an imaginary conductivity $1/\mu_0\omega\lambda^2$, and normal currents represented by σ_{nf}. These equations should be considered as *master equations*, since the actual dependence of both ρ_1 and ρ_2 as a function of temperature, magnetic field and frequency is dictated by the functional dependence of λ, δ_{nf} and ρ_v with respect to the same parameters. In particular, λ and δ_{nf} (through σ_{nf}) may vary as a function of temperature and field in different ways depending, e.g., on the order parameter symmetry. On the other hand, ρ_v may depend on temperature, magnetic field and frequency in several different ways, depending on the pinning strength, inter-vortices interactions and periodicity of the pinning potential. This will in general result in rather different dependencies for both the real and the imaginary part of the resistivity.

In order to catch some general feature out of Eqs.(4), we first consider the vortex motion contribution in the specific case of frequencies not too small, so that the long range order of the pinning potential becomes irrelevant. This should be in general the case for microwave frequencies. In this case, one may use the specific expression of ρ_v obtained for periodic pinning potentials [88, 89]:

$$\mathrm{Re}[\rho_v] = \rho_{ff}\frac{\epsilon + (\nu/\nu_0)^2}{1 + (\nu/\nu_0)^2}$$

$$\mathrm{Im}[\rho_v] = \rho_{ff}\frac{1-\epsilon}{1 + (\nu/\nu_0)^2}\frac{\nu}{\nu_0}$$

(5)

where ϵ is a creep factor that ranges from $\epsilon = 0$ (no flux creep) to $\epsilon = 1$ (free vortex motion), and ν_0 is a characteristic frequency which, in absence of creep phenomena, corresponds to the depinning frequency $\nu_p = \kappa_p/2\pi\eta$. It is worth to stress that, although Eqs.(5) have been obtained within a specific assumption, similar behaviors as a function of frequency can be reasonably assumed even in more general cases: for vortices interacting with pinning centers, in fact, it is always possible to define a characteristic frequency ν_0 which separates a low frequency regime, in which the vortices cannot follow the alternate Lorentz force acting on them, from a high frequency regime where vortices oscillate in-phase around their equilibrium positions even when strongly pinned. If the frequency is swept across

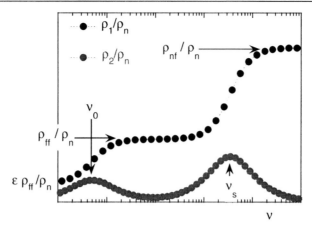

Figure 4: Exemplification of $\rho_1(\nu)$, $\rho_2(\nu)$ curves (frequency in a logarithmic scale) according to Eqs.(6). Also shown, the role of the main parameters.

ν_0, the observed Re $[\rho_v]$ increases from a low frequency value to the free flow value, while Im $[\rho_v]$ presents a maximum at ν_0. This is qualitatively the behavior predicted by Eqs.(5).

In addition, if it is further verified that $\omega\tau_{qp} \ll 1$, σ_{nf} is real and does not depend on frequency. One may write $2(\lambda/\delta_{nf})^2 = \nu/\nu_s$, with $\nu_s = 1/2\pi\mu_0\sigma_{nf}\lambda^2$.

With these substitutions Eqs.(4) can be explicitly rewritten as follows:

$$\rho_1 = \frac{1}{1+(\nu/\nu_s)^2}\left[\rho_{ff}\frac{\epsilon+(\nu/\nu_0)^2}{1+(\nu/\nu_0)^2}+\frac{\nu}{\nu_s}\left(\frac{\nu}{\nu_s\sigma_{nf}}+\rho_{ff}\frac{1-\epsilon}{1+(\nu/\nu_0)^2}\frac{\nu}{\nu_0}\right)\right]$$

$$\rho_2 = \frac{1}{1+(\nu/\nu_s)^2}\left[\frac{\nu}{\nu_s\sigma_{nf}}+\rho_{ff}\frac{1-\epsilon}{1+(\nu/\nu_0)^2}\frac{\nu}{\nu_0}-\frac{\nu}{\nu_s}\left(\rho_{ff}\frac{\epsilon+(\nu/\nu_0)^2}{1+(\nu/\nu_0)^2}\right)\right]$$

(6)

It is interesting to discuss some useful limits of Eqs.(6).

The conventional low temperature limit. At temperature low enough, so that creep phenomena are not relevant ($\epsilon = 0$ and $\nu_0 = \nu_p$) and the normal fluid conductivity can be neglected, one has $\sigma_{nf} \to 0$ and $\nu_s \to \infty$, so that

$$\rho_1 = \rho_{ff}\frac{1}{1+(\nu_p/\nu)^2}$$

$$\rho_2 = \mu_0\omega\lambda^2 + \rho_{ff}\frac{\nu_p/\nu}{1+(\nu_p/\nu)^2}$$

(7)

which are equivalent to the conventional Gittleman-Rosenblum expressions for the microwave resistivity in the mixed state, Eqs.(1), apart from the first term in the imaginary part, representing the zero-field imaginary conductivity.

The two-fluid limit. As a second example, it is easily seen that for $B \to 0$ one has $\rho_{ff} = 0$ and Eqs.(6) reduce to the two-fluid conductivity, Eqs.(2).

The high frequency limit: free flow. When the measuring frequency is much larger that the vortex characteristic frequency, $\nu \gg \nu_0$, Eq.(6) reduces to

$$\rho_1 \simeq \frac{1}{1 + (\nu/\nu_s)^2} \left[\rho_{ff} + \frac{1}{\sigma_{nf}} \left(\frac{\nu}{\nu_s} \right)^2 \right]$$

$$\rho_2 \simeq \frac{\nu/\nu_s}{1 + (\nu/\nu_s)^2} \left(\frac{1}{\sigma_{nf}} - \rho_{ff} \right)$$

(8)

In this limit vortices oscillate in phase making very short displacements from their equilibrium position. Despite the possibly finite pinning, as could be determined by, e.g., dc resistivity or magnetization, the vortex response coincides with the free flux flow. In this case the response depends on intrinsic physical quantities only: ν_s, ρ_{ff} and σ_{nf}. The analysis of the data in this limit is more stringent, since only three parameters are involved (and, in addition, ν_s and σ_{nf} are related one to each other). It should be noted that this limit might apply to various measurements at the high edge of the microwave spectrum.

Up to now we have mainly considered the vortex motion. However, both superfluid and quasiparticle densities are affected by the magnetic field. In fully gapped superconductors, such as conventional superconductors, the quasiparticle density of states (DOS) at Fermi level is nonzero only within the vortex core. As a consequence, the total DOS is proportional to the area occupied by the cores and the field-dependent depletion of the superfluid fractional density is simply $\Delta x_s \propto \frac{\xi^2}{R^2} \approx B/B_{c2}$, where the coherence length ξ gives the dimension of the vortex and $R \sim \sqrt{\Phi/B}$ is the intervortex spacing, with Φ the total magnetic flux through the sample.

The situation is rather different in superconductors with lines of nodes in the gap. In this case, it has been shown that extended states *outside* of the vortex core have the most relevant weight [95, 96]. However, such states have zero DOS only at the Fermi level, and any finite energy brings a variation of the quasiparticle fractional density. As an example, circulating supercurrents around vortices give rise to a Doppler shift [97, 95] of the quasiparticle energy. This results in a depletion of the superfluid fractional density which is proportional to the spatial range of the circulating supercurrents, $min\{\lambda, R\} \approx R$ for fields not too low, and to the number of vortices $\sim B/\Phi$. As a result, one has $\Delta x_s \approx \sqrt{B/B_{pb}}$ where B_{pb} is a characteristic pair-breaking field, that can depend on the gap gradient at the nodes, on impurity scattering and temperature, but it can be assumed (as a first approximation) to be $B_{pb} \approx cB_{c2}$ [96], with $c \sim o(1)$.

Thus, in general one can write:

$$x_s(T, B) = x_{s0}(T) - \Delta x_s(T, B) \approx x_{s0}(T) \left[1 - \left(\frac{B}{cB_{c2}} \right)^\alpha \right]$$

(9)

where α depends on the symmetry of the gap and on impurities [98], and in clean cases $\alpha = 1$ or $\alpha = \frac{1}{2}$ in fully gapped superconductors and in superconductors with lines of nodes, respectively. More accurate treatments [99, 100] do not change the qualitative result of a significant reduction of the superfluid density with the application of a magnetic field.

It is useful to plot some illustrative curve of the theoretical predictions, in order to elucidate the role of the different parameters. We will also make use of the limits above discussed.

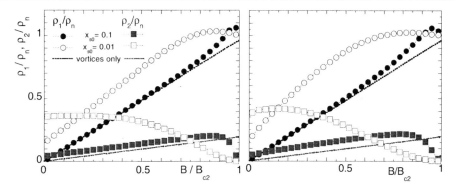

Figure 5: Reduced field dependence of the vortex state complex resistivity according to the CC model [88], Eq.(4), for $\nu/\nu_0 = 5$. The curves for small (full symbols) and very small (open symbols) superfluid concentration are reported, and compared to the vortex motion contribution alone ($\mathrm{Re}[\rho_v]$, $\mathrm{Im}[\rho_v]$ as in Eq.(5), dashed lines). Parameters are reported in the text. Left panel, fully gapped superconductors. Right panel, superconductor with lines of nodes in the gap (see text).

We present first schematic shapes of the curves $\rho_1(\nu)$ and $\rho_2(\nu)$ as obtained from the general expression, Eqs.(6). We stress that when plotting resistivity vs. frequency it is not necessary to assume any specific temperature or field dependence for the quantities appearing in the equations. As shown in Fig.(4), the shape of the two curves $\rho_1(\nu)$ and $\rho_2(\nu)$ is affected in different ways by the various parameters (ρ_{ff}, ϵ and ν_0 determine the vortex motion, σ_{nf} and ν_s the properties of the quasiparticles). In particular, it is seen that even a rather simple model can give significantly different curves for $\rho_1(\nu)$ and $\rho_2(\nu)$. Since the variation range of the involved parameters is very wide[2], it is evident that a wide measuring frequency range can prove especially useful in the interpretation of the data.

In this Chapter we will report also data for the complex resistivity at a single frequency (resonant cavity technique) and as a function of the field, so that it is appropriate to present and discuss some theoretical curve at a fixed frequency. In order to plot the microwave resistivity *vs.* the magnetic field, it is necessary to make some additional hypotheses with respect to the frequency plots. In particular, it is necessary to assume the field dependence of the parameters defining the frequency dependence of the resistivity. In order to limit the number of the assumptions, we restrict ourselves to the case of sufficiently high frequency ($\nu \gg \nu_0$) so that the field variations of ν_0 (and, to a larger extent, of ϵ) become almost irrelevant. We also assume that the vortex viscosity η does not depend on field. For the quasiparticle and superfluid response, we assume that $\sigma_{nf} = x_n(T, B)\sigma_n$ and $\lambda^2 = \lambda_0^2/x_s(T, B)$, with x_n being the normal fractional density, related to x_s through the relation $x_n + x_s = 1$. We momentarily neglect in this illustrative case possible differences between the scattering times of quasiparticles and normal electrons.   We plot the calculated curves at the fixed frequency $\nu = 50$ GHz as a function of the reduced applied magnetic field in two cases:

---

[2]All these parameters are expected to vary as a function of temperature and magnetic field. In particular, $\nu_s \to 0$ and $\sigma_{nf} \to \sigma_n$ as the superconductor enters the normal state by increasing the field or temperature; for fields not too low, $\rho_{ff} \sim B$; $\epsilon \sim 0$ at low temperatures and $\epsilon \to 1$ at higher T. Finally, the behavior of ν_0 might be extremely dependent on the pinning characteristics of the specific sample.

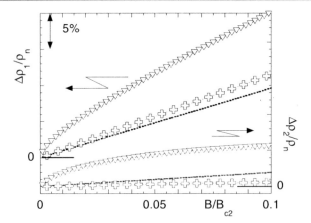

Figure 6: Plot of the field variation of the complex resistivity, at low reduced fields. Parameters as in Fig.5, with $x_{s0} = 0.01$. Blue symbols, left scale: $\Delta\rho_1/\rho_n = [\rho_1(B) - \rho_1(0)]/\rho_n$. Red symbols, right scale: $\Delta\rho_2/\rho_n = [\rho_2(B) - \rho_2(0)]/\rho_n$. Dotted lines, vortex motion only. Crosses, fully gapped superconductor. Triangles, superconductor with lines of nodes in the gap. A sublinear field dependence is clear when the superconducting gap has lines of nodes.

fully gapped superconductor and superconductor with lines of nodes in the gap, in the left and right panels of Fig.5, respectively. Curves are plotted for different values of the zero field superfluid density x_{s0} (which is equivalent to different temperatures). We use typical values for high T_c superconductors, $\lambda_0 = 1000$ Å and $\sigma_n = 10^6$ S·m^{-1}. We also fix $\nu_0 = 10$ GHz as reported for HTCS at low temperature [34, 43, 134] and $\epsilon = 0$, both independent on field. In both cases at relatively large superfluid concentration ($x_{s0} = 0.1$) the real part of the resistivity is substantially coincident with $\text{Re}[\rho_v]$ at low fields (ρ_v is the resistivity due to the vortex motion, and it does not depend, by definition, on the superfluid density x_s). This is due to the fact that with the values used for λ_0 and σ_n, $\nu_s \gg 50$ GHz at low temperatures and fields, so that, as discussed previously, Eqs.(6) reduce to the Gittleman-Rosenblum expression, Eq.(1). However, this is no longer the case at very small superfluid density (i.e., at higher temperatures close to T_c or with strong pair breaking), $x_{s0} = 0.01$, where neglecting the quasiparticle contribution results in a rather large error. Interestingly, as reported in Fig.6, in the case of lines of nodes in the superconducting gap, at low fields the variation of the imaginary part of the resistivity is clearly found to increase as $\sim \sqrt{B}$ even in presence of flux flow (linear in the applied field), while for fully gapped superconductors this increase turns out to be linear with B. Moreover, a closer inspection of the $\rho_1(B)$ curve shows that a partial $\sqrt{B}$ dependence is also present in the real part of the resistivity, for superconductors with lines of nodes in the gap, at sufficiently low superfluid concentration. It is important to stress that the role of quasiparticles is relevant whenever ν_s is not much larger than the measuring frequency ν. Due to the parameters used, in the case above exemplified ν_s is always much larger that ν (50 GHz) unless $x_s \to 0$, that is for $T \simeq T_c$ (or $B \simeq B_{c2}$).

Before commenting experimental results concerning different superconductors, we now briefly discuss which are the main possible differences between the relatively simple scenario depicted up to now and the real world. First of all, we have neglected any creep phenomena for vortices ($\epsilon = 0$), which is unlikely to be correct at high temperatures. A non-vanishing ϵ will strongly influence the low frequency ($\nu \lesssim \nu_0$) resistivity. More important, since $\nu_0 = \kappa_p/2\pi\eta$ only for $\epsilon = 0$, an increasing ϵ could introduce rather strong temperature and/or field dependencies in ν_0. In particular, since ν_0 is expected to grow rather fast with increasing ϵ, in a measurement at fixed frequency one might have $\nu > \nu_0$ at low temperatures or fields and $\nu < \nu_0$ at higher temperatures or fields. This will result in unpredictable shapes for the curves $\rho(H)$ or $\rho(T)$, depending on the field/temperature variation of ν_0.

The second main hypothesis which is not necessarily verified concerns the quasiparticle conductivity. Smaller values of ν_s can substantially change the quasiparticle contribution to the resistivity. As an example, quasiparticle relaxation rates shorter than the normal state values, as reported by several authors [64, 65, 66], would tend to decrease substantially ν_s. One might have as a consequence large quasiparticle contributions even at low temperatures and fields.

Finally, it must be noticed that Eqs.5 for the vortex resistivity are obtained in the rather special case of periodic pinning potential. Moreover, they do not take into account the resistivity due to Josephson or Abrikosov-Josephson vortices, whose dynamics might be rather different with respect to standard vortices. The expression for the vortex resistivity ρ_v might then be rather different, in the most general case. However, we will show in the following that most of the main features observed in the microwave data of several superconducting materials can be quantitatively explained using the theoretical expressions cited up to now. We will come back to specific points during the discussion of the data in Sec.4

3   Experimental Section

3.1   Samples

All measurements here presented were performed in thin, high-quality superconducting films. YBa$_2$Cu$_3$O$_{7-\delta}$, SmBa$_2$Cu$_3$O$_{7-\delta}$, and MgB$_2$ were investigated extensively. Samples were squares, of side l and thickness d. Substrates have been carefully chosen for microwave measurements. The crystal structure was investigated by X-ray $\Theta - 2\Theta$ diffraction. The c-axis orientation was assessed by measuring the full-width-half-maximum (FWHM) of the rocking curve of an appropriate peak. In-plane X-ray $\Phi-$scan was measured in samples Y1, Y2, Y3, Sm1, Sm2, and M1, showing excellent in-plane epitaxiality. Surface roughness was investigated by AFM over typical 1 μm $\times$ 1 μm area. All cuprates sample were nearly optimally doped or slightly overdoped. T_c and the resistivity above T_c, ρ_0, were estimated by electrical transport methods. Depending on the sample, we used dc or microwave resistivity. In this latter case T_c was estimated from the inflection point of the temperature-dependent real part of the microwave resistivity (this temperature is found to coincide with the temperature where the real and imaginary parts of the microwave fluctuational conductivity equal one to the other, which is an accurate evaluation of the mean-field critical temperature [101]), and ρ_0 from the measured real part ρ_1 in zero magnetic field. Typical ± 0.5 K uncertainties of these methods are inessential for the purposes of the present

Table 1: Data for the structural and electrical characterization of the samples investigated. $\rho_0 = \rho_1(100 \text{ K}, 48 \text{ GHz})$ and $\rho_{dc}(40 \text{ K})$ in RE-BCO and MgB$_2$, respectively. N.A.: not available.

Material	YBCO	YBCO	YBCO	YBCO	SmBCO	SmBCO	MgB$_2$
Sample no.	Y1	Y2	Y3	Y4	Sm1	Sm2	M1
Substrate	LaAlO$_3$	LaAlO$_3$	LaAlO$_3$	CeO$_2$/YSZ	LaAlO$_3$	LaAlO$_3$	sapphire
Thickness (nm)	220	220	220	200	220	220	100
Lateral dimension (mm)	10	10	10	10	10	10	5
$\Theta - 2\Theta$ FWHM	0.1°	0.1°	0.1°	0.2°	0.2°	0.2°	N.A.
Surface roughness (nm)	2	2	2	N.A.	3	3	N.A.
T_c (K)	89.5	89.5	90	88	87	87	36
ρ_0 ($\mu\Omega\cdot$cm)	130	140	130	250	300	300	5
References	[102, 103]	[102, 103]	[102, 103]	[104]	[105, 106]	[105, 106]	[107]

paper. Material parameters and appropriate references are reported in Table 1. More details on sample preparation and characterization are reported in the References (see Table).

3.2 Microwave Response in Thin Films

Since in all the measured samples the thickness is of the order of, or smaller than, the commonly reported values for the penetration depth, it is appropriate to shortly describe the electromagnetic response of an electromagnetically thin (super)conducting film.

In the case of bulk samples, for which the sample thickness is much greater than the electromagnetic field penetration depth (of the order of $\min(\lambda, \delta_{nf})$), the surface impedance is given by the usual expression already mentioned: $Z_s = (i\omega\mu_0\tilde{\rho})^{1/2}$.

When the sample thickness $d \lesssim \min(\lambda, \delta_{nf})$, the electromagnetic field is transmitted through the film and reaches the substrate and any supporting layer, usually a metallic back-plate. The field is therefore determined by the interaction with both the underlying layers and the film of finite thickness. In this situation the simple expression for Z_s no longer holds, being substituted by an effective surface impedance Z_s^{eff} which can be derived by means of standard impedance transformation relations [108]:

$$Z_s^{eff} = Z_s \frac{Z_d^{eff} + iZ_s \tan(k_s d)}{Z_s + iZ_d^{eff} \tan(k_s d)} \tag{10}$$

where $k_s = \mu_0\omega/Z_s$ is the HTCS propagation constant and Z_d^{eff} is the effective surface impedance of the substrate. The full expression of Eq.(10) can be significantly simplified when two main conditions are met: $|Z_d^{eff}| \gg |Z_s|$, meaning that the substrate contribution can be neglected, and $d \ll \min(\lambda, \delta)$, which is usually the case for epitaxially grown high T_c superconductor films. In this case the so-called thin-film approximation [109, 110] is obtained:

$$Z_s^{eff} \simeq \frac{\tilde{\rho}}{d} \tag{11}$$

The applicability of this equation heavily depends on the (possibly temperature-dependent) properties of the substrate. Metallic [111] and semiconducting [112, 113] substrates strongly affect Z_s^{eff} and they impose the use of the full expression, Eq.(10). On

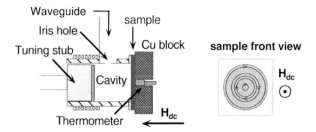

Figure 7: Left: straight section of the resonant cavity. Right: current patterns on the square sample.

the other hand dielectric substrates with backing metallic plate have often impedances high enough so that Eq.(11) can safely be used[3].

3.3   Cavity Measurements

The microwave response at high frequency was measured in RE-BCO by the end-wall cavity technique [118] at 48.2 GHz. The cylindrical cavity, of 8.2 mm diameter, was inserted in a liquid/solid Nitrogen cryostat, so that temperatures in the range 60-150 K could be reached. Cryogenic waveguides were used in order to couple the one-port cavity to the external microwave source. The temperature of the entire cavity, including the sample, could be stabilized for hours within ± 10 mK. A magnetic field H was applied along the c axis and supplied by a conventional electromagnet. The maximum attainable field in this setup was $\mu_0 H \leq 0.8$ T. In the following, we will assume that inside the sample $B \simeq \mu_0 H$[4].

For the measurements, undercoupling of the cavity and the TE$_{011}$ resonant mode were chosen. In this mode the currents flow along circular paths on the end-wall occupied by the sample, have zeroes at the center of the end walls and at the joints between the end walls and the body of the cavity, and maxima at half the radius of the end wall. This configuration allows to neglect the losses due to imperfect electric contact between the sample and the cavity; moreover, it makes possible the use of a mechanical stub to tune the cavity at the desired resonant frequency (in our case, in the range 48.0 ± 0.5 GHz). The degenerate TM mode, strongly suppressed by electrical isolation between the end walls and the body of the cavity, was further shifted in frequency by appropriate mode traps. Fig.7 reports a sketch of the cavity setup and of the pattern of the currents on a typical 10 mm $\times$ 10 mm sample.

Measurements of the field and/or temperature induced changes of the quality factor Q and resonance frequency f_0 yielded changes in the effective surface impedance through the well known expression $\Delta Z_s^{eff}(H,T) = G\Delta\left[\frac{1}{Q(H,T)} - 2\mathrm{i}f_0(H,T)\right]$, where G is an appropriate geometrical factor [108]. Separate calibration of the response of the cavity allowed for the determination of the absolute value of R_s^{eff}. In most of the measurements

---

[3]Noticeable exceptions are dielectrics with strong temperature dependent permittivity [110, 114, 115, 116] or with an accidentally unfavorable combination of thickness, permittivity and operating frequency [117].

[4]We note that demagnetization effects determine a penetration field much smaller than H_{c1} in thin films [119], and field inhomogeneities inside the sample are expected (and also directly found [120]) to be irrelevant in thin films for fields greater than a few mT in our temperature range.

here reported, Q and f_0 were measured as a function of the magnetic field at various fixed temperatures in the range 60 K - T_c. Measurements were performed by sweeping the field from zero up to 0.8 T either after zero-field-cooling (ZFC) to the desired temperature, or by increasing the temperature after each field sweep. We did not observe hysteresis, apart a 10% effect at temperatures below 70 K in some of the samples. In some cases, to check the relevance of the hysteresis, full magnetic cycles were performed. Measurements here reported refer only to the cases where hysteresis is absent, or well below 10% of the total response.

From the field induced change of Q and f_0 the field induced change of the effective surface impedance $\Delta Z_s^{eff} = Z_s^{eff}(H,T) - Z_s^{eff}(0,T)$ could be obtained [121]. Due to the small thickness of the films, one has $Z_s^{eff} = \tilde{\rho}/d = \rho_1/d + i\rho_2/d$, where $\tilde{\rho}$ is the complex resistivity.

The main experimental errors are due to the following reasons: the absolute value of Z_s^{eff} is affected by errors in the calibration of the cavity and in the evaluation of the geometrical factors [122, 123, 124]. Additionally, $\tilde{\rho}$ has an intrinsic uncertainty due to the evaluation of the film thickness (typically 10%). By contrast, calibration of the cavity does not affect the field variation of $\tilde{\rho}$ at fixed temperature, and geometrical factors and film thickness give only a possible overall scale factor. All these sources of errors are strongly reduced (if not eliminated at all) when working with reduced complex resistivity changes, $\Delta\tilde{\rho}/\rho_0 = \Delta\rho_1/\rho_0 + i\Delta\rho_2/\rho_0$, with $\rho_0 = \rho(100\,\text{K})$ in YBCO and SmBCO. In the following the data will be reported in one of the mentioned formats.

Finally, some useful features of the present setup should be stressed. First, the microwave currents and fields lie in the (a, b) planes, thus avoiding any c axis contribution to the measured response. Second, the dc magnetic field is perpendicular to the microwave currents (maximum Lorentz force configuration). Third, the microwave currents essentially probe only an annular region of 4.1 mm mean diameter and ~ 2 mm width, centered on the sample. For the subsequent discussion, it is relevant that flux lines are not forced to cross the border of the sample: the oscillation induced by the microwave currents are limited to the annular region above mentioned. In particular, no edge effects are relevant in our configuration.

3.4 Corbino Disk Measurements

The frequency dependence of the vortex-state microwave response was measured in YBCO and MgB$_2$ films. Measurements are obtained through a Corbino disk geometry [34]: a swept frequency microwave radiation is generated by a vector network analyzer (VNA) and guided to the sample under study through a coaxial cable. The sample, placed inside the cryomagnetic apparatus, shortcircuits the coaxial cable. A double spring is used to obtain a good electrical contact between the end cable connector and the sample. To further improve the contact, a thin annular indium foil is placed between the sample and the external part of the coaxial cable. Due to the limitations discussed below, data are reported for the frequency range 2-20 GHz, well inside the capabilities of the VNA (45 MHz-50 GHz) and of the cables (cutoff frequency 50 GHz). A magnetic field up to 10 T was applied along the c axis. The temperature ranges 70 K - 100 K and 5 K - 40 K were explored, in YBCO and MgB$_2$, respectively. In Fig.8 we show a sketch of the Corbino disk cell and of the current pattern

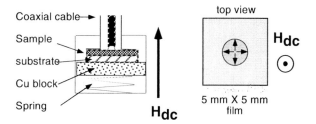

Figure 8: Left: straight section of the Corbino disk cell. Right: current patterns on the square sample.

on the sample.

In principle, a measurement of the complex reflection coefficient Γ_0 at the sample surface directly yields the complex effective surface impedance Z_{eff} of the sample by the standard relation

$$\Gamma_0 = \frac{Z_s^{eff} - Z_0}{Z_s^{eff} + Z_0}$$

where Z_0 is the impedance of the cable [108]. In our case, however, the measured quantity is the reflection coefficient at the instrument location Γ_m, which contains, besides the response of the sample, reflections and attenuation due to the cable line between the sample and the VNA. Due to the long line, necessary to place the VNA far from the stray field of the magnet, the contribution of the sample to the overall Γ_m is rather small. Further, the response of the part of the line inside the cryostat depends on the temperature profile across the cable, that varies in general during the measurement. Thus, full calibration of the cable and, consequently, direct measurements of absolute values of the impedance of the sample are not feasible. To overcome this problem, we developed a custom calibration procedure, through which we could obtain the variations of the effective surface impedance with the temperature or with the field, $\Delta Z_s^{eff}(\nu, H, T)$. The description of the calibration procedure is rather cumbersome, and it has been extensively described elsewhere [125, 126]. Here we recall only the major hypotheses necessary to obtain the impedance data from the measured complex reflection coefficient.

Reliable measurements of the *temperature* variations of the effective surface impedance, $\Delta Z_s^{eff}(\nu, H, T) = Z_s^{eff}(\nu, H, T) - Z_s^{eff}(\nu, H, T_{ref})$, require *(i)* that the temperature variation of the complex reflection coefficient are not dominated by the change in the response of the cable, and *(ii)* that at least at sufficiently low temperature $\Gamma_0(\nu) \simeq -1$ (or, equivalently, $R_s^{eff}, X_s^{eff} \ll Z_0$) [125]. In particular, variations of R_s^{eff} mainly reflect on variations in the modulus of Γ_0, while variations of X_s^{eff} have as a main effect a phase change of Γ_0. The accuracy and reliability of the measurement of ΔR_s^{eff} and ΔX_s^{eff} depend then on relative variations of modulus and phase of Γ_0 with respect to corresponding variations (due to, e.g., the change of the cable characteristics) of Γ_m, respectively. It turns out that for the measurements of R_s^{eff} the requirements are fulfilled in the temperature ranges 5 K - 40 K and 70 K - 100 K in MgB$_2$ and YBCO, respectively. Unfortunately, the temperature variations of the phase signal due to the sample are of the same order of magnitude of the temperature variations of the phase due to the cable, so that a reliable measurement of

X_s^{eff} is not feasible with this kind of measurement. No such problems arise at fixed temperature and with sweeping field: in this case, the response of the cable is almost exactly constant (we checked that the line has a very small magnetic response): *field* variations $\Delta Z_s^{eff}(\nu, H, T_0) = Z_s^{eff}(\nu, H, T_0) - Z_s^{eff}(\nu, 0, T_0)$ can be reliably obtained.

These measurements can be converted to absolute values of R_s^{eff} and/or X_s^{eff} by assuming some known $R_s^{eff}(\nu)$ or $X_s^{eff}(\nu)$ at some specific temperature or field. In general, one can take $R_s^{eff}(\nu) \simeq 0$ at sufficiently low temperatures and fields, while X_s^{eff} can be obtained from normal state measurements (at high enough temperatures or fields).

There are a few additional remark on the different behavior of the two materials investigated with the Corbino disk. First, the quality of the electrical contact between the coaxial connector and the film is much better in MgB_2 than in YBCO. This results in reliable measurements in the ranges 2 - 20 GHz and 5 - 20 Ghz in MgB_2 and YBCO, respectively. Second, with the available magnetic field it was possible to reach the normal state in MgB_2, so that absolute measurements of R_s^{eff} and X_s^{eff} can be obtained by increasing the field above the upper critical field. In YBCO measurements were taken with field-cooling the sample (thus lowering the temperature), and only the real part of Z_s^{eff} could be obtained. Finally, even after the calibration procedure, there still remain detectable oscillations in the frequency-sweeps. Those oscillations, together with uncertainties in geometrical factors, are eliminated by reporting the data normalized to the values measured above T_c.

As for the cavity measurements, the thin film approximation holds [110] and one has $Z_{eff} = \tilde{\rho}/d$. Most important, like and more than in the cavity measurements, in the Corbino disk geometry there are no contributions at all from vortex entry or exit: the area probed by the microwave currents is a small circle, of radius $\sim$2 mm, at the center of the sample, and currents flow along linear paths between the inner contact and the outer conductor, so that vortices are forced to oscillate along circular orbits. Again, no edge effects affect our measurements.

4   Results

In this Section, we present and discuss the results obtained on the various superconducting materials under study. We present separately the results for the different materials.

4.1 $YBa_2Cu_3O_{7-\delta}$

$YBa_2Cu_3O_{7-\delta}$ is a somewhat paradigmatic case: as it will be shown in the following, the microwave properties in the vortex state follow quite closely the simple models summarized in Sec.2, allowing for a rather detailed discussion of the various dynamical regimes as a function of frequency, temperature and magnetic field. By means of combined wideband (5-20 GHz) and high-frequency (48 GHz) measurements we will show (a) that it exists a relatively wide temperature range below the critical temperature T_c in which the resistivity, while clearly lower than the normal state value, is at the same time independent on frequency, indicating that the effect of pinning (if any) is not relevant in this T range, (b) that vortex motion follows closely the predictions of the mean-field, Coffey-Clem theory, and (c) that approaching the critical temperature the field dependence of the microwave resis-

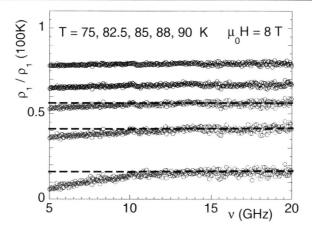

Figure 9: Normalized real resistivity in YBCO vs. frequency at various temperatures and $\mu_o H = 8$ T. A detectable frequency dependence only appears for $T < 85$ K (red data curves). Temperature increases from bottom to top.

tivity points to a relevant role of the field-dependent superfluid depletion. We also obtain the temperature dependence of the vortex parameters η and ν_0.

We begin with the frequency dependence of the microwave real resistivity ρ_1 measured in sample Y3. In Fig.9 we report the frequency dependence of ρ_1/ρ_0 measured with the Corbino disk at various temperatures and at a fixed field $\mu_0 H = 8$ T. As it can be seen, while ρ_1 is sensibly lower than ρ_0 for $T < 90$ K, a frequency dependence is observed only below $T = 85$ K$< T_c$. This is a first indication that in the superconducting state there is a region of the (H, T) phase diagram where the microwave response in the frequency region here investigated is independent (or very weakly dependent) on frequency. This finding restricts the range of physical phenomena that can give rise to the measured response. More insight can be gained by representing the data at fixed frequency as a function of the temperature, as depicted in Fig.10, where we compare the normalized real resistivity at two fixed frequencies (6 and 20 GHz) as a function of temperature, for $\mu_0 H = 2$ T and 8 T. At each field the data at two different frequencies lie on a single curve in a temperature range extending from above to below T_c, while they are markedly different at lower temperatures, below an easily estimated crossover temperature $T_f(H)$. This abrupt change in the frequency dependence across T_f marks the boundary between two different dynamic regimes.

The nature of these two regimes can be understood considering the interplay of vortex and quasiparticle response: at low temperature flux motion contribution is dominant. Near the transition temperature T_c quasiparticles play a significant role in the electromagnetic response (see Fig. 5). Very close and above T_c thermal fluctuations set in. In absence of disorder, all those mechanisms merge smoothly one into the other: in fact, the low temperature limit of the fluctuations-dominated dc resistivity coincides with the free flux flow [127], and a nearly frequency-independent ρ_1 results for both the fluctuation induced dissipation [128, 129, 130, 131, 132, 133] and for flux (free) flow resistivity (as can be easily seen from Eqs.(5), considering that the free flow limit corresponds to the case $\epsilon = 1$).

This framework drastically changes in presence of disorder. In that case vortices are

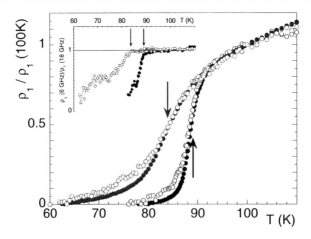

Figure 10: Normalized real resistivity in YBCO vs. temperature at 6 GHz (full symbols) and 20 GHz (open symbols) and $\mu_o H = 2$ T (black symbols) and 8 T (red symbols). It is apparent that at each field there is no frequency dependence down to a typical temperature $T_f < T_c$, depicted by the arrows. The same feature is more evident in the inset, where the ratios $\rho_1(6\,\text{GHz}, H, T)/\rho_1(20\,\text{GHz}, H, T)$ have a clear departure from 1 at $T_f(H)$.

more or less pinned to defects, and the overall dc resistivity becomes smaller than the free flow expression (viscous/plastic regime), eventually reaching a zero value if the interactions are strong enough to completely lock the flux lines (frozen regime). For what concerns the frequency response, the viscous/plastic as well as the frozen regime are characterized by a value of the creep factor ϵ (see Eqs.(5)) lower than unity (zero in the frozen regime), thus resulting in a frequency dependent real resistivity. In particular, $\text{Re}[\rho_v]$ should grow from ρ_{dc} to the free flow value ρ_{ff} when the frequency increases across the depinning frequency ν_p. This pinning frequency is estimated to be $\nu_p \simeq 10$ GHz [34, 43, 134] in YBCO, well within the frequency range here explored. As a result, the large frequency dependence of our data below T_f is explained with the increased effect of disorder. We now turn to the analysis of the data deep in the strongly frequency-dependent regime, that is $T < T_f$. We report in Fig.11 typical frequency sweeps for the normalized real resistivity in sample Y3 at low temperature and at various magnetic fields. It is immediately apparent that $\text{Re}[\rho_1]$ increases as a function of frequency, eventually reaching a saturation value at high frequencies. This is exactly the behavior expected from mean-field theories of the vortex state at sufficiently low temperatures. Within this framework, the field variation of the real resistivity is entirely given by the vortex motion, so that we fitted the data with the first of Eq.(5), using ϵ, ν_0 and $\rho_{ff}/\rho_n = \Phi_0 B/\eta\rho_n$ as fitting parameters. As it can be seen (thick lines), good fits are obtained. Similar results are obtained at different temperatures and fields. It should be mentioned that at low fields, and/or at high enough temperatures, the experimental curves become nearly featureless, and fitting is less reliable. We report here the vortex parameters obtained in the most reliable temperature and field ranges, that correspond to the region close to the frozen regime ($\epsilon \simeq 0$, $\nu_0 \simeq \nu_p$). It is found that ν_p smoothly depends on temperature and magnetic field, as reported in Fig.(12), and decreases for both increasing field and temperature.

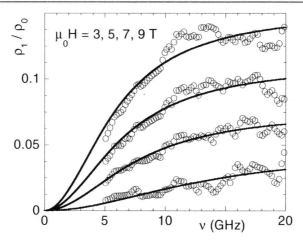

Figure 11: Typical swept-frequency measurements at different fields and $T = 70$ K. A marked frequency dependence is observed, denoting the change of vortex response. Continuous curves are fits with Eq.(5).

The vortex viscosity, as obtained from swept frequency measurements in the low temperature region is reported in Fig.(13). It is remarkable that no significant magnetic field dependence is detected, thus indicating that the flux flow follows a magnetic-field linear dependence, $\rho_{ff} \sim B$.

To gain more insight in the electromagnetic response at microwave frequencies, we present now measurements of the complex resistivity obtained with the cavity resonator at a much higher frequency, $\nu = \omega/2\pi = 48$ GHz, as a function of the applied magnetic field on YBCO samples from the same batch. Measurements span the temperature range 65 K $-T_c$, and the field is limited to 0.8 T. In Fig.(14) we show typical field-sweeps of the variation of the complex resistivity at some significant temperature. It is seen that the real resistivity ρ_1 has an almost linear variation with the applied field, apart possibly the low field region, and that this linear dependence changes to a concave downward behavior approaching closely T_c. At the same time, the variation of the imaginary part $\Delta\rho_2$ with the field is nearly absent at low temperatures, and is *negative* at higher temperatures, close to T_c.

This behavior is easily recognized as an essentially free flux flow, with a possibly significant contribution of the field-induced superfluid depletion at high temperatures. This is completely consistent with the results from Corbino disk measurements, if one considers that ν_0 was found (at low temperature) below 15 GHz. As a consequence $(\nu/\nu_0)^2 \gg 1$ and, following Eq.(5), the vortex contribution reduces, as a first approximation, to the real flux flow term. Then, not too close to T_c, where the field dependence is essentially linear, $\rho_1 \simeq \rho_{ff} = \Phi_0 B/\eta$ and $\rho_2 \simeq 0$, from which the vortex viscosity is readily obtained. Such data points are reported in Fig.(13) and compared to the Corbino disk results. The agreement is excellent: field sweeps and frequency sweeps give exactly the same vortex viscosity.

At higher temperatures both the downward curvature of $\Delta\rho_1(H)$ and the pronounced, negative $\Delta\rho_2(H)$ can be accounted for by the field dependence of the superfluid and quasi-

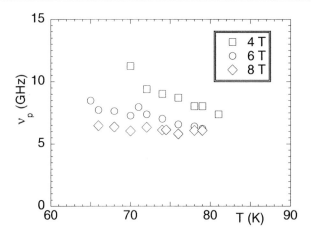

Figure 12: Temperature dependence at various fields of the depinning frequency ν_p in the temperature region below T_f. It is seen that $\nu_p < 15$ GHz, at all fields and temperatures.

particle concentration. As discussed in Sec.2, in this case one has to use the full expression of the vortex state resistivity, Eq.(4), with the noticeable simplification that $\tilde{\rho}_v \simeq \rho_{ff}$ due to the high operating frequency. Fitting necessarily add other parameters: the zero field, temperature-dependent superfluid fraction $x_{s0}(T)$, the zero-temperature London penetration depth λ_0, the exponent α and the factor c (See Eq.(9)). The upper critical field is not a different fitting parameter, assuming $\eta = \Phi_0 B_{c2}/\rho_n$. We were not able to fit the pairs of curves $\Delta\rho_1(B)$, $\Delta\rho_2(B)$ with the exponent $\alpha=1$ typical of fully-gapped wavefunction. Instead, by using $\alpha = \frac{1}{2}$ typical of a wavefunction with lines of nodes in the gap [46, 95, 96], we obtained fits as reported in Fig.(15) with the reasonable choice $\lambda_0=160$ nm, and $c \simeq 0.15$ (we found $c \simeq 1$ in a different sample, with the analysis of the real part only [42]). The resulting η connects smoothly to the experimental data obtained at lower temperatures, and $x_{s0}(T) \sim \left[1 - \left(\frac{T}{T_c}\right)^2\right]$. While this is not a direct evidence for unconventional superconducting gap, we note that by taking into account the possible existence of lines of nodes (and the consequent increased weakness of the superfluid stiffness), the vortex state microwave response is fully described in its temperature, frequency and field dependence from well below up to very close to T_c. It is interesting to notice that the small quasiparticle contribution observed (apart very close to T_c) is consistent with $\tau_{qp} \sim \tau_n$ (see Fig.5). Our results suggests then that in our YBCO films $\tau_{qp}/\tau_n \simeq 1$, closer to the results of [67] rather than to those of [65]. In this case YBCO is a paradigmatic case, and the simplest to understand. We will see in the following that all the other materials below reported present additional physics, that prevent from a straightforward application of the CC model. Additional mechanisms are revealed by measurements of the microwave response.

4.2   MgB$_2$

It is somehow expected that MgB$_2$, due to its firmly established two-gap nature, can hardly be described by oversimplified models. In order to didactically describe the difficulties

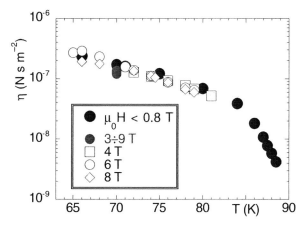

Figure 13: Temperature dependence of the vortex viscosity η in YBCO, obtained from Corbino disk swept frequency method (open symbols and red full dots, fields depicted in the figure) and from field-sweeps, resonant cavity technique (blue full dots). All the sets of data coincide, indicating that the intrinsic viscosity is measured. Moreover, no significant magnetic field dependence is detected.

hidden in the analysis of the data in this two-gap compound, we first present in Fig.(16) the real resistivity as a function of the frequency at fixed temperature and various magnetic fields, in complete analogy with the data for YBCO reported in Fig.(9). The reported data are qualitatively similar to those measured on YBCO, showing at each field a (real) resistivity increasing as a function of frequency, reaching a plateau value ρ_{pl} at high frequencies. Moreover, it is shown that, rather surprisingly, the simple, single-gap CC model quantitatively fit the data. One might then be led to the conclusion that simple vortex motion, as given by Eq.(5), captures the physics involved in microwave response in MgB$_2$. However, despite the quality of the fits, the behavior of the parameters is not easily understood within simple models for vortex motion: as an example, we report in Fig.(17) the ratio $\Phi_0 B / \rho_{ff,fit}$ that should coincide, in this oversimplified model, to the vortex viscosity η. It is found that $\Phi_0 B / \rho_{ff,fit}$ depends very strongly on the applied field (as opposed to the behavior theoretically expected and experimentally found in YBCO), while its temperature dependence changes with the field, becoming less and less temperature dependent as the field increases.

More puzzling results can be revealed by reporting the normalized complex resistivity as a function of magnetic field, at fixed temperature and different frequencies. In Fig.(18) we report these data at $T = 15$ K, but similar results are obtained at all temperatures [134]. From this figure, it is immediately apparent that the straightforward application of a single-gap model, such as the CC model in its original formulation, leads to considerable contradictions. First of all, should one identify the initial, B-linear part of ρ_1 at the highest frequency with the flux flow resistivity ρ_{ff}, in analogy with the field-sweeps in YBCO, the resulting H_{c2} as obtained from the linear extrapolation depicted in Fig.(18) would be a factor 1.5-2 lower than H_{c2} independently measured on the same sample by dc resistivity [135]. Second, at low fields and high frequency there is a steep increase in $\rho_2(H)$, up to a smoothly temperature-dependent field H_1, as reported in Fig.20. Taking the vortex motion as the

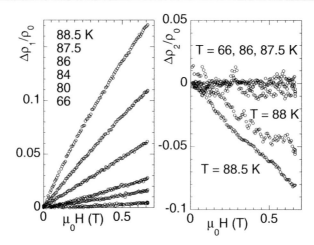

Figure 14: Typical magnetic field dependence of the variation of the complex resistivity in YBCO, measured at 48 GHz. Left panel, real resitivity. Right panel, imaginary resistivity. Note at high temperature the downward curvature in $\Delta\rho_1(H)$, and the decrease of $\Delta\rho_2(H)$.

only source of response, this is possible only with the assumption that $\nu_0 \approx 15$ GHz, and strongly field-dependent. However, if this were true, one should observe a large variation (as a function of frequency) of $\rho_1(H)$ in the same field and frequency ranges. There is no indication of such a strong difference between the experimental curves $\rho_1(H)$ measured at different frequencies for $\nu > 10$ GHz, indicating that typical vortex frequencies are not placed in this range. The steep increase of the imaginary response, followed by the more gradual decrease, is then more likely ascribed to an unconventional increase of the screening length, as it could be given by the fast suppression of the superfluid fraction in the weak π band [136]. We finally note that, above a crossover field H_2, ρ_2 is nearly independent on frequency: there are at present no indications on the possible origins for this feature, so that in the following we confine the discussion to fields $H < H_2$.

We can thus conclude that swept-frequency measurements reveal a clearly richer physics of the vortex-state in MgB$_2$ than, e.g., in YBCO. Accordingly, the interpretation of the data must be based on more complex models.

In this perspective, it is interesting to compare the measured behaviors of $\rho_1(H)$ and $\rho_2(H)$ with the predictions of the model that takes into account both vortices and quasiparticles, Eqs.(6) and Fig.5, left panel (we assume that, according to literature, MgB$_2$ is an s-wave superconductor [137]). By comparing the measured data with the predicted curves, it is clear that the motion of vortices cannot take into account the whole observed high frequency behavior. Indeed, a rather large quasiparticle contribution (curves with small x_{s0}) is needed to obtain behaviors similar to those observed: it is seen from Fig.5, left panel, that (at high frequency: in Fig.5 it is assumed $\nu \gg \nu_0$) $\rho_1(H)$ should increase approximately linearly with field while $\rho_2(H)$ should be more or less constant at low fields and then decrease approaching the upper critical field. Apart from the low field region ($H < H_1$), this behavior is rather similar to the measured one. We can then conclude that, provided a large quasiparticle contribution is taken into account, the behaviors observed could possibly be

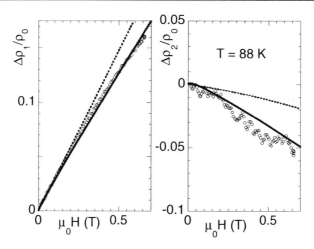

Figure 15: Typical fits of the magnetic field dependence of the variation of the complex resistivity in YBCO, measured at 48 GHz, with the flow expression, Eq.(8). Continuous lines: superconductor with lines of nodes, dashed lines: fully gapped superconductor.

described in terms of rather simple one-band models for $H > H_1$. This result is fully consistent with spectroscopic measurements [81, 82] which show that the superfluid density of the π band decreases strongly with the magnetic field, nearly vanishing at a characteristic field H^*: the microwave response in the vortex state at low fields, where both gaps play a role, cannot be described by a singe gap model, while the higher field data can be nicely described by that model. This picture is reinforced by noting that both the temperature dependence and the numerical values of H_1 agree very well with those reported for H^* [81, 82, 83, 138].

We now show that the microwave resistivity above $H_1(T)$ quantitatively coincides with a model for a single-band superconductor, Eqs.(6), if the conductivity of the π band is taken as purely real.

We make some rather crude approximations in order to reduce the number of fitting parameters. First of all, we assume that above H_1 the residual $\pi-$band contribution to the superfluid can be neglected, so that the superfluid fraction $x_s(T, B)$ is related to the superfluid fraction of the σ band only, $x_{s,\sigma}(T, B) = N_{s,\sigma}/N_\sigma$ (that is, the superfluid volume density divided by the total volume density of electrons in the σ band), through the relation $x_s(T, B) = Kx_{s,\sigma}(T, B)$, with $K = N_\sigma/(N_\sigma + N_\pi)$, being N_π the volume density of electrons in the π band. The penetration depth above H_1 becomes then $\lambda^2(T, B) = \lambda^2_{0,\sigma}/x_{s,\sigma}(T, B)$.

For what concerns the normal fluid resistivity $\rho_{nf} = 1/\sigma_{nf}$, we assume that, since the σ and π bands interact very weakly with each other [139], above H_1 one can write $\sigma_{nf} \simeq \sigma_{n,\pi} + [1 - x_{s,\sigma}(T, B)]\sigma_{n,\sigma}$. Using the independently measured [107, 140], temperature dependent H_{c2} anisotropy of our sample, the Gurevich model [141] yields $\sigma_{n,\sigma}/\sigma_{n,\pi} \simeq 0.25$. Onc then obtains, to a good approximation, $\rho_{nf}/\rho_n - 1$ in all the field and temperature region $H > H_1(T)$, and the T and B variations of $\nu_s(T, B)$ are then entirely determined by $x_s(T, B) = x_{s0}(T)(1 - b)$, where the last equality

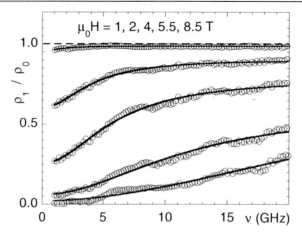

Figure 16: Frequency dependence of ρ_1 in MgB$_2$, at $T = 10$ K and different magnetic fields (analogous to Fig.9). Remarkably, a blind fit with Eq.(5) gives good agreement (thick continuous lines). However the resulting parameters exhibit very contradictory field and temperature dependences, see text and Fig.17.

comes from the single-gap-like behavior above H_1, $b = B/B_{c2}(T) \simeq H/H_{c2}(T)$ and $x_{s0}(T) = Kx_{s,\sigma}(B = 0, T)$ is the (extrapolated) zero field value[5] of x_s. Once this general frame has been established, it is possible to extract from the data, by means of a selfconsistent procedure reported in the Appendix, the temperature dependent σ-band superfluid fraction $x_{s0}(T)$, the temperature dependent upper critical field $H_{c2}(T)$, the vortex motion resistivity $\rho_v(\nu)$ and, consequently, the characteristic frequency ν_0. We now discuss those quantities. In Fig.19 we report the behavior of $x_{s0}(T)$, normalized to the lowest temperature value $x_{s0}(T = 5$ K$)$. We compare the T dependence of $x_{s0} = Kx_{s,\sigma}(B = 0, T)$, with $x_{s,\sigma}(B = 0, T)$ as obtained using the expressions of the BCS theory [11], and the values of the gap $\Delta_\sigma(T)$ measured by point contact spectroscopy [142]. The theoretical calculation and data points from our microwave measurements agree very well, giving a strong support to the consistency of the analysis here performed, and adding evidence that the temperature dependence of the zero-field superfluid fraction in the σ band follows a rather conventional behavior.

Similarly, we obtain values of the upper critical field H_{c2} (Fig. 20) in nearly perfect agreement with values obtained independently from the dc measurements [136], giving further confirmation on the reliability and consistency of the underlying model. We remark that this is a nontrivial result: previous analysis have never been able to describe the microwave data using the same parameters obtained from low frequency measurements. In particular, the upper critical field was found to be field dependent [143] or anomalous field dependencies for the vortex viscosity were invoked [76].

Finally we report in Fig.21 the vortex motion complex resistivity, isolated from the experimental data by means of the selfconsistent procedure. We remark that now those data are not flawed, as the raw data, by the presence of the strong contribution of the π band,

---

[5]We remark that $x_{s0}(T)$ is different from the value that is obtained through experiments that measure the superfluid density at $B = 0$, being in that case $x_s^{meas}(B = 0, T) = x_{s,\sigma}(B = 0, T) + x_{s,\pi}(B = 0, T)$

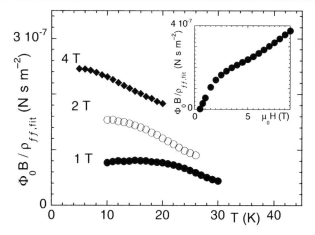

Figure 17: Field and temperature dependence of the ratio $\Phi_0 B/\rho_{ff,fit}$ as obtained from the MgB$_2$ data fitted through Eq.(5). This ratio should equals η. It is readily seen that both temperature and magnetic field dependences present very exotic behavior.

or of the overall field dependence of the superfluid concentration (see the Appendix). We then fitted all the pairs of curves $\mathrm{Re}[\tilde{\rho}_v]/\rho_n(\nu)$, $\mathrm{Im}[\tilde{\rho}_v]/\rho_n(\nu)$ to Eqs.(5) at all given temperatures and fields, using ϵ and ν_0 as fitting parameters ($\rho_{ff}/\rho_n = H/H_{c2}$ is determined by the value of H_{c2} obtained previously). A typical fit for $T = 15$ K and $\mu_0 H = 3$ T is reported in Fig.(21). We find, consistently with the indication given by the small frequency dependence of ρ_1, that ν_0 never exceeds 10 GHz, for any temperature and field. In addition, we find a rather strong field dependence of ν_0 at different temperatures, as reported in Fig.(22). This is an indication, in agreement with previous findings [143], of the collective nature of the pinning forces in this material. This conclusion is also supported by the low temperatures collapsing of all curves $\nu_0(b)$, with $b = B/B_{c2}(T)$.

Summarizing, we have shown that multifrequency measurements are a key factor to elucidate the role of quasiparticle, superfluid and vortex motion in the overall transport properties of MgB$_2$ in the vortex state. By suppressing the π band contribution with a sufficiently strong magnetic field it has been possible to evaluate the σ band superfluid density, the upper critical field and the characteristic vortex frequency.

4.3   SmBa$_2$Cu$_3$O$_{7-\delta}$

The case of SmBa$_2$Cu$_3$O$_{7-\delta}$ (SmBCO) is somehow similar to the case of MgB$_2$, in that an additional, strong contribution different from vortex motion affects the complex resistivity. In this Subsection we report measurements of the complex resistivity for temperatures down to 65 K in moderate fields $\mu_0 H < 0.8$ T and at the high operating frequency $\omega/2\pi = 48$ GHz. Most of the measurements are thus taken below the irreversibility line of SmBCO [144, 145]. Typical field sweeps for the variation of the complex resistivity at various sample temperatures are reported in Fig.(23). The data here reported are representative of the behavior observed in other SmBCO films measured under the same conditions and in similar temperature and field ranges. We list the most relevant experimental features that

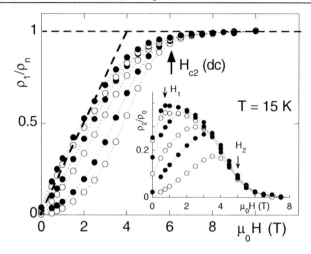

Figure 18: Field dependence of the complex resistivity at fixed frequencies ν =2, 5, 9, 12, 15, 18 GHz (frequency increases from the bottom curve to the top one) and $T = 15$ K in MgB$_2$. Main panel: ρ_1/ρ_0. The intersection of the dashed lines indicates the upper critical field as it would result from a naive application of a vortex motion model. Inset: ρ_2/ρ_0. Arrows indicate the field H_1 where the high-frequency imaginary resistivity reaches a maximum and the field H_2 where there is no more a clear frequency dependence in ρ_2.

will determine the discussion of the data, and we specifically compare them to the behavior of the parent compound YBCO.

At all temperatures, $\Delta\rho_1/\rho_0$ exhibits a pronounced downward curvature at low fields, followed by an approximately linear increase at higher fields. By contrast, $\Delta\rho_2/\rho_0$ never shows a linear variation. Moreover, $\Delta\rho_2/\rho_0$ changes from positive to negative as the temperature increases, but without changing the curvature of the data. Those considerations can be put on more quantitative grounds by plotting the data as a function of $\sqrt{H}$, as reported in Fig.24. It is seen that $\Delta\rho_2/\rho_0$ is well approximated by a straight line, which corresponds to a $\sim \sqrt{H}$ dependence, while upward curvature in $\Delta\rho_1/\rho_n$ vs. $\sqrt{H}$ indicates the presence of both a square root and a linear term in the H dependence. Summarizing, to the best of our experimental accuracy, the complex resistivity in SmBCO can be described by $\Delta(\rho_1 + i\rho_2)/\rho_0 \simeq [a_1(T) + ia_2(T)] \sqrt{\mu_0 H} + b_1(T)\mu_0 H$.

With respect to YBCO we then encounter two main differences: the field dependence, which has a strong sublinear component in the entire temperature range explored, and the very relevant increase of the imaginary part in even moderate fields. In order to discuss the data, in analogy to the discussion on the other materials here investigated, we first tentatively ascribe the observed behavior to the vortex motion alone. For the purpose of the discussion, it is sufficient to focus on the data at low enough temperature, that is below the irreversibility line, where creep can be neglected. We then consider for the preliminary discussion Eqs.(7). In the conventional data analysis the superfluid density (or, which is the same, the London penetration depth) does not change appreciably with the field, and pinning affects mostly the imaginary part of the response. One might then be tempted to assign the difference between YBCO and SmBCO simply to a much stronger, or different,

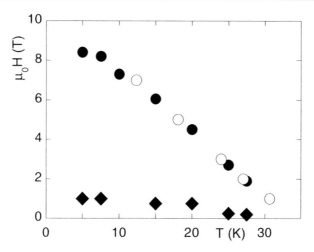

Figure 19: Full dots: temperature dependence of the zero-field σ−band superfluid fractional density obtained from the selfconsistent fitting procedure. Open symbols are calculations with the standard BCS relation [11] using gap values from [142].

pinning in SmBCO. Making use of the vortex-motion-only expressions, Eqs.(7), one would directly calculate the pinning frequency from the data[6] as: $\nu_p^{calc} = \nu\frac{\Delta\rho_2}{\Delta\rho_1}$. Proceeding further within the same framework, once the tentative ν_p has been calculated the viscosity follows immediately by calculating $\frac{\Phi_0 B}{\Delta\rho_1}\frac{1}{1+(\nu_p/\nu)^2}$, see Eqs.(7). However, within this framework we would obtain a field-dependent vortex viscosity, in particular $\sim \sqrt{B}$, at odds with the measured vortex viscosity in YBCO. The procedure is illustrated in Fig.25, where the as-calculated vortex parameters are reported at one temperature.

Then, it would appear that both the pinning mechanism (implicit in ν_p^{calc}) and the electronic state (implicit in the calculated vortex viscosity η^{calc}) are very different in SmBCO and YBCO. This conclusion does not appear very reasonable, due to the structural, electrical and superconducting similarities between the two compounds. In particular, while pinning may well be sample-dependent, the much different field behavior of the vortex viscosity put doubts on the correctness of the simple model used.

We now propose a possible alternative explanation for our data. Since, as a first approximation, the vortex motion resistivity at low fields should be proportional to the number of flux lines, that is to the induction $B \simeq \mu_0 H$, we tentatively assume that only the linear part of our data is due to the vortex motion. In this case, we deduce from the data that there is almost no vortex motion contribution to the imaginary resistivity. This is exactly the behavior as exhibited by YBCO at the same frequency, that shows free flux flow at 48 GHz.

We have now to identify the physical mechanism responsible for the $\sim \sqrt{H}$ part of the real and imaginary resistivity and its peculiar features (in particular, the change of sign of $\Delta\rho_2(H)$ with increasing temperature). As shown in Fig. 5, right panel, the change of sign of $\Delta\rho_2(H)$ with increasing temperature is a signature of the field-dependent superfluid

---

[6]This kind of analysis has been widely used in the past in various HTCS, as reviewed in [35].

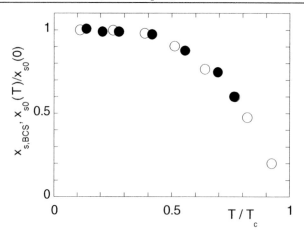

Figure 20: $H - T$ phase diagram in MgB_2. Open circles: H_{c2} from dc measurements. Full dots, H_{c2} from the selfconsistent fitting procedure. Diamonds, H_1.

depletion, while the $\sim \sqrt{H}$ dependence points to a specific electronic state, namely a super-conducting gap with lines of nodes. Thus, it is possible to compare theory and experiment. Making use of Eq.(6), we can expand for low fields (small $H/H_{c2} \simeq B/B_{c2}$) and we obtain explicit expressions for the coefficients $a_1(T)$ and $a_2(T)$. Those expressions contain several temperature-dependent quantities, for which we believe it is appropriate to use the simplest possible model: we take the superfluid fraction $x_{s0}(T) = (1 - t^2)$, which is a recognized approximation in a very wide temperature range for HTCS [64], and the pair breaking field $B_{pb} \propto B_{c2}(T)$, as theoretically suggested [95, 96], so that $B_{pb} = B_{pb0}(1 - t^2)$, with $t = T/T_c$. In order to gain qualitative information, we do not attempt to insert some temperature dependence in the quasiparticle scattering time τ_{qp}, that we use as a free parameter. As can be seen in Fig.26, the fits reproduce the shape, height, width of the curve given by the experimental data, including the temperature of change of sign for a_2. The resulting $\tau_{qp} \simeq 0.7$ ps compares well to the highest values reported in YBCO, e.g., 0.2 ps at ≈ 80 K as obtained from microwave measurements crystals [64] and to 0.5 ps at ≈ 80 K as obtained from millimeter-wave interferometry in YBCO film [65]. Even if the model can certainly be improved (e.g., by considering a temperature-dependent τ_{qp}), the substantial agreement with the data led us to conclude that the microwave resistivity in the vortex state observed in our SmBCO samples is strongly affected by the field-induced superfluid depletion, and that the differences between different samples of YBCO and SmBCO are determined by mere quantitative differences in τ_{qp}.

Coming back to the fluxon dynamics, once the temperature dependence of the carrier conductivity has been assessed from the above described fit, one can extract the vortex vis-cosity in a wide temperature range[7]. The so-obtained data points for the vortex viscosity are reported in Fig.27. Both numerical values and temperature dependences compare favorably to similar data in YBCO (Fig. 13), adding evidence that the vortex motion contribution has been correctly identified.

---

[7]At low temperature the slope b_1 directly yields η, but approaching T_c the quasiparticle screening length comes into play and a correction is needed, see Eq.(4).

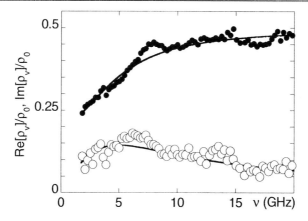

Figure 21: Contribution of the vortex motion to the overall field-dependent complex resistivity in MgB$_2$, isolated as described in the text. Full dots, Re$[\rho_v]$; open circles, Im$[\rho_v]$. Continuous lines are simultaneous fits by Eqs.(5).

4.4   General Remarks

As we have shown in this study, the physics of superconductors in the vortex state can take advantage from investigations at microwave frequencies. However, in order to assess with some confidence the various mechanisms responsible for the microwave response, it is necessary to combine temperature, field and frequency dependent measurements. This is especially needed in the two-gap superconductor MgB$_2$, where single frequency measurements are not able to assess the nature of the strong anomaly in the low-field response.

By the combined measurements here reported, it has been shown that, in all cases here investigated, the correctly identified vortex motion contribution follows to a great accuracy the conventional models. In particular, the high frequency regime always coincides with conventional flux flow (above H^* in MgB$_2$): this is a remarkably noticeable point, since it implies that, in the field and temperature ranges studied, the dominant excitations inside the vortex cores in the materials examined are conventional quasiparticle excitations, as indicated by the temperature dependence and field independence of the vortex viscosity. In order to emphasize this point, we plot in Fig. 27 the vortex viscosity as measured by us in YBCO and SmBCO thin films and the depinning frequency obtained by us in YBCO films, together with data obtained in YBCO crystals by multifrequency cavity measurements [43]. It is immediately seen that the vortex viscosity has the same behavior in YBCO and SmBCO, films and crystals. Moreover, the data for different materials scale one onto the other with mere numerical factors of order unity, consistently with the viscosity given by the standard expression $\eta = \Phi_0 B_{c2}/\rho_n$. Correctly, the depinning frequency in YBCO appears to be different in films and in crystals, indicating that defects in samples (or size effects) play a role. We especially emphasize this point, in comparison to earlier studies (see the early review in [35]) that reported the very anomalous sample independent depinning frequency and sample dependent vortex viscosity, quite the opposite of the expected behavior.

In fact, a second important indication that emerges from our measurements is that it is in general not appropriate to neglect the field dependence of the superfluid density and quasi-

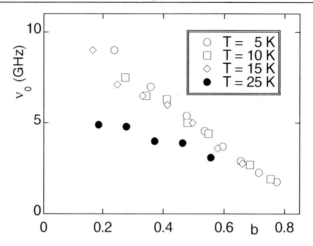

Figure 22: Reduced field dependence of the vortex characteristic frequency ν_0 in MgB$_2$ at various temperatures. The reduced field $b = B/B_{c2}$.

particle density of states in the materials under study: albeit coming from different physical origins, the superfluid density results to be much weaker than in conventional superconductors upon the application of an external magnetic field. This latter feature seems to be governed by the (sample-dependent) quasiparticle scattering time in RE-BCO. In MgB$_2$ the low fields behavior is only qualitatively understood, so that this material appears to be most interesting for future studies. In particular, there is at present no consensus on a representation of the vortex dynamics when both gaps are effective, which is then a promising field for investigation.

5 Conclusion

We have extensively investigated the experimental microwave response of innovative superconductors (cuprates and MgB$_2$) combining resonant and swept-frequency techniques. We have shown that this combination allows for the identification of the major differences between those superconductors and the conventional, metallic superconductors. The main difference resides in the weakness of the superfluid density with respect to an applied magnetic field both in cuprates and MgB$_2$, coming however from different physics. By contrast, the vortex motion contribution appears to be well described by conventional models (above H^* in MgB$_2$).

Appendix A: Selfconsistent Fitting Procedure in MgB$_2$

Here we describe the selfconsistent fitting procedure for the swept-frequency data in MgB$_2$. We refer to the general expressions, Eqs.(4).

The values of x_{s0} and H_{c2} for each temperature can be obtained as follows. As a first step, we notice that, if $\frac{\nu}{\nu_s(B,T)} = 2 \left(\frac{\lambda(B,T)}{\delta_{nf}(B,T)} \right)^2$ is known at any fixed temperature and field,

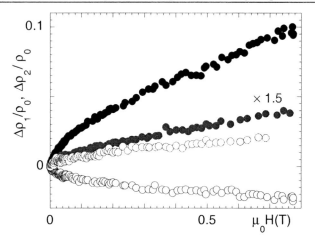

Figure 23: Field dependence of the complex resistivity changes in SmBCO (sample Sm2) at two temperatures. Black symbols, $T = 82$ K. Red symbols, $T = 71$ K. A sublinear component is evident both in $\Delta\rho_1/\rho_0$ (full dots) and in $\Delta\rho_2/\rho_0$ (open circles). Note the large values and the sign change of $\Delta\rho_2/\rho_0$ with increasing temperature. $\Delta\rho_1/\rho_0$ at 71 K has been scaled to avoid crowding.

$r_1(\nu, B, T)$ and $r_2(\nu, B, T)$ can be obtained from the measured ρ_1/ρ_n and ρ_2/ρ_n by inverting Eqs.(4). Although the frequency, temperature and field dependence of r_1 and r_2 are not known *a priori*, the high frequency limit of r_1 is known: at high enough frequency Eq.(5) gives $\mathrm{Re}[\rho_v]/\rho_n \rightarrow \rho_{ff}/\rho_n = b$, with $b = B/B_{c2}(T) \simeq H/H_{c2}(T)$, having assumed the validity of the Bardeen Stephen expression [13] for the flux flow resistivity. We then choose a given temperature T_0 and calculate, for any value of H, a value of $\nu_s(B, T)$ using tentative values of $x_{s0}(T_0)$ and $H_{c2}(T_0)$ and $x_s(H, T_0) = x_{s0}(T_0)(1 - b)$. We then obtain for all fields $r_1(\nu, B, T_0)$ inverting Eqs.(4) and we check that they approach a constant value at high frequency. The values of $x_{s0}(T_0)$ and $H_{c2}(T_0)$ are then changed until, for all fields, the high frequency value for r_1 is equal to $b = H/H_{c2}(T_0)$. Thus, *having assumed only the high frequency limit of r_1*, $r_1(\nu)$ and $r_2(\nu)$ at various fields and temperatures are obtained. Since $\frac{\lambda(B,T_0)}{\delta_{nf}(B,T_0)}$ has already been selfconsistently determined at each temperature T_0, we can isolate the *pure vortex motion complex resistivity* $\mathrm{Re}[\rho_v(\nu)]$ and $\mathrm{Im}[\rho_v(\nu)]$ at various fields and temperatures. Those curves can then be fitted to Eq.(5) to get the vortex parameters.

Acknowledgments

This work has been partially supported by INFM under the national projects PRA-H.O.P and PRA-U.M.B.R.A., and by MIUR under a FIRB project "Strutture Semiconduttore/Superconduttore per l'elettronica integrata". We thank V. Ferrando and C. Ferdeghini for supplying the MgB$_2$ sample, C. Camerlingo for YBCO sample Y4 and M. Boffa and A.M. Cucolo for the remaining YBCO and SmBCO samples. We acknowledge useful discussions with S. Anlage, M. Ausloos, M.W. Coffey, R. Fastampa, M. Giura, J. Halbritter, A. Maeda, R. Marcon, D. Neri, D. Oates, R. Wördenweber.

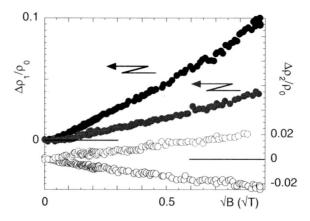

Figure 24: Same data of Fig.(23), replotted *vs.* $\sqrt{H}$. Left scale, $\Delta\rho_1/\rho_0$. Right scale, $\Delta\rho_2/\rho_0$. Symbols as in Fig.(23). It clearly appears that $\Delta\rho_2/\rho_0 \propto \sqrt{H}$.

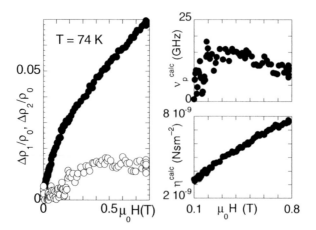

Figure 25: Left panel: complex resistivity changes at 74 K in sample Sm1. Right panels: field dependence of the calculated vortex parameters ν_p^{calc} and η^{calc}, as they would result from the conventional framework for vortex-state complex resistivity. The calculated viscosity η^{calc} presents a strong field dependence, at odds with models and with experimental data in YBCO.

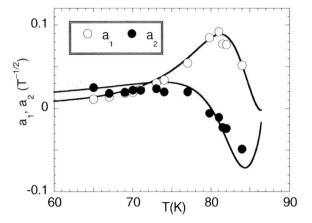

Figure 26: Plot of the coefficients $a_1(T)$ and $a_1(T)$. Continuous lines are simultaneous fits with the pair breaking expression described in the text. Details are reported in [146].

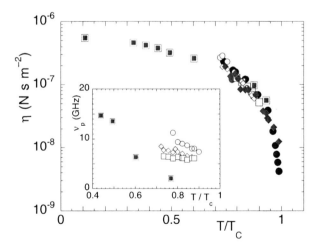

Figure 27: Temperature dependence of the vortex viscosity η obtained in this study in YBCO (data and symbols as in Fig.13) and SmBCO films (green filled diamonds), compared to data obtained in crystals at 19.1 GHz (red squares, from [43]). The data in SmBCO have been multiplied by 2. All the data collapse on the same curve, indicating that the same electronic mechanisms take place in the vortex core. Inset: depinning frequency in YBCO films (sample Y3, same data and symbols as in Fig.12) compared to measurements at 19.1 GHz in single crystals (red squares, from [43]).

References

[1] Biondi, M. A.; Forrester, A. T.; Garfunkel, M. P.; Satterthwaite, C. B. *Rev. Mod. Phys.* 1958, 30 1109.

[2] Biondi, M. A.; Garfunkel, M. P. *Phys. Rev.* 1959, 116 853.

[3] D'Aiello, R.V.; Freedman, S.J. *Phys. Rev. Lett.* 1969, 22, 515.

[4] Lehoczky, S.L.; Briscoe, C.V. *Phys. Rev. Lett.* 1969, 23, 695.

[5] Lehoczky, S.L.; Briscoe, C.V. *Phys. Rev. B* 1971, 4, 3938.

[6] Pippard, A. B. *Proc. R. Soc.* 1947, A191, 399.

[7] Pippard, A. B. *Proc. R. Soc.* 1950, 203, 98.

[8] Waldram, J. R. *Adv. Phys.* 1964 49, 1

[9] Rosenblum, B.; Cardona, M. *Phys. Lett.* 1964, 9, 220.

[10] Rosenblum, B.; Cardona, M. *Phys. Rev. Lett.* 1964, 12, 657.

[11] Tinkham, M. *Introduction to Superconductivity 2nd edition* ; McGraw-Hill: New York, US, 1996.

[12] see, e.g., Abrikosov, A. A. *Fundamentals of the Theory of Metals* ; North-Holland, 1988.

[13] Bardeen, J.; Stephen, M. J. *Phys. Rev.* 1965 140, A1197.

[14] de Gennes, P. G. *Superconductivity of Metals and Alloys* ; Addison-Wesley.

[15] Kim, Y. B. ; Stephen, M. J. In *Superconductivity*; Parks, R. D., Ed., Marcel Dekker, Inc. : New York, US, 1969; pp 1107.

[16] Campbell, A. M.; Evetts, J. E. *Adv. Phys.* 1972 21, 199

[17] Suhl, H. *Phys. Rev. Lett.* 1965, 14, 226.

[18] Matsuda, Y.; Ong, N. P.; Yang, Y. F.; Harris, J. M.; Peterson, J. B. *Phys. Rev. B* 1994, 49, 4380.

[19] Kopnin, N. B. ; Vinokur, V. M. ; *Phys. Rev. Lett.* 1998, 81, 3952.

[20] Sonin, E. B. *Phys. Rev. B* 2001, 63, 054527.

[21] Han, J. H. ; Kim, J. S. ; Kim, M. J.; Ping Ao *Phys. Rev. B* 2005, 71, 125108.

[22] Gittleman, J. I.; Rosenblum, B. *Phys. Rev. Lett.* 1966, 16, 734.

[23] Wu, D. H.; Shridar, S. *Phys. Rev. Lett.* 1990, 65, 2074.

[24] Yeh, N.-C. *Phys. Rev. B* 1991, 43, 523.

[25] Golosovsky, M.; Naveh, Y.; Davidov, D. *Phys. Rev. B* 1992, 45, 7495.

[26] Huang, M. X.; Bhagat, S. M.; Findicoglu, A. T.; Venkatesan, T.; Manheimer, M. A.; Tyagi, S. *Physica C* 1992, 193, 421.

[27] Owliaei, J. ; Sridhar, S. ; Talvacchio, J. *Phys. Rev. Lett.* 1992, 69, 3366.

[28] Pambianchi, M. S. ; Wu, D. H. ; Ganapathi, L. ; Anlage, S. M. *IEEE Trans. Appl. Supercond.* 1993, 3, 2774.

[29] Silva, E.; Marcon, R.; Matacotta, F. C. *Physica C* 1993, 218, 109.

[30] Golosovsky, M.; Tsindlekht, M.; Chayet, H.; Davidov, D. *Phys. Rev. B* 1994, 50, 470; *ibidem* 1995, 51, 12062.

[31] Revenaz, S.; Oates, D. E.; Labbé-Lavigne, D.; Dresselhaus, G.; Dresselhaus, M. S. *Phys. Rev. B* 1994, 50, 1178.

[32] Morgan, D. C. ; Zhang Kuan ; Bonn D. A. , Liang Ruixing ; Hardy, W. N. ; Kallin, C. ; Berlinsky, A. J. *Physica C* 1994, 235-240, 2015.

[33] Parks, B.; Spielman, S.; Orenstein, J.; Nemeth, D. T.; Ludwig, F.; Clarke, J.; Marchant, P.; Lew, D. J. *Phys. Rev. Lett.* 1995, 74, 3265.

[34] Wu, D. H.; Booth, J. C.; Anlage, S. M. *Phys. Rev. Lett.* 1995, 75, 525.

[35] Golosovsky, M.; Tsindlekht, M.; Davidov, D. *Supercond. Sci. Technol.* 1996, 9, 1 and references therein.

[36] Belk, N.; Oates, D. E.; Feld, D. A.; Dresselhaus, G.; Dresselhaus, M. S. *Phys. Rev. B* 1996, 53, 3459.

[37] Belk, N.; Oates, D. E.; Feld, D. A.; Dresselhaus, G.; Dresselhaus, M. G. *Phys. Rev. B* 1997, 56, 11966.

[38] Lütke-Entrup, N.; Plaçais, B.; Mathieu P.; Simon, Y. *Phys. Rev. Lett.* 1997, 79, 2538.

[39] Ong, N. P.; Wu, H. *Phys. Rev. B* 1997, 56, 458.

[40] Hanaguri, T.; Tsuboi, T.; Tsuchiya, Y.; Sasaki, K.; Maeda, A. *Phys. Rev. Lett.* 1999, 82, 1273.

[41] Rogai, R.; Marcon, R.; Silva, E.; Fastampa, R.; Giura, M.; Sarti, S.; Boffa, M.; Cucolo, A. M. *Int. J. Mod. Phys. B* 2000, 14, 2828.

[42] Silva, E.; Fastampa, R.; Giura, M.; Marcon, R.; Neri, D.; Sarti, S. *Supercond. Sci. Technol.* 2000, 13, 1186.

[43] Tsuchiya, Y.; Iwaya, K.; Kinoshita, K.; Hanaguri, T.; Kitano, H.; Maeda, A.; Shibata, K.; Nishizaki, T.; Kobayashi, N. *Phys. Rev. B* 2001, 63, 184517.

[44] Silva, E.; Marcon, R.; Muzzi, L.; Pompeo, N. ; Fastampa, R.; Giura, M.; Sarti, S.; Boffa, M.; Cucolo, A. M.; Cucolo, M. C. *Physica C* 2004, 404, 350.

[45] Parks, B.; Orenstein, J.; Mallozzi, R.; Nemeth, D. T.; Ludwig, F.; Clarke, J.; Merchant, P.; Lew, D. J.; Bozovic, I.; Eckstein, J. N. *J. Phys. Chem. Solids* 1995, 56, 1815.

[46] Mallozzi, R.; Orenstein, J.; Eckstein, J. N.; Bozovic, I. *Phys. Rev. Lett.* 1998, 81, 1485.

[47] Blatter G.; Feigel'man M. V.; Geshkenbein V. B.; Larkin A. I.; Vinokur V. M. *Rev. Mod. Phys.* 1994, 66, 1125.

[48] Brandt, E. H. *Rep. Prog. Phys.* 1995, 58, 1465.

[49] Pakulis, E. J.; Osada, T. *Phys. Rev. B* 1988, 37, 5940.

[50] Marcon, R.; Fastampa, R.; Giura, M.; Matacotta, C. *Phys. Rev. B* 1989, 39, 2796.

[51] Giura, M.; Marcon, R.; Fastampa, R. *Phys. Rev. B* 1989, 40, 4437.

[52] Giura, M.; Fastampa, R.; Marcon, R.; Silva, E. *Phys. Rev. B* 1990, 42, 6228.

[53] Wosik, J.; Kranenburg, R. A.; Wolfe, J. C.; Selvamanickam, V.; Salama, K. *J. Appl. Phys.* 1991, 69, 874.

[54] Wosik, J.; Xie, L. M.; Chau, R.; Samaan, A.; Wolfe, J. C. *Phys. Rev. B* 1993, 47, 8968.

[55] Fastampa, R.; Giura, M.; Marcon, R.; Silva, E. In *Studies of High Temperature Superconductors*; Narlikar, A. V.; Ed., Nova Science: New York, US, 1996; Vol. 17, pp 115-145.

[56] Halbritter, J. *J. Supercond.* 1995, 8, 691.

[57] Gurevich, A. *Phys. Rev. B* 1992, 46, 3187.

[58] Gurevich, A. *Phys. Rev. B* 2002, 65, 214531.

[59] Xin, H.; Oates, D. E.; Dresselhaus, G.; Dresselhaus, M. S. *Phys. Rev. B* 1990, 65, 214533.

[60] Lee, Sheng-Chiang ; Lee, Su-Young ; Anlage, Steven M. *Phys. Rev. B* 2005, 72, 024527.

[61] Gaganidze, E.; Heidinger, R.; Halbritter, J.; Shevchun, A.; Trunin, M.; Schneidewind, H. *J. Appl. Phys.* 2003, 93, 4049.

[62] Hardy, W. N.; Bonn, D. A.; Morgan, D. C.; Liang, Ruixing; Zhang, Kuan *Phys. Rev. Lett.* 1993, 70, 3999.

[63] Mao, Jian ; Wu, D. H.; Peng, J. L.; Greene, R. L.; Anlage, S. M. *Phys. Rev. B* 1995, 51, 3316.

[64] Bonn, D. A.; Liang, R.; Riseman, T. M.; Baar, D. J.; Morgan, D. C.; Zhang, K.; Dosanjh, P.; Duty, T. L.; MacFarlane, A.; Morris, G. D.; Brewer, J. H.; Hardy, W. N.; Kallin, C.; Berlinsky, A. J. *Phys. Rev. B* 1993, 47, 11314.

[65] Nagashima, T.; Hangyo, M.; Nakashima, S.; Murakami, Y. In *Adv. in Superconductivity VI*; Fujita, T.; Shiohara, Y.; Eds.; Springer-Verlag: Tokyo, JP, 1994; pp 209-212.

[66] Jakobs, T.; Sridhar, S.; Rieck, C. T.; Scharnberg, K.; Wolf, T.; Halbritter, J. *J. Phys. Chem. Solids* 1995, 56, 1945.

[67] Gao, F. ; Carr, G. L. ; Porter, C. D. ; Tanner, B. D. ; Williams, G. P. ; Hirschmugl, C. J. ; Dutta, B. ; Wu, X. D. ; Etemad, S. *Phys. Rev. B* 1996, 54, 700.

[68] Nagamatsu, J.; Nakagawa, N.; Muranaka, T.; Zenitani, Y.; Akimitsu, J. *Nature* (London) 2001, 410, 63.

[69] Bouquet, F.; Fisher, R. A.; Phillips, N. E. ; Hinks, D. G.; Jorgensen, J. D. *Phys. Rev. Lett.* 2001, 87, 047001.

[70] Iavarone, M.; Karapetrov, G.; Koshelev, A. E.; Kwok, W. K.; Crabtree, G. W.; Hinks, D. G.; Kang, W. N.; Choi, Eun-Mi; Hyun Jung Kim; Kim, Hyeong-Jin ; Lee, S. I. *Phys. Rev. Lett.* 2002, 89, 187002.

[71] Gonnelli, R. S.; Daghero, D.; Ummarino, G. A.; Stepanov, V. A.; Jun, J.; Kazakov, S. M.; Karpinski, J. *Phys. Rev. Lett.* 2002, 89, 247004.

[72] Kim, Mun-Seong ; Skinta, John A.; Lemberger, Thomas R.; Kang, W. N.; Kim, Hyeong-Jin ; Choi, Eun-Mi ; Lee, Sung-Ik *Phys. Rev. B* 2002, 66, 064511.

[73] Jin, B. B.; Klein, N.; Kang, W. N.; Kim, Hyeong-Jin; Choi, Eun-Mi; Lee, Sung-Ik; Dahm, T.; Maki, K. *Phys. Rev. B* 2002, 66, 104521.

[74] Ghigo, G.; Botta, D.; Chiodoni, A.; Gozzelino, L.; Gerbaldo, R.; Laviano, F.; Mezzetti, E.; Monticone, E.; Portesi, C. *Phys. Rev. B* 2005, 71, 214522.

[75] Dulčić, A.; Požek, M.; Paar, D.; Choi, Eun-Mi ; Kim, Hyun-Jung; Kang, W. N.; Lee, Sung-Ik *Phys. Rev. B* 2003, 67, 020507(R).

[76] Shibata, A.; Matsumoto, M.; Izawa, K.; Matsuda, Y.; Lee, S.; Tajima, S. *Phys. Rev. B* 2003, 68, 060501(R).

[77] Golubov, A. A.; Brinkman, A.; Dolgov, O. V. ; Kortus, J.; Jepsen, O. *Phys. Rev. B* 2002, 66, 054524.

[78] Ohishi, K.; Muranaka, T.; Akimitsu, J.; Koda, A.; Higemoto, W.; Kadono, R. *J. Phys. Soc. Japan* 2003, 72, 29

[79] Serventi, S.; Allodi, G.; De Renzi, R.; Guidi, G.; Romanò L.; Manfrinetti, P.; Palenzona, A.; Niedermayer, Ch.; Amato, A.; Baines, Ch. *Phys. Rev. Lett.* 2004, 93, 217003.

[80] Babaev, E. *Phys. Rev. Lett.* 2002, 89, 067001.

[81] Eskildsen, M. R.; Kugler, M.; Tanaka, S.; Jun, J.; Kazakov, S. M.; Karpinski, J.; Fischer, Ø. *Phys. Rev. Lett.* 2002, 89, 187003.

[82] Gonnelli, R. S.; Daghero, D.; Calzolari, A. ; Ummarino, G. A.; Dellarocca, V.; Stepanov, V. A.; Jun, J.; Kazakov, S. M.; Karpinski, J. *Phys. Rev. B* 2004, 69, 100504(R).

[83] Cubitt, R.; Eskildsen, M. R.; Dewhurst, C. D. ; Jun, J.; Kazakov, S. M.; Karpinski, J. *Phys. Rev. Lett.* 2003, 91, 047002.

[84] Jackson, J. D. *Classical Electrodynamics*; Wiley: New York, US, 1962.

[85] Portis, A. M.; WBlazey, K.; Muller, K. A.; Bednorz, J. G. *Europhys. Lett.* 1988, 5, 467.

[86] Sonin, E. B.; Tagantsev, A. K. *Zh. Eksp. Teor. Fiz.* 1989, 95, 994 [*Sov. Phys.-JETP* 1989, 68, 572].

[87] Marcon, R.; Fastampa, R.; Giura, M.; Silva, E. *Phys. Rev. B* 1991, 43, 2940.

[88] Coffey, M. W.; Clem, J. R. *Phys. Rev. Lett.* 1991, 67, 386.

[89] Brandt, E. H. *Phys. Rev. Lett.* 1991, 67, 2219.

[90] Sonin, E. B.; Tagantsev, A. K.; Traito, K. B. *Phys. Rev. B* 1992, 46, R5830.

[91] Plaçais, B.; Máthieu, P.; Simon, Y.; Sonin, E. B.; Traito, K. B. *Phys. Rev. B* 1996, 54, 13083.

[92] Coffey, M. W.; Clem, J. R. *Phys. Rev. B* 1992, 45, 10527.

[93] Coffey, M. W.; Clem, J. R. *Phys. Rev. B* 1992, 46, 11757.

[94] Coffey, M. W.; Clem, J. R. *Phys. Rev. B* 1993, 48, 342.

[95] Volovik, G. E. *JETP Lett.* 1993, 58, 469.

[96] Won, H.; Maki, K. *Phys. Rev. B* 1996, 53, 5927.

[97] Yip, S. K.; Sauls, J. A. *Phys. Rev. Lett.* 1992, 69, 2264.

[98] Nakai, N ; Miranoviç, P.; Ichioka, M.; Machida, K. *Phys. Rev. B* 2004, 70, 100503(R).

[99] Dahm, T.; Graser, S.; Iniotakis, C.; Schopohl, N. *Phys. Rev. B* 2002, 66, 144515.

[100] Laiho, R.; Lähderanta, E.; Safonchik, M.; Traito, K. B. *Phys. Rev. B* 2004, 69, 094508.

[101] Silva, E.; Marcon, R.; Sarti, S.; Fastampa, R.; Giura, M.; Boffa, M.; Cucolo A. M. *Eur. Phys. J. B* 2004, 37, 277.

[102] Beneduce, C.; Bobba F.; Boffa M.; Cucolo A. M.; Cucolo M. C.; Andreone A.; Aruta C.; Iavarone M.; Palomba F.; Pica G.; Salluzzo M.; Vaglio R. *Int. J. Mod. Phys. B* 1999, 13, 1333.

[103] Neri, D.; Marcon, R.; Rogai, R.; Silva, E.; Fastampa, R.; Giura, M.; Sarti, S.; Cucolo, A. M.; Beneduce, C.; Bobba, F.; Boffa, M ; Cucolo, M.C. *Physica C* 2000, 341-348 2679.

[104] Camerlingo, C.; Lissitski, M. P.; Russo, M.; Salvato, M.; presented at *INF Meeting, June 2002, Bari, Italy* (unpublished)

[105] Boffa, M. A.; Bobba, F.; Cucolo, A. M.; Monaco, R. *Int. J. Mod. Phys. B* 2003, 17, 768.

[106] Boffa, M.; Cucolo, M. C.; Monaco, R.; Cucolo, A. M. *Physica C* 2003 384, 419.

[107] Ferrando, V.; Amoruso, S.; Bellingeri, E. ; Bruzzese, R.; Manfrinetti, P.; Marré, D.; Velotta, R.; Wang, X.; Ferdeghini, C. *Supercond. Sci. Technol.* 2003, 16, 241.

[108] Collin, R. E. *Foundation for microwave engineering 2nd edition* ; Electrical Engineering Series; McGraw-Hill International Series: Singapore, 1992.

[109] Sridhar, S. *J. Appl. Phys.* 1988, 63, 159.

[110] Silva, E.; Lanucara, M.; Marcon, R. *Supercond. Sci. Technol.* 1996, 9, 934.

[111] Beeli, P. *Physica C* 2000, 333, 65 and references therein.

[112] Pompeo, N.; Marcon, R.; Silva, E. In *Applied Superconductivity 2003 - Proc. of VI European Conference on Applied Superconductivity, Sorrento, Italy, 14-18/9/2003* ; Andreone, A.; Pepe, G. P.; Cristiano, R.; Masullo, G.; Eds.; Institute of Physics, Conference Series 181; 2004; pp 2629-2634.

[113] Pompeo, N.; Marcon, R.; Méchin, L.; Silva, E. *Supercond. Sci. Technol.* 2005, 18, 531.

[114] Klein. M ; Chaloupka, H.; Müller, G.; Orbach, S.; Piel, H.; Roas, B.; Schultz, L.; Klein, U.; Peiniger, M. *J. Appl. Phys.* 1990, 67, 6940.

[115] Hartemann, P. *IEEE Trans. Appl. Supercond* 1992, 2, 228.

[116] Hein, A. M.; Strupp, M.; Piel, H.; Portis, A. M.; Gross, R. *J. Appl. Phys.* 1994, 75, 4581.

[117] Silva, E.; Lanucara, M.; Marcon, R. *Physica C* 1997, 276, 84.

[118] Poole, C. P. *Electron Spin Resonance. A Comprehensive Treatise on Experimental Techniques*; Interscience: New York, US, 1967.

[119] Clem, J. R. ; Sanchez, A. *Phys. Rev. B* 1994, 50, 9355.

[120] Xing, X. ; Heinrich, B. ; Zhou, Hu ; Fife, A. A. ; Cragg, A. R. *J. Appl. Phys.* 1994, 76, 4244.

[121] Silva, E.; Lezzerini, A.; Lanucara, M.; Sarti, S.; Marcon, R. *Meas. Sci. Technol.* 1998, 9, 275.

[122] Ceremuga, J.; Krupka, J.; Kosciuk, T. *J. Supercond.* 1995, 8, 681.

[123] Mazierska, J. *J. Supercond.* 1997, 10, 73.

[124] Silva, E.; in *Superconducting Materials: Advances in Technology and Applications* , ed by A. Tampieri and G. Celotti, World Scientific, pp. 279-306 (2000)

[125] Sarti, S.; Amabile, C.; Silva, E. (2004) COND-MAT/0406313

[126] Amabile, C.; Fastampa, R.; Giura, M.; Sarti, S.; Silva, E.; Ferrando, V.; Tarantini, C.; Ferdeghini, C. In *Applied Superconductivity 2003 - Proc. of VI European Conference on Applied Superconductivity, Sorrento, Italy, 14-18/9/2003* ; Andreone, A.; Pepe, G. P.; Cristiano, R.; Masullo, G.; Eds.; Institute of Physics, Conference Series 181; 2004; pp 1281-1288.

[127] Ikeda, R.; Ohmi, T.; Tsuneto, T. *J. Phys. Soc. Jpn.* 1991, 60, 1051.

[128] Mikeska, H.-J.; Schmidt, H. *Z. Physik* 1970 230, 239.

[129] Schmidt, H. *Z. Phys* 1968, 216, 336. Ibidem 1970, 232, 443.

[130] Skocpol, W. J.; Tinkham, M. *Rep. Prog. Phys.* 1975, 38, 1049.

[131] Klemm, R. A. *J. Low Temp. Phys.* 1974, 16, 381.

[132] Dorsey, A. T. *Phys. Rev. B* 1991 43, 7575.

[133] Silva, E. *Eur. Phys. J. B* 2002, 27, 497.

[134] Sarti, S.; Silva, E.; Amabile, C.; Fastampa, R.; Giura, M. *Physica C* 2004, 404, 330.

[135] Ferrando, V.; Manfrinetti, P.; Marré, D.; Putti, M.; Sheikin, I.; Tarantini, C.; Ferdeghini, C. *Phys. Rev. B* 2003, 68, 94517.

[136] Sarti, S.; Amabile, C.; Silva, E.; Giura, M.; Fastampa, R.; Ferdeghini, C.; Ferrando, V.; Tarantini, C. *Phys. Rev. B* 2005, 72, 024542.

[137] Seneor, P.; Chen, C.-T.; Yeh, N.-C.; Vasquez, R. P.; Bell, L. D.; Jung, C. U.; Park, Min-Seok ; Kim, Heon-Jung ; Kang, W. N.; Lee, Sung-Ik *Phys. Rev. B* 2001, 65, 012505.

[138] Koshelev, A. E.; Golubov, A. A. *Phys. Rev. Lett.* 2003, 90, 177002.

[139] Mazin, I. I.; Andersen, O. K.; Jepsen, O.; Dolgov, O. V.; Kortus, J.; Golubov, A. A.; Kuz'menko, A. B.; van der Marel, D. *Phys. Rev. Lett.* 2002, 89, 107002.

[140] Ferrando, V.; Tarantini, C.; Manfrinetti, P. ; Marré, D.; Putti, M.; Tumino A.; Ferdeghini, C. In *Applied Superconductivity 2003 - Proc. of VI European Conference on Applied Superconductivity, Sorrento, Italy, 14-18/9/2003* ; Andreone, A.; Pepe, G. P.; Cristiano, R.; Masullo, G.; Eds.; Institute of Physics, Conference Series 181; 2004; pp 1263-1270.

[141] Gurevich, A. *Phys. Rev. B* 2003, 67, 184515.

[142] Szabó, P.; Samuely, P.; Kacmarcik, J.; Klein, T.; Marcus, J.; Fruchart, D.; Miraglia, S.; Marcenat, C.; Jansen, A. G. M. *Phys. Rev. Lett.* 2001, 87, 137005.

[143] Dulčić, A.; Paar, D.; Požek, M.; Williams, G. V. M.; Krämer, S.; Jung, C. U.; Park, Min-Seok ; Lee, Sung-Ik *Phys. Rev. B* 2002, 66, 014505.

[144] Murakami, M.; Sakai, N.; Higuchi, T.; Yoo, S. I. *Supercond. Sci. Technol.* 1996, 9, 1015.

[145] Küpfer, H.; Wolf, Th.; Zhukov, A. A.; Meier-Hirmer, R. *Phys. Rev. B* 1999, 60, 7631.

[146] Silva, E.; Pompeo, N. ; Muzzi, L. ; Marcon, R.; Sarti, S. ; Boffa, M. ; Cucolo, A. M. COND-MAT/0405324 (2004).

In: Recent Developments in Superconductivity Research   ISBN 978-1-60021-462-2
Editor: Barry P. Martins, pp. 245-274   © 2007 Nova Science Publishers, Inc.

Chapter 8

A CENTURY OF SUPERCONDUCTIVITY – UPDATES OF THE PERIODIC TABLE OF SUPERCONDUCTING ELEMENTS

Cristina Buzea

Physics Department, Queen's University, Kingston K7L 3N6, Canada

Abstract

We present the updates of the superconductivity in simple elements, as reported at the turn of the twenty first century. After almost a hundred years of superconductivity research, the non-superconducting gaps in the periodic table of elements are shrinking while the maximum superconducting temperatures achieved by simple elements is raising to values unforeseen several decades ago. The research of superconductivity in simple elements was revived by the recent developments of high pressure diamond anvil cells together with the discovery of superconductivity in magnesium diboride at a remarkable high transition temperature (40 K). If an element is not superconducting down to very low temperatures, there area several methods to transform it into a superconductor. Among the most used method to probe superconductivity of elements is by subjecting them to high pressure, irradiation, charge doping. On the other side, quenched condensed, templated, and very thin films allow amorphous, structural phase, or proximity induced superconductivity in non-superconducting elements. This review offers the readers a comprehensive picture of superconductivity, its correlations and trends for simple elements.

1   Introduction

After almost a century from the discovery of zero resistance at low temperatures in mercury, the knowledge of superconductivity shown by simple elements is scattered in various journals and conference proceedings along the years. Some of these journals are well known, easy to access, while other are with a limited circulation, however the information they contain is very important and sometimes not published in more reachable journals. Each journal's accessibility is usually limited by the institution or personal subscription. We tried to be as comprehensive as possible, to cite the first reports announcing superconductivity of one element, the articles containing the most relevant information, and articles that announce important discoveries, despite the fact that sometimes the reports are not confirmed. We would like to acknowledge that this review was made possible by the new era of electronic information storage and online information access [1] as well as by the fine subscription list of Queen's University library.

At the end of nineteenth century there was no widely accepted theory for the behaviour of free electrons in metals, particularly at low temperatures (T). It was known that the resistivity of metals decreases with temperature, but it was not known what limiting value the resistance would approach, if the temperature was reduced very close to 0 K. The opinions were divided. Some scientists believed the resistivity of a pure metal tends to zero as the temperature is decreased to absolute zero; others predicted that the metallic resistivity would decrease to a minimum level at low T and then increase for lower temperatures. It seemed that the first investigator to solve the cryogenic problem would settle the questions about the electrical resistance near absolute zero. And this was Kamerlingh Onnes who succeeded in liquefying helium (boiling point at 4.2 K) in 1908. Kamerlingh Onnes investigated the resistivity of metals at low temperatures, assisted by his graduate student, Gilles Holst, who made the actual measurements of resistance. It was thought natural to measure metals like gold and platinum, which have low resistance at room temperature. Nobody expected that the metals with poor conductivity will provide the most spectacular behaviour. And definitely no one expected zero dc resistance to occur above 0 K. The first experiments with platinum showed a gradually diminished resistance followed by a levelling-off at 4.3 K, in agreement with one of the theories. But the same behaviour could be obtained in the presence of impurities. Experiments on gold showed similar results, making uneasy a precise answer to the resistivity behaviour problem. Onnes decided to use mercury, as it was liquid at room temperature and therefore easier to purify than platinum or gold. The measurement results for mercury were completely unexpected. The resistance of mercury suddenly disappeared at 4.2 K. Onnes called this newly discovered state, superconductivity [2].

After the discovery of mercury superconductivity, all the simple elements were tested in order to know if they were superconductors. Surprisingly, the best electricity carrier metals (such as copper, gold and silver) presented no superconductivity at all. Then, not long after, a list of the superconductors made of these simple elements was established. Among these simple elements, the niobium has the highest T_c (9.2 K).

Since 1986, most of superconductivity research was directed towards high temperature superconductors (HTSC), with much higher critical temperatures of up to 133 K for a mercury based cuprate $HgBa_2Ca_2Cu_3O_8$, and 164 K under pressure [3]. Only recently the applications of HTSCs (such as yttrium and bismuth based cuprates) are moving from research and development to scale-up [4]. After the discovery of magnesium diboride in 2001 [5] at an unprecedented high critical temperature of 40K for a binary compound, the interest has shifted toward simpler materials. Some of these materials were recently reported to superconduct at, surprisingly high critical temperatures - 20 K for lithium under pressure - the highest T_c for a simple element [6]. During the last few years, superconductivity has been observed in many new elements, some of them under pressure, such as sulphur 17 K [7], oxygen 0.5 K [8], carbon in nanotube 15 K [9] and diamond forms 4K [10], a non-magnetic state of iron 1 K [11], and the light elements lithium 20 K [6] and boron 11 K [12] (Figs. 1 and 2). Despite the fact that thousands of papers have been published on the topic, our understanding of superconductivity after almost one hundred years of research is limited to a phenomenological level, its microscopic origin still eluding us.

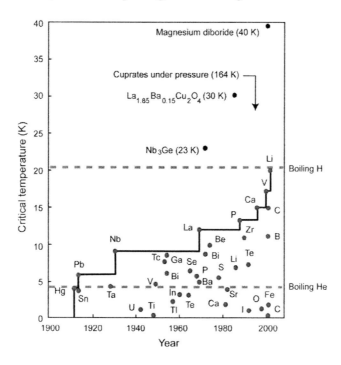

Figure 1. Historical development of the critical temperature of simple elements. Reference points for Nb3Ge cuprates, and magnesium diboride are given for comparison.

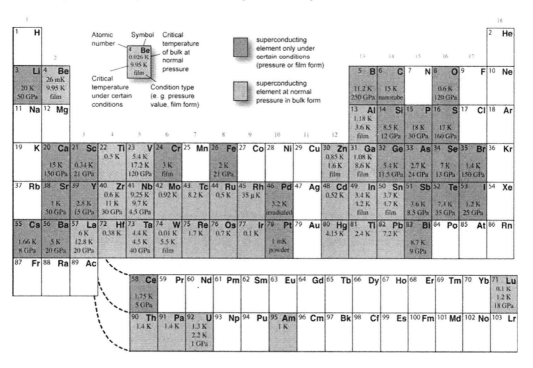

Figure 2. Periodic table of superconducting elements

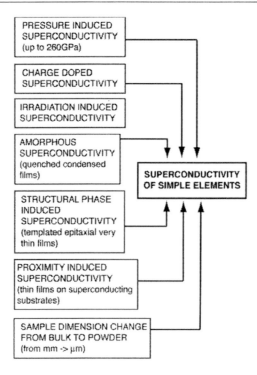

Figure 3. Techniques used to transform normal elements into superconductors

2   Inducing Superconductivity

Once thought a rare anomaly, it now appears that superconducting states can be formed in most materials, including thousands of inorganic compounds, most of the natural elements, alloys, oxides, and even organic materials and, significantly, the double-stranded molecules of DNA [13].

If a given element is not superconducting at normal pressure, there are several ways of causing it to superconduct, illustrated schematically in Fig. 3. Transport of electron pairs from a superconductor in close proximity creates a small superconducting interface layer in some materials, a phenomenon called proximity induced superconductivity. Crystal strain or non-equilibrium crystal textures can also induce superconductivity, through applied pressure, quenched growth on a cold substrate, or epitaxial templating in the form of thin films. Chemical doping that affects electron occupancy (charge doping) produces superconductivity in many materials, including diamond. Finally, irradiation induced lattice disorder can also lead to superconductivity through the suppression of spin fluctuations.

The atomic structure of a material determines its superconducting characteristics with superconductivity arising as a collective transport behavior through the electronic potential of the crystal. An illustration of this is the existence of different critical temperature values for polymorphic forms of certain elements, such as lanthanum, which superconducts at T_c = 4.8 K when double hexagonal closed packed, dhcp, compared to T_c = 6 K when face centered cubic, fcc [14].

2.1   Proximity Effect

The superconducting proximity effect is a phenomenon that appears between a superconductor and a normal metal, where the superconducting wave function varies smoothly across the interface causing a suppression of the pair amplitude in the superconductor and an enhancement of pairing on the normal side [15]. The characteristic distance in which superconductivity "penetrates" into the normal region is the normal state coherence length, $\xi_N = \sqrt{\hbar D/2\pi kT}$ where D is the diffusivity. The length scale at which superconductivity is suppressed in the superconducting material is the superconducting coherence length ξ.

2.2   Superconductivity under Pressure

Applying pressure is often the simplest way of drastically altering the interatomic distances and the crystalline structure of a material, where the material can undergo structural phase transitions, each structure being characterized by a unique electronic configuration. Phase transformations occur in many elements under compression and some high-pressure phases exhibit superconductivity.

The renaissance in high-pressure research on superconductivity is primarily the result of recent development of high-pressure diamond anvil cells [16]. While the highest attainable pressure in 1968 was approximately 25 GPa employing Bridgman type anvils at liquid helium temperatures [17], today measurements can be performed to a maximum pressure of 260 GPa [12]. This made possible the surprise observation of higher-pressure emergent phases of simple elements that exhibit superconducting properties. Of particular interest is the research of the light elements, as they are predicted to superconduct at record high temperatures [18]. In addition, low atomic number (Z) materials are highly compressible, having a large variation in interatomic distances. In addition, high sensitivity electric and magnetic susceptibility techniques have been developed to examine superconductivity at megabar pressures [19]. Superconductivity under pressure is seen in many elements, as is summarized in the periodic table of Fig. 2. Some elements transformed to a metallic state without showing signs of superconductivity, e.g. xenon [20], others to semiconductors, e.g. nitrogen [21].

2.3   Quenched-Condensed Films

The pioneering work of Buckel and Hilsch [22] showed that films prepared on substrates held at cryogenic temperatures has superconducting T_c's very different from the bulk transition temperature. For some materials T_c was raised, and the rise was correlated with the Debye temperature. Additional studies showed that the particle size increases as the substrate temperature was raised, along with a decrease in T_c toward the bulk value. It was experimentally observed that T_c of superconductors in contact with dielectric barriers, evaporated on cryogenic substrates can have much higher transition temperatures than their bulk counterparts [23].

A multitude of reports on superconductivity of quenched-condensed films, some of them unconfirmed, were published several decades ago, from which many are of a particular author

- C. Reale. One must emphasize that unconfirmed results should be considered with reluctance. During the seventies and eighties, Reale reports the measurement of superconductivity of various elements [24-30], describing deposition of films by thermal evaporation in vacuum on various substrates held at 4.2K, in-situ measurements of resistivity and critical fields, variation of transition temperature with annealing temperature and film thickness. He reports that after film fabrication and in-situ assessment of superconductivity, the films were cyclically annealed at increasing temperatures until superconductivity was irreversible quenched. His results show a decrease of the critical temperature T_c with the annealing temperature as well as with the film thickness. For many of the elements already reported to be superconductive, the critical temperature was in concordance with the results obtained by other authors in that time (1970-1980), while recent experiments demonstrated a higher critical temperatures for those elements. However, Reale reported superconductivity of several elements, some of which were later confirmed to be superconductive under pressure and some still unconfirmed.

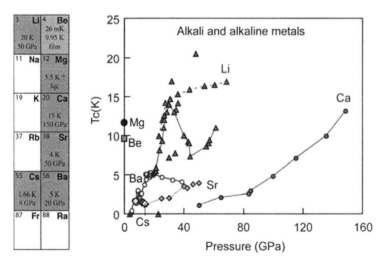

Figure 4. Critical temperature dependence on pressure for alkali and alkaline metals, Li [6, 31, 32], Ca [33], Ba [34-40], Sr [34].

3   Superconducting Elements

3.1   Alkali and Alkaline Earth Metals

Among the alkali metals (Li, Na, K, Rb, Cs, Fr) only Li and Cs were found to superconduct under pressure, while most of the alkaline earth metals superconduct under pressure (Ca, Sr, Ba) or film form (Be). An unconfirmed report authored by Reale [30] stated superconductivity of Mg as a quenched-condensed film.

Hydrogen, the most common element in the Universe, has a simple structure - only one proton and one electron. If cooled below 20 K becomes a liquid, and a solid below 14 K, in both states an insulator. At the beginning of the last century, it was predicted that under extreme pressures H would become a metal [41], its molecules dissociating into atoms. Recent calculations [42-44] predicted that solid hydrogen should become a molecular metal above 300 GPa pressure, before transforming into an alkali metal. In 1 960s, Aschroft [18] predicted that H would conduct electricity without resistance, and possibly have a critical temperature around room temperature. These prediction stimulated high pressure studies of hydrogen and search of its superconductivity. The main technological problem encountered in the study of H, the smallest of atoms, is its high diffusivity and chemical reactivity, which make it the most difficult material to contain at high pressure, tending to embrittle and weaken gaskets and anvils. Despite these difficulties, the study of H has generated several interesting results, among which is its transformation into a metallic fluid at 140 GPa [45], and the modification of its optical properties (darkening) at higher pressures (320 GPa) [46]. Hydrogen remains in its molecular state up to 316 GPa at least all attempts at demonstrating its superconductivity, however, have had negative results.

The first report, however unconfirmed, on possible lithium superconductivity as a quenched-condensed film with T_c=2.4 K was published in 1975 [25]. Later measurements of Li resistivity under pressure (1986) revealed a phase transition accompanied by a sudden drop of resistivity around 7 K, attributed to possible superconductivity recent calculations (1999) predicted superconductivity of Li at high pressure [48] with a T_c of up to 80 K [49]. A clear demonstration of Li superconductivity was published in 2002 [6, 31], with T_c increasing with pressure and reaching an unprecedented 20 K for a simple element, under 50 GPa [6]. A plot of its critical temperature dependence on pressure is shown in Fig. 4 together with the behavior of superconducting alkaline earth metals. The experimental behavior of Li is not adequately resolved, and agreement has not been reached on its T_c dependence on pressure [6, 31, 32]. Data measured under almost hydrostatic pressures [32] (compressed in an anvil while surrounded by helium to provide uniform pressure without shear) shows an increase then decrease of T_c with pressure, evidence that superconductivity competes with symmetry breaking structural phase transitions which occur near 20, 30 and 62 GPa. The sudden appearance of superconductivity with a T_c of 5.5 K is likely to be due to a transition from the low temperature phase Rh6 to the fcc structure [32]. T_c increases monotonically to 14 K at 30 GPa, above this pressure the derivative dP/dT changing - a clear sign of structural phase transition (probably to a cI16 or intermedite hR1 phase). Deemyad and Schilling [32] found that critical temperature decreases rapidly in a dc magnetic field, signature of a type I superconductor, in contrast with the results of Shimizu et al. with $H_c(0)$ values of 30000 Oe, pointing toward a type II superconductivity in Li [6].

Cesium under pressure has a much lower critical temperature than Li, with T_c decreasing monotonically from 1.66 K for increasing pressure [50]. Qunched-condensed films of Cs on RbI, KI, and RbBr substrates were reported to have transition temperatures between 1.52 and 2.15 K [24, 25]. In bulk form Cs does not show superconductivity.

Beryllium, has a modest T_c of 26 mK in bulk form [51], but attains a T_c of 9.95 K when deposited as a quenched-condensed film [52].

For the other alkalis, preliminary calculations suggest that Na [53], K, and Rb [54] might become superconducting under pressure with transition temperatures of up to 10 K, but has yet to be experimentally observed.

Reale published in 1975 data showing Mg superconductivity at 5.5 K as a quenched-condensed film [30], however measurements performed by Belzons et al. [55] under similar conditions contradict this finding. Reale reported that Mg superconductivity vanishes at an annealing temperature of 100 K, concluding that the bulk is not superconducting [30].

The heavier alkaline earth metals, Ca, Ba, and Sr transform into superconductors only under pressure [56]. Calcium [33, 57], and strontium [34] have a positive dT_c/dp curve, while the critical temperature of Ba increases and then decreases [34-39], as seen in Fig. 4.

Calcium was reported to pass into superconducting state around 2 K under pressures of 44 GPa by Dunn and Bundy [57], superconductivity confirmed later by Okada et al. [33], with a T_c reaching 15 K under pressures of 150 GPa[33].

In addition to these reports, Reale [24, 30] claims superconductivity of alkaline earth metals Be, Mg, Ca, Sr, Ba as vapor-quenched films with thickness from 10-200 nm [30]. He reported critical temperatures for films of Ca, Sr, Ba of 4.2 K, 3.7 K [30], and 4.25 K [24], respectively.

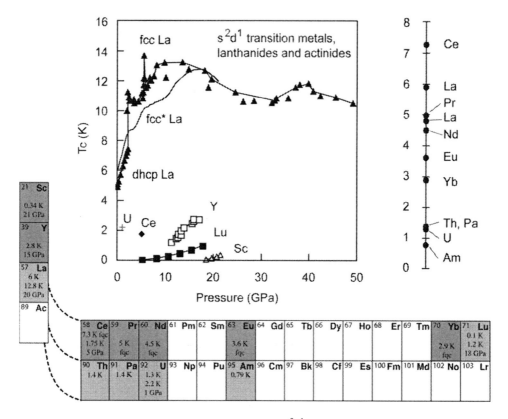

Figure 5. Critical temperature dependence on pressure for s^2d^1 transition metals, La [14, 58] Lu [59], Sc [59], Y [50], Ce [24], Pr, Nd, Eu, and Yb [60], Th [61], Pa [62], Am [63], U [64-66].

3.2 s^2d^1 Transition Metals

Scandium is the lightest among the s^2d^1 transition metals Sc, Y, La, and Lu. It was suggested that antiferromagnetic spin fluctuations may be responsible for the absence of superconductivity in Sc at normal pressure [59].

Yttrium [50], as its related s^2d^1 transition metals Sc, and Lu [59], have a positive dT_c/dp slope, as shown in Fig. 5, being almost superconductor at normal pressure.

Lanthanum has two superconducting phases at normal pressure in bulk form, dhcp at 4.8 K and fcc at 6 K [58]. La also shows a more complicated critical temperature dependence on pressure above 2 GPa, as well as phase transitions [14, 58]. Reale claimed superconductivity of 9.8 K of quenched condensed films fabricated at liquid helium temperature [60]. T_c is reported to decrease with annealing temperature and with increasing film thickness.

Cerium is shown to superconduct at 1.75 K under 5 GPa pressure [50]. Reale also reports of its superconductivity as quenched-condensed film of about 10 to 210 nm thick on NaF substrates with a maximum transition temperature of 7.3 K [24] and on alkali-zinc borosilicate substrates at 5.8 K [60]. Its T_c decreases with annealing temperature and with film thickness.

Unverified reports of Pr, Nd, Eu, and Yb superconductivity as 10-210 nm thin films quenched-condensed on substrates held at 4.2 K and measured in situ claim critical temperatures of 5, 4.5, 3.6, and 2.9 K, respectively, shown to decrease with annealing temperature and increasing film thickness [60]. Among them, Pr and Nd are suggested to superconduct in bulk form at lower temperatures.

Lu [59] has a positive dT_c/dp slope, as shown in Fig. 5.

Thorium is superconductor below 1.4 K [61], his properties being very well described by the BCS theory: a weakly coupled type I superconductor, exhibiting a complete Meissner effect, with a critical field curve, energy gap, and electronic specific heat in agreement with the BCS theory [64].

Pa and Am become superconducting at 1.4 K [61, 62] and 0.79 K [63], respectively.

A metal whose superconductivity has attracted much attention since its discovery (1942) is uranium, with several superconducting phases. In its room temperature α-U phase shows filamentary superconductivity. Uranium is among very few elemental type-II superconductors and has an unique room temperature structure. Its superconductivity is also unique because it was the first example of a metal that becomes a superconductor under pressure without a crystallographic transition [66]. The T_c of α-U (orthorombic) increases from 1.3 K to 2 K under 1 GPa, saturating and slowly decreasing for higher pressures [64]. From heat capacity studies it was concluded that α-U was filamentary in nature unless subjected to about 1 GPa, when its transformed to bulk superconductivity. A series of three low-temperature charge-density-wave (CDW) structural phase changes in the normal state occur at 43, 38, and 22 K, and the superconducting transition shows a large positive isotope effect when the CDW is suppressed by 1 GPa of pressure [66]. Recent measurements of transition temperature of α-U in single crystal form demonstrated superconductivity below 0.8 K in filamentary form [65]. The higher T_c in polycrystalline bulk form may be due to strain at grain boundaries, giving rise to a similar bulk effect as induced at high pressures. A possible explanation for the filamentary nature of U superconductivity is that distortions in the crystal lattice due to the charge-density-wave CDW are somehow responsible for causing superconducting filaments [65]. The critical temperature of γ-u is almost identical with that of α-U under pressure [64].

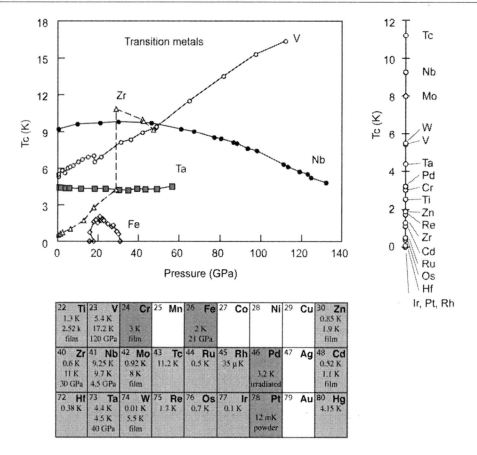

Figure 6. Critical temperature dependence on pressure for transition metals, Zr [67, 68], V [69-71], Nb [72], Ta [72, 73], Fe [11], Ti [74, 75], Cr [76], Zn [23], Zr [77], Mo [78], Tc [79], Ru [80], Rh [81], Pd [82], Cd [83], W [75], Re [84], Os [85], Hf [86], Pt [87].

3.3   Transition Metals

Most of the transition metals are superconducting at normal pressure, the highest critical temperatures being for elements in the fifth group: Nb at 9.25 K [72], V at 5.4 K [88], [89] and Ta at 4.4 K [90].

Early measurements performed by Meissner on the resistance of titanium single crystals showed a drop 80% at 4 K, while a zero resistance was reached at 1.3 K [74]. When deposited by ion-beam sputtering in noble gasses, Ti had a considerable increase compared to its vacuum deposited thin film form critical temperature [75]. The films deposited with Xe have a T_c of 2.52 K compared to the vacuum deposited with a T_c of 0.41 K [75]. Recent experiments gave convincing evidence that pure Ti is an intrinsic type-II superconductor with the lowest T_c up to now due to its very low renormalized Fermi velocity ($H_{c2}(0)$=0.46 kG) [91].

Unlike niobium and tantalum, vanadium has a large positive pressure coefficient of the critical temperature [69], its critical temperature under pressure reaching 17.2 K at 120 GPa, among the highest for simple elements [70]. The increase of T_c is thought to be due to suppression of electron spin fluctuations through broadening of the d - band width [71]. Vanadium epitaxial films superconduct at 5.4K [89].

Chromium, normally an antiferromagnetic metal, nonsuperconducting when vacuum evaporatred, was found to superconducts when ion-beam sputtered with noble gases [75, 92]. The films deposited with noble gases of higher mass have higher Tcs than those with noble gases with lower mass, more exactly T_c reaches 0.6 K, 0.9 K, and 1.5 K, for Ar, Kr, Xe gases, respectively [75]. It was suggested that the disturbance of the electronic structure by noble gas incorporation suppresses the antiferromagnetism normally observed in chromium. Superconductivity was also observed in fcc Cr films epitaxially sandwiched between gold layers Au/Cr/Au below 3 K [76].

The correlation between superconductivity and magnetism makes the study of ferromagnetic metals (Fe, Co, Ni, Gd) particularly important. Iron is the classic magnetic metal, being strongly ferromagnetic at standard conditions in its bcc phase. In this ferromagnetic form iron does not superconduct to the lowest temperatures attainable. At pressures above 10 GPa iron assumes an hcp structure, believed to be non-magnetic. The superconductivity of the iron hcp phase was predicted [93] and was recently confirmed experimentally [11]. A small but definite drop in the electrical resistivity was observed at temperatures below 2 K at pressures above 16 GPa [11]. The critical field of iron is 0.2 T.

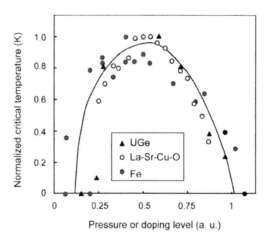

Figure 7. Normalized critical temperature variation with pressure and doping level for materials with ferromagnetic or antiferromagnetic behaviour. Data are taken from references [94, 95] and [11] for Uge2, La-Sr-Cu-O, and Fe, respectivly

As illustrated in Fig. 6, the $T_c(p)$ diagram for iron has a specific bell-shaped curve. Interestingly, this phase diagram for iron [11] shows a striking resemblance to the critical temperature versus doping or carrier concentration of superconductors with antiferromagnetic parent compounds - $La_{2-x}Sr_xCu_2O_4$ [94], or UGe_2 [95], as illustrated in Fig. 7. While their relationship is not fully understood, it was believed that ferromagnetism and superconductivity were competing mechanisms, and the superconductivity of iron came as a surprise. Perhaps the explanation lies in theoretical model of magnetism mediated superconductivity [95]. The validity of this model was questioned until recently, when superconductivity was observed in $_{UGe2,}$ on the border of ferromagnetism within the same electrons that produce band magnetism [95]. These recent discoveries of superconductivity in iron [11], as well as in cobalt compounds [96, 97], re-focuses attention on magnetic mediated superconductivity as an alternative to phonon mediated.

Zinc attains 1.9 K in ultrathin superconducting films sandwiched between insulating layers [23], 1.52 K as quenched condensed thin films [98], while in bulk form has only 0.85 K.

A material less studied since its superconductivity discovery [67, 99] is zirconium, with two superconducting phases, hcp and bcc (body center cubic), reaching a critical temperature of 11 K at 30 GPa in its bcc form [68]. In bulk form Zr attains a modest 0.6 K, while zirconium thin films with thickness from 200 to 1200 nm attain 1.3 K [77]. If sputtered with Xe, zirconium films reach 4 K [75].

Niobium also holds the record for the highest critical temperature of an element at normal pressure. No structural phase transitions are observed in Nb, so changes in T_c are thought to be the result of changes in electronic structure alone, creating a $T_c(p)$ that is nearly constant from 10 Gpa to 70 GPa [72](Fig. 6).

A great deal of research on superconducting properties of refractory metals was performed in the sixties and seventies, when high purity refractory metals could be synthesized. Molybdenum, in bulk form, has a T_c of 0.9 15 K as reported by Rorer et al. [100], while in quenched-condensed films on liquid-helium cooled substrates 30 nm thickness molybdenum films attain a much higher temperature of 8 K, as reported by Koepke and Bergmann [78]. When deposited by ion-beam sputtering in noble gasses, Mo had a considerable increase compared to its bulk critical temperature [75]. The films deposited with noble gases of higher mass have higher T_cs than those with noble gases with lower mass, more exactly T_c reaches 4.8 K, 6.4 K, 6.8 K, and 7.2 K, for Ne, Ar, Kr, Xe gases, respectively [75].

Technetium, named from the Greek word "teknetos", meaning artificial, has been reported to have a superconducting transition temperature onset at 11.2 K and end at 8 K in powder form [79]. The magnetic behavior of this material is characteristic of a type II superconductor with $H_{c2}(0) = 2600$ Oe and a thermodynamical critical fiels $H_c(0) = 1400$ Oe [101].

Ru has a modest critical temperature of 0.5 K [80].

The investigation of superconductivity in platinum metals (Pt, Rh, Pd) demonstrated superconductivity of bulk Rh, $T_c = 325$ K [81], irradiated Pd [102], and powder Pt [103]. All three elements show strongly exchange-enhanced paramagnetism [103]. The associated spin-density fluctuations are supposed to be detrimental to superconductivity. To this day, Rh remains the element of the periodic table with the lowest transition temperature into superconducting state. Its critical field reaches 0.49 nT.

Palladium, a metal with particularly interesting properties, is not superconducting in bulk form. From its high electronic density of states at the Fermi surface and its phonon spectrum, one would expect strong electron-phonon coupling, and therefore a reasonable high superconducting transition temperature. Experimentally, however, superconductivity does not occur down to 1.7 mK [102] due to spin fluctuations, though palladium can be transformed into a superconductor with a T_c of 3.2 K through the introduction of lattice disorder via low-temperature irradiation with He+ ions [82], in qualitative agreement with theoretical predictions for superconductivity in palladium without spin fluctuations [82].

Cd thin films exhibit a superconducting transition temperature of 1.1 K [83] compared to the bulk of 0.52 K. Early measurements of Hf superconductivity showed a transition temperature of 0.38 K [86].

The critical temperature of Ta in bulk for is 4.4 K [104] being nearly constant from ambient to pressures up to 40 GPa, as illustrated in Fig. 6 [72]. During this pressure range no structural phase transition was observed.

Bulk W is used as a temperature calibration fixed-point for very low-temperature thermometry at 15.6 mK [105]. Evaporated and sputtered W films have been fabricated with superconducting transition temperatures as high as 4.1 K, several order of magnitude higher than the bulk value [106]. It was suggested that T_c is higher as the thin film particles size decreases. For W films sputtered with noble gases it was found that their T_c increases with the mass of the noble gas used, i.e. for Ne, Zr, Kr, and Xe the films of tungsten become superconductive below 3 K, 3.7 K, 4.2 K and 5.5 K. [75].

Rhenium hcp single crystals are found to have a superconducting transition temperature T_c of 1.7 K [84]. Osmium and iridium have modest critical temperatures of 0.7 K [80] and 0.1 K [85], respectively.

Platinum shown superconductivity as compacted powder, its bulk being nonsuperconductive. Pt, a fcc metal, has a strong electron-phonon coupling, expected to favour superconductivity. Conversely, its strong electron-electron exchange interaction and resulting high paramagnetic Pauli susceptibility ($\chi_{exp}[T{\rightarrow}0]=S\chi_{Pauli}$, with a Stoner enhancement factor S = 3.9) brings this metal close to a ferromagnetic instability [103]. As a consequence, on one side, the 5d conduction electrons show strong spin fluctuations, that tend to suppress superconductivity, and on the other side, the effective moments of magnetic 3d impurities like Fe, Mn, and Co in platinum are enhanced due to the polarization of the neighbouring 5d conduction electrons of the host metal. The investigations of superconductivity on platinum performed on bulk samples with dimensions of a few millimeters were unsuccessful. However, recent studies of the magnetic behaviour of compacted platinum powder with a typical grain size of 2 μm showed a very different magnetic behavior from the bulk. The Stoner enhancement of Pauli susceptibility in compact powders shows a much weaker temperature dependence of the dynamic susceptibility in the mili and microkelvin range [103]. In contrast to the spin glass behavior observed in bulk platinum with small amount of impurities, the susceptibility of the compact powders shows hardly any temperature dependence down to 1.5 mK. At lower temperatures the susceptibility drops, being accompanied by a step in the electrical resistance, showing a clear transition into superconducting state. A fundamental difference between bulk and the compacted powders is a large surface to volume ratio that could lead to an appreciable lattice softening, thought to be essential in the enhancement of T_c in granular nontransition superconductors, however no T^2 contribution was observed in specific heat indicative for surface phonons [87]. Also, a relatively strong lattice strain was observed in compacted powders [87]. Indeed, the superconducting transition temperature is enhanced more than an order of magnitude (12 mK) for grains with submicrometer size, as compared to those with larger size (micrometers) [87].

Solid mercury usually crystallizes in a rhombohedral structure (α-Hg), stable to low temperatures under normal pressure. This phase was shown to be superconducting by Onnes at the beginning of the last century with a critical temperature of 4.15 K [2]. Another phase, β-Hg, was also shown to superconduct below 3.9 K [107].

There remain only two areas in the periodic system where the metals are not superconducting not magnetic: some alkali-alkaline metals and the noble (Cu, Au, Ag) metals. Despite considerable efforts, none of these remaining metallic elements indicated superconductivity, although the lowest accessible temperature has been lowered down to a few microkelvins during the past decade. The absence of superconductivity in noble metals is attributed to the weak electron-phonon interaction and low electronic specific-heat coefficients. One factor that has to be taken into account is the purity of the metals, especially regarding the magnetic impurities that can hinder the appearance of superconductivity.

3.4 Metals

Metals from groups 13-15 (Al, Ga, In, Tl, Sn, Pb, Bi) situated on the right edge of the transition metals in the periodic table show several common characteristics. First, they all superconduct at normal pressure, except bismuth, which is almost a semi-metal. Second, the T_cs of the elements decrease with increasing pressure, as shown in Fig. 8 for Al [108], Sn [110], Pb [37, 73, 112-114] and Bi [36, 37], the derivative dT_c/dp having similar values. In addition, the critical temperature seems to increase from group 13 to 15.

Aluminum superconducts below 1.18 K [108], and when in mesoscopic wire form shows a peculiar resistance peak in the superconducting state near T_c, attributed to thermal fluctuations producing phase slips of the superconducting order parameter [109]. Aluminum thin films show a much higher T_c than the one of bulk Al [116], which was observed to increase for smaller film thickness. The highest T_c in ultrathin film form sandwiched between insulating layers is 5.7 K [23].

Gallium has a critical temperature of 1.08 K in bulk form, and in amorphous film form, quench-condensed on liquid-helium cooled substrates, attains 8.6 K [117].

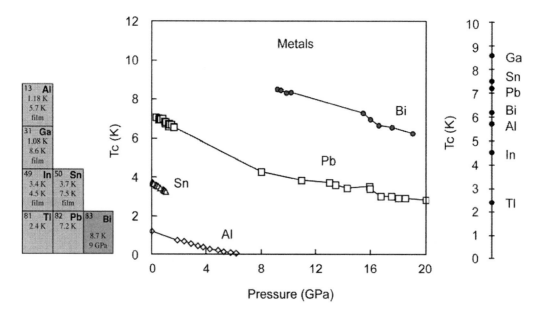

Figure 8. Critical temperature dependence on pressure for metals, Al [108, 109], Sn [110, 111], Pb [37, 73, 112-114], Bi [36, 37], In [23], Tl [115].

Indium superconducts in bulk form below 3.4 K [118]. As ultrathin films sandwiched between insulating layers quenched on cryogenic substrates, In exhibits higher T_c of 4.5 K [23].

Sn was among the first metals to be tested for superconductivity in 1911 by K. Onnes. Its superconductivity in bulk form was later confirmed to be at 3.72 K [119]. Indium superconducts in bulk form below 3.4 K [118]. As ultrathin films sandwiched between insulating layers quenched on cryogenic substrates, Sn exhibit much higher T_c of 6 K [23]. Sn critical temperature seems to vary with the substrate type [111]. Reale reported a fcc phase of Sn obtained by vapour-quenching tin on Rh, Pd, and Ag substrates that superconducts at 7.5, 6.7, and 5.5 K respectively [111]. Above a critical thickness, T_c falls to a thickness-independent limiting value. Recent research reports show that single crystal superconducting tin nanowires with diameters of 40-60 nm have a transition temperature closed to the bulk value [120]. Magnetization measurements indicate that the critical field of the nanowires increases significantly with decreasing diameter to 0.3 T, being almost an order of magnitude larger than in bulk values, consistent with similar data in Pb wires [121].

Thallium, a poisonous metal, has a critical temperature of 2.4 K [115].

Pb was among the first metals to be tested for superconductivity in 1911 by K. Onnes and discovered to be superconductive at 7.2 K. Under pressures its critical temperature decreases monotonically with pressure [37, 73, 112-114]. Recent experiments on superconductivity in single crystal nanowires [122] revealed the fact that their superconducting transition is closed to the bulk one, but significantly suppressed in polycrystalline wires. The critical field was found to increase monotonically with decreasing nanowire diameter [121].

Bismuth is an example of those elements that do not display superconductivity under ordinary circumstances, but undergo superconducting phase transitions in a fcc phase [123] after being subjected to hydrostatic pressure [36], [37]. Bi superconducts with a T_c of 8.7 K at 9 GPa [37], with a T_c = 4 K [123] in thin film form on ferromagnetic Ni substrates, and with T_c = 6.2 K [124] when amorphous.

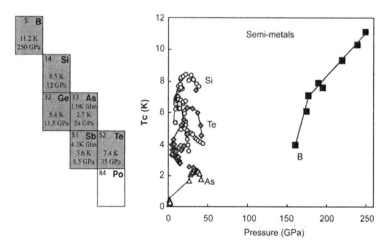

Figure 9. Critical temperature dependence on pressure for semi-metals, B [12], Si [125-127], As [27, 128, 129], Te [130-132], Sb [27].

3.5   Semi-metals

A general characteristic of semi-metals is that they do not show superconductivity at normal pressure, but superconduct when compressed. Most of them have structural phase transitions within the studied pressure range, and exhibit different T_cs for different phases. As a common feature of Si [125-127], As [128, 129], and Te [130-132], the critical temperature increases with pressure, followed by saturation and a decline (Fig. 9).

The recent research of superconductivity in the lightest semi-metal - boron, probably triggered by the announcement of MgB_2 superconductivity at nearly 40 K [5], spurred interest in superconducting physics on boron-related compounds [133]. Boron superconductivity was observed with a maximum T_c of 11.2 K at 250 GPa [12], as shown in Fig. 9. Under ambient conditions, boron has a molecular-like structure based on 12-atom icosahedral clusters, feature that gives rise to complex polymorphism [12]. Mailhiot et al. [134] predicted a transition from icosahedral boron to a body-centered tetragonal (bct) structure around 210 GPa. The pressure of metallization is a little lower than the theory, around 180 GPa [12]. The magnitude of T_c appears to be consistent with an electron-phonon coupling origin of superconductivity.

Silicon is the raw material most often used in integrated circuit fabrication. It is the second most abundant substance on the earth. At pressures above 9 GPa silicon, which is not a superconductor under normal conditions, acquires superconducting properties [112, 125, 127] .For amourphous Si, the T_c of β-Sn phase is about 6.8 K and is relatively independent of pressure [127]. The T_c of sh phase increases from 7.4 K at 15 GPa to a maximum around 8.5 K at about 21 GPa, then decreases monotonically up to 38 GPa [127]. For crystalline Si the critical temperature of the high pressure phases is about 3 K lower than that of the corresponding phases of amorphous Si [127]. The transition temperature also depends slightly on the type of conduction (n- or p-type) and the initial carrier density [125].
There are less data on the superconductivity of germanium [112].

Aside the pressure studies in semi-metals, unconfirmed reports on superconductivity in quenched-condensed films of As and Sb on LiF and MnO substrates, claimed superconductivity of these materials at 1.9 K and 4.3 K, respectively [27].

Te suffers different phase transformations under pressures up to 42 GPa [130, 132], showing the following sequence of pressure-induced structural phase transitions: hexagonal→monoclinic (4 GPa) → orthorhombic (6 GPa) → β-Po (11 GPa) → bcc (27 GPa). A semiconductor-metal transition occurs at 4 Pa, corresponding to the structural transition to monoclinic phase. The T_c of Te exhibits a significant enhancement at 35 GPa, resultant from the structural transition from β-Po type to bcc phase, reaching a maximum of 7.4 K, followed by a sharp decline to less than 5 K for pressures of up to 42 GPa [130].

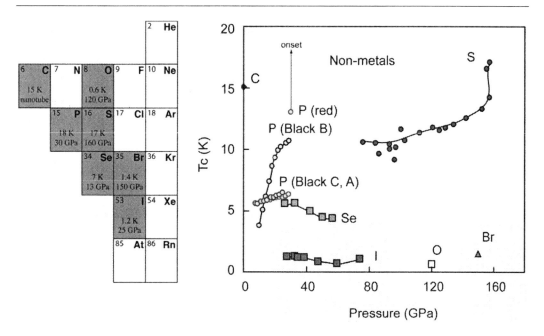

Figure 10. Critical temperature dependence on pressure for non-metals, P [135, 136], S [7, 137], Se [130], Br [137], I [138].

3.6 Non-metals

Among non-metals, carbon (in the form of diamond and nanotubes), oxygen, bromine, and iodine have recently been transformed into superconductors, while the superconductivity under pressure of phosphorus, sulfur, and selenium was observed much earlier (Fig. 10). Nitrogen was transformed into a semiconductor under pressure without exhibiting superconductivity [21]. The different pressure dependencies of their transition temperature indicate that the superconductivity in non-metals is controlled by different factors.

Carbon likely occurs in the widest variety of elemental forms, the most known allotropes being graphite, diamond, C60 carbon spheres (Buckminster-fullerenes), and carbon nanotubes derived from curved graphene sheets. The possibility of superconductivity in carbon nanotubes was predicted in 1995 with T_c expected to be inversely proportional to the nanowire diameter, an effect related to electron-phonon interactions [139]. In 2001 Kociak et al. [140] reported superconductivity below 0.55 K in ropes of single-walled carbon nanotubes with low-resistance contacts to non-superconducting metallic pads. Later the same year Tang et al. [9] succeeded in observing superconductivity in one-dimensional 0.4nm diameter single-walled carbon nanotubes encapsulated in channels of zeolite crystals with a record high T_c of 15 K. The latest surprise from superconductivity in carbon-based materials came from boron doped diamond at 4 K [10]. The discovery of superconductivity in a diamond-type structure suggests that Si and Ge, also having a diamond crystal structure, might superconduct under appropriate doping conditions.

Diamond, the king of gems, has always adorned as a jewel. It is the hardest known material, has the highest thermal conductivity, being an electrical insulator. However, when doped with boron it becomes a p-type semiconductor, or when heavily doped becomes metallic [141]. To achieve superconductivity, it is apparently crucial to realize a carrier concentration sufficiently high to induce an insulator-to-metal transition. Onset critical temperatures of 3.2 K and 7.4 K were reported for bulk diamond synthesized by high pressure sintering [10] or in film form by chemical vapor deposition (CVD) [141]. The higher T_c for CVD diamond is attributed to higher concentrations of boron compared to bulk diamond synthesized at high pressures. The critical field for CVD diamond approaches 10.4 T [141]. The superconductivity in bulk diamond doped with boron was reported to appear in samples fabricated at high pressure, near 100,000 atmospheres, and temperatures of about 2500-2800 K, and with boron concentrations of 4×10^{21} cm^{-3} [10]. Diamond is shown to be a type-II superconductor, with a $H_{C2}(0)$ of 3.4 T and a coherence length of = 10 nm. The application of pressure decreases the superconducting transition temperature without any significant broadening of the transition. The slow decrease of T_c with pressure $dT_c/dP = -6.4 \times 10^{-2}$ K/GPa contrasts with positive $dT_c/dP=0.05$ K/GPa reported for elemental boron for pressures higher than 180 GPa, ruling out the possibility that superconductivity in doped diamond is due to free boron [10]. The strong diamagnetic response demonstrates that superconductivity is not filamentary.

In recent years extraordinary progress on superconductivity of electron or hole-doped fullerenes has been reported by scientists at Bell labs, with the highest transition temperature of up to 117 K. However, the experiments could not be reproduced and the validity of the research started to be questioned by other scientists. The suspected fraud conducted to the retraction of several papers in prominent journals: Science [142], Nature [143-145] and Applied Physics Letters, which questioned the peer-review flaws [146]. A review committee cleared scientists involved in the research of the above mentioned paper, except Jan Hendrik Schon, found to fabricate or alter experimental data. Therefore, until experimental confirmation, a large question mark remains on the superconductivity of charge doped fullerenes.

Oxygen is known to show magnetic properties at low temperatures, so was not expected to superconduct. At pressures exceeding 95 GPa, solid molecular oxygen undergoes a phase transition and becomes metallic. At around 100 GPa solid oxygen becomes superconducting below 0.6 K, as revealed by resistive and magnetic measurements [8, 137]. It seems that the oxygen superconductivity is present even in the molecular metallic state, in contrast with iodine and bromine. The transition temperature of oxygen is rather low compared to other elements in the 16th series: S, Se, and Te which superconduct at 17, 7, and 7.4 K, respectively, in spite of all expectations from both the structural sequence and the increase of T_c for this group. Its critical field is about 2000 G.

Since the discovery of superconductivity in phosphorous [147] at 5.8 K under 17 GPa its critical temperature was increased under 30 GPa pressure to 18 K [148]. Simultaneous measurements of electrical resistivity and X-ray diffraction of black phosphorus carried out at high pressures [149] revealed the first-stage phase transition from the orthorhombic to arsenic-type structure begin to occur at about 4. 0 GPa. These two phases coexist between about 4. 0 and 6. 5 GPa. No distinct anomaly in resistivity change is observed in these pressure regions. The arsenic-type phase was found to transform to a simple cubic phase at around 9. 5 GPa. This phase was very stable in the wide range of pressure. Further phase transformation was not observed at least up to 60 GPa.

The superconducting transition temperature of phosphorus under pressure depends on the path in the pressure-temperature diagram [135, 136, 148, 150]. Black insulating phosphorous with an orthorhombic structure is the most stable form at ambient conditions. The three phases of orthorhombic, rhombohedral and simple cubic in black P show all superconductivity with critical temperatures between 2-4, 5-6, 6-10 K, respectively [151]. (Path A) When pressure is applied at room temperature, black-P is converted into simple cubic metallic phase [136]; after subsequent cooling down to liquid helium temperature under constant pressure is found to be superconductive with T_c of about 6 K. For increasing pressures (path A) T_c increases slightly with pressure [136]. (Path B) If black-P is suddenly cooled down to liquid helium temperature at atmospheric pressure, then the pressure is continuosly increased at this temperature, he superconductivity appears around 4K at 11 GPa [136]. T_c increases steeply with pressure up to almost 11 K, with the critical temperature onset at 13 K [136, 150, 152]. (Path C) After black-P is transformed into semi-metallic rhombohedral phase by applying pressure up to 8.7 GPa, the sample is cooled down to liquid helium temperature keeping constant the pressure, and superconductivity is observed about 5.7 K at 9 GPa [136].

If red phosphorus is used as the starting material and pressurized at 4.2 K, the T_c increases up to 13 K with the onset of transition at 18 K under 30 GPa [135]. The resistance transition versus temperature exhibits a drop with a finite temperature width, with onset at 18 K and zero resistance at 8 K. A new simple bcc structure has been observed above 262 GPa [153], giving hope that a similar or higher T_c could be achieved in this phase. It was suggested that this bcc phase of phosphorous might be stabilized at ambient conditions using suitable templates such as V(100), Fe(100), or Cr(100) substrates [154]. This could lead to a breakthrough in technological spintronic applications, where the combined spin and superconducting degrees of freedom will provide a new level of functionality for microelectronic devices. An unconfirmed report claims superconductivity of quenched-condensed P films at 9.2 K [26].

Under ambient conditions, S forms a molecular solid consisting of eight-membered rings. In situ high-pressure diffraction studies on the metallic phase of S shows a base-centred orthorhombic (b.c.o.) structure above 83 GPa, that transforms to a β-Po structure at 162 GPa, isostructural with the high-pressure metallic phases of Se and Te [155]. The first studies of superconductivity of sulfur were performed in 1978 with reports of transitions into the superconducting state at 5.7 K [156] and 9.8 K [157]. Recent studies of compressed sulfur [7, 137, 158] show that the element transforms directly from an insulator into a superconductor at 93 GPa with $T_c = 10$ K, with T_c jumping to 17 K at 160 Gpa, the transition to the β-Po structure phase (Fig. 10), supposed to arise from a combination of a moderate mass enhancement parameter and a large overall phonon energy scale [159]. This is the third highest reported transition temperature for an elemental solid. This behavior contrasts with the negative dT_c/dp observed at much lower pressures in the heavier superconducting chalcogens Se and Te [130].

Density-functional calculations predict that, with increasing compression, sulfur favors more open structures: β-Po $\rightarrow$ sc $\rightarrow$ bcc [159]. The superconducting temperature of these higher pressure phases is calculated to decrease with pressure for the β-Po and simple cubic phases, while the bcc phase will show a jump to about 17 K at about 550 GPa [159].

An unconfirmed report claims superconductivity of quenched-condensed S films at 4.9 K [26].

Selenium exists in several different forms: the thermodynamically stable chain hexagonal structure, metastable monoclinic, and amorphous [160]. The search for superconductivity in the three Se forms indicate that different forms of Se behave differently when pressurized [160]. The following structural sequence of phase transitions is suggested tro occur in Se: hexagonal $\rightarrow$ SeII (14 GPa) $\rightarrow$ monoclinic (23 GPa) $\rightarrow$ orthorhombic (28 GPa) $\rightarrow\beta$-Po (60 GPa) $\rightarrow$ bcc (140 GPa) [130]. Experimental measurements under pressure for pressures up to 60 GPa demonstrate that critical temperature of selenium decreases with pressure [130]. A semiconductor-metal transition occurs at 23 GPa, corresponding to the structural transition to monoclinic phase [130]. The pressure dependence of T_c shows an anomaly at about 30 GPa, corresponding to a structural transition from monoclinic to orthorhombic phase. The T_c of monoclinic phase is 5.6 K at 26 GPa, and is independent of pressure. Density functional calculations predicts that selenium will show an enhancement in T_c up to 11 K at the transition into the bcc phase, fact still to be experimentally confirmed [159].

The successful searches for superconductivity in the halogen group were iodine [138] and bromine [137]. In the case of bromine under pressure, a molecular to monoatomic phase transition takes place at 80 GPa. At pressures higher than 90 GPa bromine becomes a superconductor with a T_c of about 1.4 K and a critical field of 1 KOe (at 150 GPa) [137, 161]. Quenched condensed bromine films with an fcc structure were reported to become metallic when cooled down to 0.1K, however not superconducting [28], unlike fcc iodine phases quench-deposited on CaO or SrO films [29].

Molecular crystals of iodine at ambient pressure are insulating, while under pressure change to a metal before the molecular dissociation [137, 138]. Superconductivity appears in monoatomic phases and not in its molecular metallic phase. Iodine seems to be among the first example of a hole metal superconductor. Measurements of the Hall effect reveals that the carrier density changes at the boundary of the molecular dissociation, the absence of superconductivity being attributed to the low carrier density before the molecular dissociation [137].

None of the noble gases (He, Ne, Ar, Kr, Xe, Rn) are known to superconduct. A report on xenon indicates its transforms into a metallic state but without showing signs of superconductivity [20].

4   Critical Temperature Correlations

Critical temperature correlations with various physical properties of simple elements were addressed by Hirsch [162] from a statistical point of view, however, the sampling of the studied elements and the updates in the periodic table of superconducting elements would make the respective research outdated. Hirsch [162] concluded that there is a relationship between the existence of superconductivity and the following normal-state properties: bulk modulus, melting temperature, work function, and Hall coefficient. Namely, larger bulk modulus, melting temperature, work functions, and Hall coefficients are associated with superconductivity. He also found that superconductivity is independent of ionic mass, electronic specific heat, magnetic susceptibility, electrical and thermal conductivity, Debye temperature, ionization potential and atomic volume. The magnitude of T_c is positively correlated with electronic specific heat, magnetic susceptibility, atomic volume and negatively correlated to electrical and thermal conductivity and Debye temperature. The magnitude of T_c is uncorrelated with ionic mass, ionization potential, work function, melting temperature and Hall coefficient.

As a general trend, the maximum critical temperature of simple elements scales with the atomic number Z, the highest T_c's being achieved for low Z (Fig. 11), implying that light elements are the best candidates for the utmost critical temperatures. Other correlations of T_c with normal state properties [163], such as bulk modulus, work function, Hall coefficient, Debye temperature, etc. should be re-evaluated in the light of the most recent findings in the critical temperature of simple elements. An interesting path would be a search for possible correlation of T_c with Hall coefficient changes under pressure.

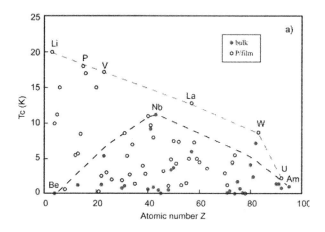

Figure 11. Critical temperature correlation with atomic number Z for bulk and under pressure or/and film form simple elements.

As future directions, further theoretical and experimental studies of superconductivity at pressures exceeding 50 GPa for most of the elements are needed.

While applying pressure to a material is an established way of exploring new crystalline structures, recent thin film fabrication methods have shown great utility in growing unique and non-equilibrium crystals through sandwiched epitaxy [76] or control of geometrical effects in vapor aggregation [164, 165]. These methods should be further employed in the search for new superconducting phases.

The study of elemental superconductors has an enormous impact on our understanding of superconductivity. Current work is laying the foundations of elemental superconductor nanowires applications and further elucidation of the low dimensionality effect on their properties. Advances in materials and nanomaterials fabrication and characterization will certainly lead to observation of superconductivity in elements at higher temperatures, creating new opportunities for applications such as spintronics [154] or DNA integrated molecular electronics [166]. As suggested by the fast pace of discoveries in this field of superconductivity, it seems the periodic table of superconducting elements will likely need updating in the near future.

Acknowledgments

We thank I. Pacheco and K. Robbie for simulating discutions and invaluable suggestions.

References

[1]   see for example: Engineering Village, http://www.engineeringvillage2.org/; ISI Web of Science, http://isi6.isiknowledge.com/; SPIN Web Search, http://scitation.aip.org/ spinweb/.

[2]   Onnes, H. K. (1911) The superconductivity of mercury, *Comm. Phys. Lab. Univ. Leiden:* **122**, 124.

[3]   Gao, L., Xue, Y. Y., Chen, F., Xiong, Q., Meng, R. L., Ramirez, D., Chu, C. W., Eggert, J. H. & Mao , H. K. (1994) Superconductivity up to 164 K in $HgBa_2Cam-1CumO_{2m+2+d}$ (m= 1, 2, and 3) under quasihydrostatic pressures, *Phys. Rev. B:* **50**, 4260–4263

[4]   Scanlan, R. M., Malozemoff, A. P. & Larbalestier, D. C. (2004) Superconducting materials for large scale applications, *Proc. IEEE:* **92**, 1 639-1 654.

[5]   Nagamatsu, J., Nakagawa, N., Muranaka, T., Zenitani, Y. & Akimitsu, J. (2001) Superconductivity at 39 K in magnesium diboride. *Nature:* **410**, 63-64.

[6]   Shimizu, K., Ishikawa, H., Takao, D., Yagi, T. & Amaya, K. Superconductivity in compressed lithium at 20 K. (2002) *Nature:* **419**, 59 7-599.

[7]   Struzhkin, V. V., Hemley, R. J., Mao, H. K. & Tomofeev, Y. A. (1997) Superconductivity at 10-17 K in compressed sulphur. *Nature:* **390**, 382-384.

[8]   Shimizu, K., Suhara, K., Ikumo, M., Eremets, M. I. & Amaya, K. (1998) Superconductivity in oxygen. *Nature:* **393**, 767-769.

[9]   Tang, Z. K., Zhang, L., Wang, N., Zhang, X. X., Wen, G. H., Li, G. D., Wang, J. N., Chan, C. T. & Sheng, P. (2001) Superconductivity in 4 angstrom single-walled carbon nanotubes. *Science:* **292**, 2462 -2465.

[10]   Ekimov, E. A., Sidorov, V. A., Bauer, E. D., Mel'nik, N. N., Curro, N. J., Thompson, J. D. & Stishov, S. M. (2004) Superconductivity in diamond, *Nature:* **428**, 542-545.

[11]   Shimizu, K., Kimura, T., Furomoto, S., Takeda, K., Kontani, K., Onuki, Y. & Amaya, K. (2001) Superconductivity in the non-magnetic state of iron under pressure, *Nature:* **412**, 316-318.

[12]   Eremets, M. I., Struzhkin, V. V., Mao, H. K. & Hemley, R. J. (2001) Superconductivity in boron, *Science:* **293**, 2 72-2 74.

[13]   Kasumov, A. Yu., Kociak, M., Gueron, S., Reulet, B., Volkov, V. T., Klinov, D. V. & Bouchiat H. (2001) Proximity-induced superconductivity in DNA, *Science:* **291**, 280-282.

[14]   Maple, M. B., Wittig, J. & Kim, K. S. (1969) Pressure-induced magnetic-nonmagnetic transition of Ce impurities in La, *Phys. Rev. Lett.:* **23**, 1375-1377.

[15]   Bourgeois, O., Frydman, A. & Dynes, R. C. (2003) Proximity effect in ultrathin Pb/Ag multilayers within the Copper limit, *Phys. Rev. B:* **68**, 092509.

[16]   Timofeev, Y. A., Struzhkin, V. V., Hemley, R. J., Mao, H. K. & Gregoryanz, E. A. (2002) Improved techniques for measurement of superconductivity in diamond anvil cells by magnetic susceptibility, *Rev. Sci. Instrum.:* **73**, 3 71-377.

[17] Brandt, N. B. & Berman, I. V. (1968) Superconductivity of phosphorus at high pressures, *JETP Lett.:* **7**, 323.

[18] Ashcroft N W 1968 Mettalic hydrogen: a high-temperature superconductor? *Phys. Rev. Lett.* 1748-1749

[19] Hemley, R. J. & Mao, H. K. (1998) New phenomena in low-Z materials at megabar pressures, J. Phys. Condens. Matter.: **10**, 11157-11167.

[20] Eremets, M. I., Gregoryanz, E. A., Struzhkin, V. V., Mao, H. K., Hemley, R. J., Mulders, N. & Zimmerman, N. M. (2000) Electrical Conductivity of Xenon at Megabar Pressures, *Phys. Rev. Lett.:* **85**, 279 7-2800.

[21] Eremets, M. I., Hemley, R. J., Mao, H. K. & Gregoryanz, E. (2001) Semiconducting non-molecular nitrogen up to 240 GPa and its low-pressure stability, *Nature:* **411**, 170-174.

[22] Buckel V. & Hilsch, R. (1952) *Z. Physik* **132**, 420.

[23] Strongin, M. & Kammerer, O. F. (1968) Superconductivitive phenomena in ultrathin films, *J. Appl. Phys.* **39**, 2509-2514 and references therein.

[24] Reale, C. (1977) A1-type superconducting quenched-condensed Cs, Ba, La and Ce films, *Vacuum* **27**: 3-6.

[25] Reale, C. (1975) Superconducting films of alkali metals and their alloys with alkaline earth metals, *Phys. Lett.* **55A**: 1 65-1 66.

[26] Reale, C. (1978) Superconducting properties of quenched-deposited A1-type phosphorous and sulfur films, *Phys. Lett.* **68A**: 453-455.

[27] Reale, C. (1978) Determination of the critical magnetic field of a superconducting A1-type As, Sb, and Bi phase, *Vacuum:* **28**, 79-81.

[28] Reale, C. (1980) New bromine crystalline modifications obtained by vapor-quenching, *Vacuum:* **30**, 315-317.

[29] Reale, C. (1979) Normal conductivity and superconductivity of a A1-type iodine phase, *Vacuum:* **29**, 245-248.

[30] Reale, C. (1975) Occurrence of superconductivity in vapour-quenched films of alkaline earth metals, *Phys. Lett.:* **51A**, 353-354.

[31] Eremets, M. I., Struzhkin, V. V., Mao, H. K. & Hemley, R. L. (2003) Exploring superconductivity in low-Z materials at megabar pressures, *Physica B:* **329**–333, 1312-1316.

[32] Deemyad, S. & Schilling, J. S. (2003) Superconducting Phase Diagram of Li Metal in Nearly Hydrostatic Pressures up to 67 GPa, *Phys. Rev. B:* **91**, 167001.

[33] Okada, S., Shimizu, K., Kobayashi, T. C., Amaya, K. & Endo, S. (1996) Superconductivity of calcium under high pressures, *J. Phys. Soc. Japan:* **65**, 1924-1926.

[34] Dunn, K. J. & Bundy, F. P. (1982) Pressure-induced superconductivity in strontium and barium, *Phys. Rev. B:* **25**, 194-19 7.

[35] Wittig, J. & Matthias, B. T. (1969) Superconductivity of barium under pressure, *Phys. Rev. Lett.:* **22**, 634-636.

[36] Il'ina, M. A. & Itskevich, E. S. (1970) Superconductivity of barium at high pressures, *JETP Lett.:* **11**, 15-17.

[37] Il'ina, M. A., Itskevich, E. S. & Dizhur, E. M. (1972) Superconductivity of bismuth, barium, and lead at pressures exceeding 100 kbar, *Sov. Phys. JETP:* **34**, 1263 -1265.

[38] Moodenbaugh, A. R. & Wittig, J. (1973) Superconductivity in the high-pressure phases of barium, *J. Low. Temp. Phys.:* **10**, 203-206.

[39]   Probst, C. & Wittig, J. (1977) Superconductivity of bcc barium under pressure, *Phys. Rev. Lett.:* **39**, 1161-1163.

[40]   Vasvari, B., Animalu, A. O. E. & Heine, V. (1967) Electronic Structure of Ca, Sr, and Ba under Pressure, *Phys. Rev. Lett.:* **154**, 535–539.

[41]   Wigner, E. & Huntington, H. B. (1935) On the possibility of a metallic modification of hydrogen, *J. Chem. Phys.:* **3**, 764.

[42]   Johnson, K. & Aschroft, N. W. (2000) Structure and band gap closure in dense hydrogen, *Nature:* **403**, 632-635.

[43]   Satdele, M. & Martin, R. M. (2000) Metallization of molecular hydrogen: prediction from exact-exchange calculations, *Phys. Rev. Lett.:* **84**, 6070-6073.

[44]   Natoli, V., Martin, R. M. & Ceperley, D. (1995) Crystal structure of molecular hydrogen at high pressures, *Phys. Rev. Lett.:* **74**, 1601-1604.

[45]   Nellis, W. J., Weir, S. T. & Mitchell, A. C. (1999) Minimum metallic conductivity of fluid hydrogen at 140GPa (1.4Mbar), *Phys. Rev. B:* **59**, 3434-3449.

[46]   Loubeyre, P., Occelli, F. & LeToullec, R. (2002) Optical studies of solid hydrogen to 320 GPa and evidence for black hydrogen, *Nature:* **416**, 613-617.

[47]   Lin, T. H. & Dunn, K. J. (1986) High-pressure and low-temperature study of electrical resistance of lithium, *Phys. Rev. B:* **33**, 80 7-811.

[48]   Neaton, J. B. & Ashcroft, N. W. (1999) Pairing in dense lithium, *Nature:* **400**, 141-145.

[49]   Christensen, N. E. & Novikov, D. L. (2001) Predicted superconductive properties of lithium under pressure, *Phys. Rev. Lett.:* **86**, 1861-1864.

[50]   Wittig, J. (1970) Pressure-induced superconductivity in cesium and yttrium, *Phys. Rev. Lett.* **24**, 812-815.

[51]   Falge, R. L. Jr. (1967) Superconductivity of hexagonal beryllium, *Phys. Lett. A:* **24**, 579-580.

[52]   Grandqvist, C. G. & Claeson, T. (1974) Superconducting transition temperatures of vapour quenched Beryllium, *Phys. Lett. A:* **47**, 97-98.

[53]   Neaton, J. B. & Ashcroft, N. W. (2001) On the constitution of sodium at higher densities, *Phys. Rev. Lett.:* **86**, 2830-2833.

[54]   Shi, L., Papaconstantopoulos, A. & Mehl, M. J. (2003) Superconductivity in compressed potassium and rubidium, *Solid State Commun.:* **127**, 13-15.

[55]   Belzons, M., Blanc, R. & Payan, R. (1976) Electrical resistance of thin magnesium layers deposited at 4.2 K, *Comptes Rendus Hebdomadaires des Seances de l'Academie des Sciences, Series B:* **283**, 241-244.

[56]   McMahan, A. K. (1986) Pressure-induced changes in the electronic structure of solids, *Physica B&C:* **139**-140, 31-41.

[57]   Dunn, K. J. & Bundy, F. P. (1981) Electrical-resistance behavior of Ca at high pressures and low temperatures, *Phys. Rev. B:* **24**, 1 643-1 650.

[58]   Tissen, V. G., Ponyatovskii, E. G., Nefedova, M. V., Porsch, F. & Holzapfelet W. B. (1996) Effect of pressure on the superconducting Tc of lanthanum, *Phys. Rev. B:* **13**, 8238-8240.

[59]   Wittig, J., Probst, C., Schmidt, F. A. & Gschneidner, K. A. Jr. (1979) Superconductivity in a new high-pressure phase of scandium, *Phys. Rev. Lett.:* **42**, 469-4 72.

[60]   Reale, C (1975) Superconductivity in vapour-quenched lanthanide films, *Thin Solid Films:* **28**, L29-L30.

[61]   Fowler, R. D., Matthias, B. T., Asprey, L. B. & Hill, H. H. (1965) Superconductivity of protactinium, *Phys. Rev. Lett.*: **15**, 860-862.

[62]   Smith, J. L., Spinlet, J. C. & Muller, W. M. (1979) Superconducting preoperties of protactinium, *Science* **205**, 188-190.

[63]   Smith, J. L. & Haire, R. G. (1978) Superconductivity of americium, *Science* **200**, 535-537.

[64]   Hill, H. H. (1971) Superconductivity in the "actinide" elements, Physica: 55, 186-206 and references therein.

[65]   O'Brien, J. L., Hamilton, A. R., Clark, R. G., Mielke, C. H., Smith, J. L., Cooley, J. C., Rickel, D. G., Starrett, R. P., Reilly, D. J., Lumpkin, N. E., Hanrahan, R. J.& Hults, W. L. (2002) Magnetic susceptibility of the normal-superconducting transition in high-purity single-crystal -uranium, *Phys. Rev. B*: **66**, 064523 and references therein.

[66]   Lashley, J. C., Lang, B. E., Boerio-Goates, J., Woodfield, B. F., Schmiedeshoff, G. M., Gay, E. C., McPheeters, C. C., Thoma, D. J., Hults, W. L., Cooley, J. C., Hanrahan, R. J. & Smith, J. L. (2001) Low-temperature specific heat and cruitical magnetic field of -uranium single crystals, *Phys. Rev. B*: **63**, 224510 and references therein.

[67]   Brandt, N. B. & Ginzburg, N. I. (1964) Effect of high pressure on the superconducting properties of zirconium, *Sov. Phys. JETP:* **19**, 823-825.

[68]   Akahama, Y., Kobayashi, M. & Kawamura, H. (1990) Superconductivity and phase transition of zirconium under high pressure up to 50 GPa, *J. Phys. Soc. Japan:* **59**, 3843-3845.

[69]   Brandt, N. B. & Zarubina, O. A. (1973) Superconductivity of vanadium at pressures up to 250 kbar, *Sov. Phys. Solid State:* **15**, 3423-3425.

[70]   Ishizuka, M., Iketani, M. & Endo, S. (2000) Pressure effect on superconductivity of vanadium at megabar pressures, *Phys. Rev. B:* **61**, R3823–R3825.

[71]   Akahama, Y., Kobayashi, M. & Kawamura, H. (1995) Pressure effect on superconductivity of V and V-Cr alloys up to 50 GPa, *J. Phys. Soc. Japan:* **64**, 4049-4050.

[72]   Struzhkin, V. V., Timofeev, Y. A., Hemley, R. J. & Mao, H. K. (1997) Superconducting Tc and electron-phonon coupling in Nb to 132 GPa: magnetic susceptibility at megabar pressures, *Phys. Rev. Lett.:* **79**, 4262-4265.

[73]   Kohnlein, D. (1968) Supraleitung von vanadium, niob und tantal unter hohem druck, *Z. Phys.:* **208**, 142.

[74]   Meissner, W. (1930) Measurements with the aid of liquid helium. VI. Transition curve to superconductivity for titanium, *Z. Phys.*: **60**, 181-183.

[75]   Schmidt, P. H., Castellano, R. N., Barz, H., Cooper, A. S. & Spencer, E. G. (1973) Variation of superconducting transition temperatures of transition-metal thin films deposited with the noble gases, *J. Appl. Phys.:* **44**, 1833-1836.

[76]   Brodsky, M. S., Marikar, P., Friddle, R. J., Singer, L. & Sowers, C. H. (1982) Superconductivity in Au/Cr/Au epitaxial metal film sandwiches (EMFS), *Solid State Commun.:* **42**, 6 75-6 78.

[77]   Friebertshauser, P. E. & McCamont, J. W. (1969) Electrical Properties of Titanium, Zirconium, and Hafnium Films from 300 °K to 1.3 °K, *J. Vac. Sci. Technol.:* **6**, 184-187.

[78]   Koepke, R. & Bergmann, G. (1976) The upper critical magnetic field Bc2(T) of amorphous molybdenum films, *Solid State Commun.:* **19**, 435-437.

[79]   Daunt, J. G. & Cobble, J. W. (1953) Superconductivity of Technetium, *Phys. Rev.* **92**, 507-508.

[80]   Hulm, J.K. & Goodman, B.B. (1957) Superconducting properties of rhenium, ruthenium, and osmium, *Phys. Rev.* **106**, 659-671.

[81]   Buchal, Ch., Pebell, F., Mueller, R. M., Kubota, M. & Owerrs-Bradley, J. R. (1983), Superconductivity of rhodium at ultralow temperatures, *Phys., Rev. Lett.:* **50**, 64-67.

[82]   Stritzker, B. (1979) Superconductivity in irradiated palladium, *Phys. Rev. Lett.:* **41**, 1769-1773.

[83]   Kubatkin, S. E. & Landau, I. L. (1989) Investigation of disordered bismuth and cadmium films, *Z. Eksp. Teor. Fiz.*: **96**, 740-756.

[84]   Haq, A. U. & Meyer, O. (1982) Electrical and superconducting properties of rhenium thin films, *Thin Solid Films,* **94** (1982) 119-132.

[85]   Raub, Ch. J. (1984) Superconductivity of the platinum metals and their alloys, *Mater. & Design*: **5**, 129-136, and references therein.

[86]   Smith T. S. & Daunt, J. G. (1952) Some properties of superconductors below 1K: III. Zr, Hf, Cd, and Ti, *Phys. Rev.*: **88**, 1172–1176.

[87]   König, R., Schindler, A., Herrmannsdörfer, T., Braun, H. F., Eska, G., Gunther, D., Meissner, M., Mertig, M., Wahl, R. & Pompe, W. (2003) Superconductivity at 20mK in compacted submicrometer platinum powders, *Physica B:* **329**-333: 142 7-1428.

[88]   Webber, R. T., Reynolds, J. M. & McGuire, T. R. (1949) Superconductors in alternating magnetic fields, *Phys. Rev.:* **76**, 293-295.

[89]   Gutsche, M., Kraus, H., Jochum, J., Kemmather, B. & Gutekunst, G. (1994) Growth and characterization of epitaxial vanadium films, *Thin Solid Films:* **248**, 18-27.

[90]   Worley, R. D., Zemansky, M. W. & Boorse, H. A. (1955) Heat capacities of vanadium and tantalum in the normal and superconducting phases, *Phys. Rev.:* **99**, 44 7-458.

[91]   Shumei, L., Dianlin, Z., Xiunian, J., Li, L., Shanlin, L., Ning, K., Xiaosong, W. & Lin, J. J. (2000) Upper critical field of Ti and alpha-TiAl alloys: Evidence of an intrinsic type-II superconductivity in pure Ti, *Phys. Rev. B*: **62**, 8695-8698.

[92]   Schmidt, P. H., Castellano, R. N., Barz, H., Matthias, B. T., Huber, J. G. & Fertig, W. A. (1972) Superconducting ion beam sputtered chromium metal thin films, *Phys. Lett.*: **41A**, 367-368.

[93]   Wohlfarth, E. P. (1979) The possibility that e-Fe is a low temperature superconductor, *Phys. Lett. A:* **75**, 141-143.

[94]   Dabrowski, B., Wang, Z., Rogacki, K., Jorgensen, J. D., Hitterman, R. L., Wagner, J. L., Hunter, B. A., Radaelli, P. G. & Hinks, D. G.(1 996) Dependence of Superconducting Transition Temperature on Doping and Structural Distortion of the CuO2 Planes in La2-xMxCuO4 (M = Nd, Ca, Sr), *Phys. Rev. Lett.:* **76**, 1348-1351.

[95]   Saxena, S. S., Agarwal, P., Ahilan, K., Grosche, F. M., Haselwimmer, R. K. W. & Steine, M. J. (2000) Superconductivity on the border of itinerant-electron ferromagnetism in UGe2, *Nature:* **406**, 588-592.

[96]   Takada, K., Sakurai, H., Takayama-Muromachi, E., Izumi, F., Dilanian R. A. & Sasaki, T. (2003) Superconductivity in two-dimensional CoO2 layers, *Nature:* **422**, 53-55.

[97]   Sarrao, J. L., Morales, L. A., Thompson, J. D., Scott, B. L., Stewart, G. R., Wastin, F., Rebizant, J., Boulet, P., Colineau, E. & Lander, G. H. (2002) Plutonium-based superconductivity with a transition temperature above 18 K, *Nature:* **420**, 297-299.

[98] Terris, B. D. & Ginsberg, D. M. (1984) Localized excited states observed in superconducting quench-condensed and annealed manganese-doped zinc films, *Phys. Rev. B*: **29**, 2503-2509.

[99] Smith, T. S. & Daunt, J. G. (1952) Some properties of superconductors below 1K. III. Zr, Hf, Cd, and Ti, *Phys. Rev.*: **88**, 1172-1176.

[100] Rorer, D. C., Onn, D. G. & Meyer, H. (1965) Thermodynamic properties of molybdenum in its superconducting and normal state, *Phys. Rev.* **138**, A1661-A1668.

[101] Sekula, S. T., Kernohan, R. H. & Love, G. R. (1967) Superconducting properties of technetium, *Phys. Rev.*: **155**, 364-369.

[102] Webb, R. A., Ketterson, J. B., Halperin, W. P., Vuillemin, J. J. & Sandesara, S. B. (1978) Very low temperature search for superconductivity in Pd, Pt, and Rh, *J. Low Temp. Phys.*: **32**, 659-664.

[103] König, R., Schindler, A. & Herrmannsdörfer, T. (1999) Superconductivity of Compacted Platinum Powder at Very Low Temperatures, *Phys. Rev. Lett.*: **82**, 4528-4532.

[104] Meissner, W. (1930) Measurements with the aid of liquid helium. VII. Transition to superconductivity in tantalum and thorium, *Z. Phys.*: **61**, 191-198.

[105] Rosembaum, R., Ben-Shlomo, M., Goldsmith, S. & Boxman, R. L. (1989) Low-temperature electronic properties of W, Mo, Ta, and Zr films, *Phys. Rev. B*: **39**, 10009-100 19.

[106] Gibson, J. W. & Hein, R. A. (1964) Superconductivity of tungsten, *Phys. Rev. Lett.*: **12**, 688.

[107] Schirber, J. E. & Swenson, C. A. (1959) Superconductivity of beta mercury, *Phys. Rev. Lett.*: **2**, 296-297.

[108] Gubser, D. U. & Webb, A. W. (1975) High-pressure effects on the superconducting transition temperature of aluminum, *Phys. Rev. Lett.*: **35**, 104-10 7.

[109] Moshchalkov, V. V., Gielen, L., Neuttiens, G., van Haesendonck, C. & Bruynseraede, Y. (1994) Intrinsic resistance fluctuations in mesoscopic superconducting wires, *Phys. Rev. B*: **49**, 15412-15415.

[110] Jennings, L. D. & Swenson, C. A. (1958) Effects of pressure on the superconducting transition temperatures of Sn, In, Ta, Tl, and Hg, *Phys. Rev.*: **112**, 31-43.

[111] Reale, C. (1976) A new metallic superconducting Sn modification observed in quenched-condensed films, *Phys. Lett.*: **57A**, 65-66.

[112] Wittig, J. (1966) Super conductivity of germanium and silicon at high pressure, *Z. Phys.*: **195**, 215-22 7.

[113] Wittig, J. (1966) Superconductivity of tin and lead under very high pressure, *Z. Phys.*: **195**, 228-238.

[114] Smith, T. F. & Chu, C. W. (1967) Will pressure destroy superconductivity?, *Phys. Rev.*: **159**, 353-358.

[115] Snider, J. L. & Nicol, J. (1957) Atomic heats of normal and superconducting thallium, *Phys. Rev.*: **105**, 1242–1246.

[116] Strongin, M., Kammerer, O. F. & Paskin, A. (1965) Superconducting transition temperature of thin films, *Phys. Rev. Lett.*: **14**, 949-95 1.

[117] Buckel, W., Bulow, H. & Hilsh, R. (1954) Elektronenbeugungs-aufnahmen von dünnen metallschichten bei tifen temperaturen, *Z. Phys.*: **138**, 136.

[118] Shaw, R. W., Mapother, D. E. & Hopkins, D. C. (1960) Critical fields of superconducting tin, indium, and tantalum, *Phys. Rev.*: **120**, 88-91.

[119] Rapp, O. & Sundqvist, B. (1981) Pressure dependence of the electron-phonon interaction and the normal-state resistivity, *Phys. Rev. B*: **24**, 144-154.

[120] Tian, M., Wang, J., Snyder, J., Kurtz, J., Lium Y., Schiffer, P., Mallouk, E., Chan, M. H. W. (2003) Synthesis and characterization of superconducting single-crystal Sn nanowires, *Appl. Phys. Lett.* **83**, 1620-1622.

[121] Michotte, S., Matefi-Tempfli, S. & Piraux, L. (2003) Investigation of superconducting properties of nanowires prepared by template synthesis, *Supercond. Sci. Technol.* **16**, 557-561.

[122] Yi, G. & Schwarzacher, W. (1999) Single crystal superconductor nanowires by electrodeposition, *Appl. Phys. Lett.*: **74**, 1746-1748.

[123] Moodera, J. S. & Meservey, R. (1990) Superconducting phases of Bi and Ga induced by deposition on a Ni sublayer, *Phys. Rev. B*: **42**, 179-183.

[124] Chen, T. T., Chen, J. T., Leslie, J. D. & Smith, H. J. T. (1969) Phonon Spectrum of Superconducting Amorphous Bismuth and Gallium by Electron Tunneling, *Phys. Rev. Lett.*: **22**, 526-530.

[125] Il'ina, M. A. & Itskevich, E. S. Superconductivity of high-pressure phases of silicon in the pressure range up to 14 GPa, *Sov. Phys. Solid State*: **22**, 1833-1835.

[126] Chang K. J., Dacorogna, M. M., Cohen, M. L., Mignot, J. M., Chouteau, G. & Martinez, G. (1985) Superconductivity in high-pressure metallic phases of Si, *Phys. Rev. Lett.*: **54**, 2375-2378.

[127] Lin, T. H., Dong, W. Y., Dunn, K. J., Wagner, C. N. J. & Bundy, F. P. (1986) Pressure-induced superconductivity in high-pressure phases of Si, *Phys. Rev. B*: **33**, 7820-7822.

[128] Chen, A. L., Lewis, S. P., Su, Z., Yu, P. Y. & Cohen, M. L. (1992) Superconductivity in arsenic at high pressures, *Phys. Rev. B*: **46**, 5523-5527.

[129] Bermen, I. V. & Brandt, N. B. (1969) Superconductivity of arsenic at high pressures, *JETP Lett.*: **10**, 55-57.

[130] Akahama, Y., Kobayashi, M. & Kawamura, H. (1992) Pressure induced superconductivity and phase transition in selenium and tellurium, *Solid State Commun.*: **84**, 803-806.

[131] Berman, I. V., Binzarov, Z. I. & Zhurkin, P. (1973) Study of superconductivity properties of Te under pressure up to 260 kbar, *Sov. Phys. State*: **14**, 2192-2194.

[132] Bundy, F. P. & Dunn, K. J. (1980) Pressure dependence of superconducting transition temperature of high-pressure metallic Te, *Phys. Rev. Lett.*: **44**, 1 623-1 62 6.

[133] for a review see Buzea, C. & Yamashita, T. (2001) Review of the superconducting properties of MgB2, *Supercond. Sci. Technol.*: **14**, R115–R146, and references therein.

[134] Mailhiot, C., Grant, J. B. & McMahan, A. K. (1990) High-pressure metallic phases of boron, *Phys. Rev. B*: **42**, 9033-9039.

[135] Shirotani, I., Mikami, J., Adachi, T., Katayama, Y., Tsuji, K., Kawamura, H., Shimomura, O. & Nakajima, T. (1994) Phase transitions and superconductivity of black phosphorus and phosphorus-arsenic alloys at low temperatures and high pressures, *Phys. Rev. B*: **50**, 1 62 74–1 62 78 *and references therein*.

[136] Kawamura, H., Shirotani, I. & Tachikawa, K. (1985) Anomalous superconductivity and pressure induced phase transitions in black phosphorus, *Solid State Commun.*: **54**, 775-778.

[137] Amaya, K., Shijmizu, K., Eremets, M. I., Kobayashi, T. C. & Endo, S. (1998) Observation of pressure-induced superconductivity in the megabar region *J. Phys.: Condens. Matter.:* **10**, 111179-111190.

[138] Shimizu K., Yamauchi, T., Tamitani, N., Takeshita, N., Ishizuka, M., Amaya, K. & Endo, S. (1994) The pressure induced superconductivity of iodine, *J. Superconductivity:* **7**, 921-924.

[139] Benedict, L. X., Crespi, V. H., Louie, S. G. & Cohen, M. L. (1995) Static conductivity and superconductivity of carbon nanotubes: Relations between tubes and sheets, *Phys. Rev. B:* **52**, 14935-14940.

[140] Kociak M., Kasumov, A. Yu., Guéron, S., Reulet, B., Khodos, I. I., Gorbatov, Yu. B., Volkov, V. T., Vaccarini, L. & Bouchiat, H. (2001) Superconductivity in ropes of single-walled carbon nanotubes, *Phys. Rev. Lett.:* **86**, 2416-2419.

[141] Takano, Y., Nagao, M., Sakaguchi, I., Tachiki, M., Hatano, T., Kobayashi, K., Umezawa, H. and Kawarada, H. (2004) Superconductivity in diamond thin films well above liquid helium temperature, *Appl. Phys. Lett.:* **85**, 285 1-2853.

[142] Bao, Z., Batlogg, B., Berg, S., Dodabalapur, A., Haddon, R. C., Hwang, H., Kloc, C., Meng, H. & Schön, J. H. (2002) Retraction, Science: 298, 961.

[143] Schön, J. H., Kloc, C., Siegrist, T., Steigerwald, M., Svensson, C. & Batlogg, B. (2003) Retraction: Superconductivity in single crystals of the fullerene C70, *Nature* **422**, 92-92.

[144] Schön, J. H., Meng, H. & Bao, Z. (2003) Retraction: Self-assembled monolayer organic field-effect transistors, *Nature*: **422**, 92-92.

[145] Schön, J. H., Kloc, C. & Batlogg, B. (2003) Retraction: Superconductivity at 52 K in hole-doped C60, *Nature* **422**, 93-93.

[146] Lerner, E, J. (2002) Fraud shows peer-review flaws, The Industrial Physicist December 2002/January 2003, 12-17.

[147] Wittig, J. & Matthias, B. T. (1968) Superconducting phosphorous, *Science:* **160**, 994-996.

[148] Shirotani, I., Kawamura, H., Tsuburaya, K. & Tachikawa, K. (1987) Superconductivity of phosphorus and phosphorus-arsenic alloy under high pressures, *Jpn. J. Appl. Phys. Suppl.:* **26**, 921-922.

[149] Shirotani, I., Fukizawa, A., Kawamura, H., Yagi, T., Akimoto, S., (1985) Pressure induced phase transitions in black phosphorus, *Solid State Physics Under Pressure*: Recent Advance with Anvil Devices Conference, D. Reidel Publ Co, 1985, p 207-211.

[150] Kawamura, H., Shirotani, I. & Tachikawa, K. (1984) Anomalous superconductivity in black phosphors under high pressures, *Solid State Commun.:* 49, 8 79-881.

[151] Shirotani, I., Tsuji, K. & Kawamura, H. (1991) Crystal structure and superconductivity in black phosphorus at low temperatures and high pressures, *Synthetic Metals*: **41**-43, 1947.

[152] Kawamura, H., Shirotani, I., Tachikawa, K., (1985) Anomalous superconductivity in black phosphorous under high pressure, *Solid State Physics Under Pressure*: Recent Advance with Anvil Devices Conference, D. Reidel Publ Co, 1985, p 213-216.

[153] Akahama, Y., Kawamura, H., Carlson, S., LeBihan, T. & Häuserman, D. (2000) Structural stability and equation of state of simple-hexagonal phosphorus to 280 GPa: Phase transition at 262 GPa, *Phys. Rev. B:* **61**, 3139-3142.

[154] Ostanin, S., Trubitsin, V., Staunton, J. B, & Savrasov, S. Y. (2003) Density functional study of the phase diagram and pressure-induced superconductivity in P: implications for spintronics, *Phys. Rev. Lett.:* **91**, 087002.

[155] Degtyareva, O., Gregoryanz, E., Somayazulu, M., Dera, P., Mao, H. W. & Hemley R. J., Novel chain structures in group VI elements, *Nature Mater.* **4** (2005) 152-156, and references therein.

[156] Evdokimova, V. V. & Kuzemskaya, I. G. (1978) Superconductivity in S at high pressure, *JETP Lett.:* **28**, 390-392.

[157] Yakovlev, E. N., Stepanov, G. N., Timofeev, Y. A. & Vinogradov, B. V. (1978) Superconductivity of sulfur at high pressures, *JETP Lett.:* **28**, 340-342.

[158] Kometani, S., Eremets, M. I., Shimizu, K., Kobayashi, M. & Amaya, K. (1997) Observation of pressure induced superconductivity in sulfur, *J. Phys. Soc. Japan:* **66**, 2564-2565.

[159] Rudin, S. P., Liu, A. Y., Freericks, J. K. & Quandt, A. (2001) Comparison of structural transformations and superconductivity in compressed sulfur and selenium, *Phys. Rev. B:* **63**, 224107.

[160] Moodenbaugh, A. R., Wu., C. T. & Viswanathan, R., (1973) Superconductivity and phase stability of selenium at high pressures, *Solid State Commun.* **13**, 1413-1416.

[161] Shimizu, K., Amaya, K. & Endo, S. (1995) Electrical resistance measurements of solid bromine at high pressures and low temperatures, Proc. Joint XV IRAPT and XXXIII EHPRG Int'l Conf. High pressure science and technology, Warsaw (1995), 498-450.

[162] Hirsch, J. E. Correlations between normal-state properties and superconductivity, *Phys. Rev. B:* **55** (1997) 9007-9024.

[163] Hirsch, J. E. (1997) Correlations between normal-state properties and superconductivity, *Phys. Rev. B:* **55**, 900 7-9023.

[164] Robbie, K., Beydaghyan, G., Brown, T., Dean, C., Adams, J. & Buzea, C. (2004) Ultrahigh vacuum glancing angle deposition system for thin films with controlled three-dimensional nanoscale structure, *Rev. Sci. Instrum.:* **75**, 1089-109 7 and references therein

[165] Karabacak, T., Mallikarjunan, A., Singh, J. P., Ye, D., Wang, G. C. & Lu, T. M. (2003) Beta-phase tungsten nanorod formation by oblique-angle sputter deposition, *Appl. Phys. Lett.:* **83**, 3096-3098.

[166] Shigematsu, T., Shimotani, K., Manabe, C., Watanabe, H. & Shimizu, M. (2003) Transport properties of carrier-injected DNA, *J. Chem. Phys.:* **118**, 4245-4252.

In: Recent Developments n Superconductivity Research   ISBN 978-1-60021-462-2
Editor: Barry P. Martins pp. 275-337   © 2007 Nova Science Publishers, Inc.

Chapter 9

UNIVERSAL CAUSE OF HIGH-T$_c$ SUPERCONDUCTIVITY AND ANOMALOUS BEHAVIOR OF HEAVY FERMION METALS

V.R. Shaginyan[1,2,3*], *M.Ya. Amusia*[3,4], *A.Z. Msezane*[2], *K.G. Popov*[5]
[1]Petersburg Nuclear Physics Institute,
Russian Academy of Sciences, Gatchina, 188300, Russia
[2]CTSPS, Clark Atlanta University, Atlanta, Georgia 30314, USA
[3]Racah Institute of Physics, the Hebrew University, Jerusalem 91904, Israel
[4]A.F. Ioffe Physical-Technical Institute,
Russian Academy of Sciences, St. Petersburg, 194021, Russia
[5]Komi Science Center, Ural Division,
Russian Academy of Sciences, Syktyvkar, 167982, Russia

Abstract

Unusual properties of strongly correlated liquid observed in the high-T_c superconductors and heavy-fermion (HF) metals are determined by quantum phase transitions
taking place at their critical points. Therefore, direct experimental studies of these
transitions and critical points are of crucial importance for understanding the physics
of high-T_c superconductors and HF metals. In case of high-T_c superconductors such
direct experimental studies are absent since at low temperatures corresponding critical
points are occupied by the superconductivity. Recent experimental data on the behavior of HF metals illuminate both the nature of these critical points and the nature of the
phase transitions. We show that it is of crucial importance to simultaneously carry out
studies of both the high-T_c superconductivity and the anomalous behavior of HF metals. The understanding of this fact has been problematic largely because of the absence
of theoretical guidance. The main features of the fermion condensation quantum phase
transition (FCQPT), which are distinctive in several aspects from that of conventional
quantum phase transition (CQPT), are considered. Our paper deals with these fundamental problems through studies of the behavior of quasiparticles, leading to good
quantitative agreement with experimental facts. We show that in contrast to CQPT,
whose physics in the critical region is dominated by thermal and quantum fluctuations and characterized by the absence of quasiparticles, the physics of a Fermi system

---

*E-mail address: vrshag@thd.pnpi.spb.ru

near FCQPT or undergone FCQPT is controlled by the system of Landau-type quasi-particles. However, contrary to the conventional Landau quasiparticles, the effective mass of these strongly depends on the temperature T, magnetic fields B, the number density x, etc. Our general consideration suggests that FCQPT and the emergence of novel quasiparticles near and behind FCQPT are distinctive features of strongly correlated substances such as the high-T_c superconductors and HF metals. We show that the main properties and universal behavior of the high-T_c superconductors and HF metals can be understood within the framework of presented here theory based on FCQPT. A large number of the experimental evidences in favor of the existence of FCQPT in high-T_c superconductors and HF metals is presented. We demonstrate that the essence of strongly correlated electron liquids can be controlled by both magnetic field B and temperature T. Thus, the main properties of heavy-fermion metal such as magnetoresistance, resistivity, specific heat, magnetization, volume thermal expansion, etc, are determined by its position on the $B - T$ phase diagram. The obtained results are in good agreement with recent facts and observations.

PACS numbers: 71.27.+a, 74.20.Fg, 74.25.Jb

1   Introduction

In the last two decades a new class of materials such as heavy-fermion (HF) metals and high-T_c superconductors has been found, which display a dazzling variety of physical phenomena, see e.g. [1, 2]. The high-T_c superconductors are strongly correlated metals with normal state properties that are not at all those of a normal Landau Fermi liquid (LFL). In the case of HF metals, the electronic strong correlations result in a renormalization of the effective mass of the quasiparticles, which can exceed the bare mass by a factor up to 1000 or even diverge. These non-Fermi-liquid (NFL) systems demonstrate anomalous pure power-law temperature dependences in their low temperature properties over broad temperature ranges. It is believed that in this class the basic assumption of the LFL theory that at low energies the electrons in a metal should behave as weakly interacting quasiparticles is violated. Therefore, it is generally accepted that the fundamental physics that gives rise to the high-T_c superconductivity and NFL behavior with a recovery of the LFL behavior under the application of magnetic fields observed in HF metals and high-T_c compounds is controlled by quantum phase transitions. This has made quantum phase transitions a subject of intense current interest, see e.g. [3, 2, 4, 5].

A quantum phase transition is driven by control parameters such as the composition, number density x or magnetic fields B, and takes place at a quantum critical point (QCP) when temperature $T = 0$. QCP separates an ordered phase generated by quantum phase transition from a disordered phase. It is expected that the universal behavior is only observable if the system in question is very near QCP, for example, when the correlation length is much larger than microscopic length scales. Quantum phase transitions of this sort are quite common, and we shall call them as conventional quantum phase transitions (CQPT). In the case of CQPT, the physics is dominated by thermal and quantum fluctuations of the critical state, which is characterized by the absence of quasiparticles. It is believed that the absence of quasiparticle-like excitations is the main cause of the NFL behavior and other types of critical behavior in the quantum critical region. On the base of scaling related to the divergence of the correlation length, one can construct the critical contribution to the free energy and evaluate the corresponding properties such as critical exponents, the NFL behavior, etc. [3, 2, 4, 5]. However along this way one may expect difficulties. For example, having the only critical contribution, one has to describe different types of the behavior exhibited by different HF metals, see e.g. [6, 7, 8]. Note that HF metals are three-dimensional structures, (see e.g. [6, 9, 10]) and, thus, the type of behavior cannot be related to the dimension. The critical behavior observed in measurements on HF metals takes place up to rather high temperatures comparable with the effective Fermi temperature T_k. For example, the thermal expansion coefficient $\alpha(T)$ measured on CeNi$_2$Ge$_2$ shows a $1/\sqrt{T}$ divergence over more than two orders of magnitude in temperature drop from 6 K down to at least 50 mK [6]. It is hardly possible to understand such a behavior on the base of the assumption of scaling when the correlation length has to be much larger than microscopic length scales. Obviously, such a situation can take place only at $T \to 0$. At some temperature $T \sim T_k$, this macroscopically large correlation length must be destroyed by thermal fluctuations.

The next problem is related to explanations of the recovery of the LFL behavior under applied magnetic fields B, observed in HF metals and the high-T_c compounds, see e.g. [10, 1, 11]. At $T \to 0$, the magnetic field dependence of the coefficient $A(B)$, causing an electrical resistivity contribution $\Delta\rho = A(B)T^2$, the Sommerfeld coefficient $\gamma(B)$ and $\chi(B)$ in specific heat, $C/T = \gamma(B)$, and magnetic susceptibility, $\chi(B)$, shows that $A(B) \sim \gamma^2(B)$ and $A(B) \sim \chi^2(B)$, so that the Kadowaki-Woods ratio, $K = A(B)/\gamma^2(B)$ [9], is B-independent and conserved [10]. Such a universal behavior is hardly possible to explain within the picture assuming the absence of quasiparticles which takes place near QCP of the corresponding CQPT. As a consequence, for example, these facts are in variance to the spin-density-wave scenario [10] and the renormalization group treatment of quantum criticality [12]. Moreover, striking recent measurements of the specific heat, charge and heat transport and the resistivity used to study the nature of magnetic field-induced QCP in heavy-fermion metal CeCoIn$_5$ [13, 14] certainly seem to disagree with descriptions based on CQPT.

In a system of interacting bosons at temperatures lower than the temperature of Bose-Einstein condensation [15, 16], a finite number of particles is concentrated in the lowest level. In the case of a noninteracting Bose gas at zero temperature, $T = 0$, this number is simply equal to the total number of particles in the system. In a homogeneous system of noninteracting Bosons, the lowest level is the state with zero momentum, and the ground state energy is equal to zero. For a noninteracting Fermi system such a state is impossible, and its ground state energy E_{gs} reduces to the kinetic energy and is proportional to the total number of particles. Imagine an interacting system of fermions with a pure repulsive interaction. Let us increase its interaction strength. As soon as it becomes sufficiently large and the potential energy starts to prevail over the kinetic energy, we can expect the system to undergo a phase transition at $T = 0$. This takes place when the density x tends to the critical quantum point $x \to x_{FC}$ and the kinetic energy E_k becomes frustrated, while the effective inter-electron interaction, or the Landau amplitude, being sufficiently large, starts to determine the occupation numbers of quasiparticles $n(\mathbf{p})$ which deliver the minimum value to the ground state energy $E[n(p)]$. As a result, at $x < x_{FC}$, the function $n(\mathbf{p})$ is given by the standard equation that determines the minimum of functional $E[n(\mathbf{p})]$

[17, 18, 19, 20, 21]

$$\frac{\delta E[n(\mathbf{p})]}{\delta n(\mathbf{p})} = \mu. \tag{1.1}$$

Here we deal with three dimensional (3D) case and assume that the phase transition takes place at $x < x_{FC}$. At $T = 0$ Eq. (1.1) determines the quasiparticle distribution function $n_0(\mathbf{p})$, which delivers the minimum value to the ground state energy E. The function $n_0(\mathbf{p})$ being the signature of the new state of quantum liquids [22] does not coincide with the quasiparticle distribution function of the LFL theory in the region $(p_f - p_i)$, so that $0 < n_0(\mathbf{p}) < 1$ and $p_i < p_F < p_f$, with $p_F = (3\pi^2 x)^{1/3}$ being the Fermi momentum. Such a state was called the state with fermion condensate (FC) because quasiparticles located in the region $(p_f - p_i)$ of momentum space are pinned to the chemical potential μ [19, 17, 22]. We note that the behavior obtained for the single-particle spectrum and quasiparticle distribution functions is observed within exactly solvable models [23, 24, 25]. Lowering the potential energy, FC decreases the total energy. Unlike the Bose-Einstein condensation, which occurs even in a system of noninteracting bosons, FC can take place if the coupling constant of the interaction is large, or the corresponding Landau amplitudes are large and repulsive.

We note the remarkable peculiarity of FCQPT at $T = 0$: this transition is related to the spontaneous breaking of gauge symmetry, when the superconducting order parameter $\kappa(\mathbf{p}) = \sqrt{n_0(\mathbf{p})(1 - n_0(\mathbf{p}))}$ has a nonzero value over the region occupied by FC, with the entropy $S = 0$ [21, 26]. At small values of the pairing coupling constant λ_0, the gap $\Delta(\mathbf{p})$ is linear in λ_0 and vanishes provided that $\lambda_0 \to 0$, while $\kappa(\mathbf{p})$ remains finite [20, 21]. As we shall see, this peculiarity allows to construct the theory of high- T_c superconductivity based on FCQPT. Thus, the state with FC cannot exist at any finite temperatures and is driven by the parameter x: at $x > x_{FC}$ the system is on the disordered side of FCQPT; at $x = x_{FC}$, Eq. (1.1) possesses the non-trivial solutions $n_0(\mathbf{p})$ with $p_i = p_F = p_f$; while at $x < x_{FC}$, the system is on the ordered side [26].

One of the most challenging problems of modern condensed matter physics is the structure and properties of Fermi systems with large coupling constants. The first solution to this problem was offered by the Landau theory of Fermi liquids, later called "normal", by introducing the notion of quasiparticles and so called amplitudes, which characterize the effective interaction among them [27]. The Landau theory can be viewed as the low energy effective theory in which high energy degrees of freedom are removed by introducing the effective amplitudes instead of strong inter-particle interaction. Usually, it is assumed that the stability of the ground state of Landau liquid is determined by the Pomeranchuk stability conditions: the stability is violated when even one of the Landau effective interaction parameters is negative and reaches a critical value [27, 28]. Note that the new phase, at which the stability conditions are restored, can in principle be again described within the framework of the same theory. However, it has been demonstrated rather recently [17] that the Pomeranchuk stability conditions cover not all possible instabilities: one of them is missed. It corresponds to the situation when, at the temperature $T = 0$, the effective mass, the most important characteristic of Landau quasiparticles, can become infinitely large. Such a situation, leading to profound consequences, can take place when the corresponding Landau amplitude being repulsive reaches some critical value. This leads to a completely new class of strongly correlated Fermi liquids with FC [17, 22, 19], which is separated from that of a normal Fermi liquid by the fermion condensation quantum phase transition (FCQPT) [29, 30].

In the FCQPT case we are dealing with the strong coupling limit where an absolutely reliable answer cannot be given on the bases of pure theoretical first principle foundation. Therefore, the only way to verify that FC occurs is to consider both exactly solvable models and experimental facts, which can be interpreted as confirming the existence of such a state. The exactly solvable models unambiguously demonstrate that Fermi liquids with FC do exist, see e.g. [23, 24, 25]. On the other hand, these facts are seen in features of those two-dimensional (2D) systems with interacting electrons or holes, which can be represented by doped quantum wells and high-T_c superconductors. Considering the HF metals and the 2D systems of ^{3}He, we will show that FC exist also in these systems.

The goal of our review is to describe the behavior of Fermi systems with FC and to show that the existing data on strongly correlated liquids represented by the electronic (or hole) systems of high-T_c superconductors and HF metals can be well understood within the theory of Fermi liquids based on FCQPT. In Section 2, we review the general features of Fermi liquids with FC in their normal state. Section 3 is devoted to consideration of the superconductivity in the presence of FC. We show that the superconducting state is totally transformed by the presence of FC. For instance, the maximum value Δ_1 of the superconducting gap can be as large as $\Delta_1 \sim 0.1\varepsilon_F$, while for normal superconductors one has $\Delta_1 \sim 10^{-3}\varepsilon_F$, where ε_F is the Fermi level energy. In Section 4 we describe the quasiparticle's dispersion and its lineshape and show that they strongly deviate from the case of normal Landau liquids. In Section 5 we consider the field-induced LFL in the heavy electron liquid with FC. In Section 6 we apply our theory to explain the main properties of magnetic-field induced Landau Fermi liquid in the high-T_c metals. Section 7 is devoted to the appearance of FCQPT in different Fermi liquids. In Section 8 we analyze the main properties of HF metals whose electronic system is placed on the disordered side of FCQPT. HF metals with the electronic system located on the ordered side of FCQPT are considered in Section 9. In Section 10 we show that FC manifests itself in the dissymmetry of tunnelling conductivity which can be observed in measurements on the high-T_c compounds and HF metals. Finally, in Section 11, we summarize our main results.

2   Fermi Liquids with Fermion Condensate

To study the universal behavior of the high-T_c superconductors and HF metals at low temperatures, we use the heavy electron liquid model in order to ignore the complications of the anisotropy of the lattice of solids and its microscopic inhomogeneity. It is possible since we consider the universal behavior demonstrated by these materials and processes related to the power-low divergences of observables such as the effective mass, specific heat, thermal expansion, etc. These divergences are determined by small momenta transferred as compared to momenta of the same order of magnitude as those of the reciprocal lattice cell, and contributions coming from them can be safely ignored. On the other side, we can simply use the common concept of the applicability of the LFL theory when describing electronic properties of metals [27]. Thus, we may usefully ignore the complications due to lattice and its anisotropy. As a result, we regard the medium as homogeneous heavy electron isotropic liquid.

2.1 Landau Theory of Fermi Liquid

Let us start by explaining the important points of the LFL theory [27]. The LFL theory rests on the notion of quasiparticles which represent elementary excitations of a Fermi liquid. Therefore these are appropriate excitations to describe the low temperature thermodynamic properties. In the case of an electron system, these are characterized by the electron's quantum numbers and effective mass M^*. The ground state energy of the system in question is a functional of the quasiparticle occupation numbers (or quasiparticle distribution function) $n(\mathbf{p}, T)$, just like the free energy $F[n(\mathbf{p}, T)]$, entropy $S[n(\mathbf{p}, T)]$, and other thermodynamic functions. From the condition that the free energy $F = E - TS$ should be minimal, we can find the distribution function

$$\frac{\delta(F - \mu N)}{\delta n(\mathbf{p}, T)} = \varepsilon(\mathbf{p}, T) - \mu(T) - T \ln \frac{1 - n(\mathbf{p}, T)}{n(\mathbf{p}, T)} = 0. \tag{2.1}$$

Here μ is the chemical potential, while

$$\varepsilon(\mathbf{p}, T) = \frac{\delta E[n(\mathbf{p}, T)]}{\delta n(\mathbf{p}, T)}, \tag{2.2}$$

is the quasiparticle energy. This energy is a functional of $n(\mathbf{p}, T)$ just like the total energy $E[n(\mathbf{p}, T)]$. The entropy $S[n(\mathbf{p}, T)]$ is given by the familiar expression [27]

$$S[n(\mathbf{p}, T)] = -2 \int [n(\mathbf{p}, T) \ln n(\mathbf{p}, T) + (1 - n(\mathbf{p}, T)) \ln(1 - n(\mathbf{p}, T))] \frac{d\mathbf{p}}{(2\pi)^3}, \tag{2.3}$$

which stems from purely combinatorial considerations. Equation (2.1) is usually presented as the Fermi-Dirac distribution

$$n(\mathbf{p}, T) = \left\{ 1 + \exp \left[\frac{(\varepsilon(\mathbf{p}, T) - \mu)}{T} \right] \right\}^{-1}. \tag{2.4}$$

At $T \to 0$, one gets from Eqs. (2.1) and (2.4) the standard solution $n(p, T \to 0) \to \theta(p_F - p)$, with $\theta(p_F - p)$ is the step function, $\varepsilon(p \simeq p_F) - \mu = p_F(p - p_F)/M_L^*$, where M_L^* is the Landau effective mass [27]

$$\frac{1}{M_L^*} = \frac{1}{p} \frac{d\varepsilon(p, T = 0)}{dp} \Big|_{p=p_F}. \tag{2.5}$$

It is implied that M_L^* is positive and finite at the Fermi momentum p_F. As a result, the T-dependent corrections to M_L^*, to the quasiparticle energy $\varepsilon(\mathbf{p})$, and to other quantities, start with T^2-terms. The effective mass is given by the well-known Landau equation

$$\frac{1}{M_L^*} = \frac{1}{M} + \sum_{\sigma_1} \int \frac{\mathbf{p_F p_1}}{p_F^3} F_{\sigma, \sigma_1}(\mathbf{p_F}, \mathbf{p_1}) \frac{\partial n_{\sigma_1}(\mathbf{p_1}, T)}{\partial p_1} \frac{d\mathbf{p_1}}{(2\pi)^3}. \tag{2.6}$$

Here $F_{\sigma, \sigma_1}(\mathbf{p_F}, \mathbf{p_1})$ is the Landau amplitude depending on the momenta $\mathbf{p}$, spins σ, and M is the bare mass of an electron. For the sake of simplicity, we omit the spin dependence of the effective mass since in the case of a homogeneous liquid and weak magnetic fields M_L^* does not noticeably depend on the spins.

Applying Eq. (2.6) at $T = 0$ and taking into account that $n(\mathbf{p}, T = 0)$ becomes the step function $\theta(p_F - p)$, we obtain the standard result

$$\frac{M_L^*}{M} = \frac{1}{1 - N_0 F^1(p_F, p_F)/3}.$$

Here N_0 is the density of states of the free Fermi gas and $F^1(p_F, p_F)$ is the p-wave component of the Landau interaction amplitude. Since in the LFL theory $x = p_F^3/3\pi^2$, the Landau amplitude can be written as $F^1(p_F, p_F) = F^1(x)$. Assume that at some critical point x_{FC} the denominator $(1 - N_0 F^1(p_F, p_F)/3)$ tends to zero, that is $(1 - N_0 F^1(x)/3) \propto (x - x_{FC}) + a(x - x_{FC})^2 + ... \to 0$. As a result, one obtains that $M_L^*(x)$ behaves as [31, 32]

$$\frac{M_L^*(x)}{M} \simeq A + \frac{B}{x - x_{FC}} \propto \frac{1}{r}. \tag{2.7}$$

Here A and B are constants and $r = (x - x_{FC})$ is the "distance" from the QCP of FCQPT taking place at x_{FC}. The observed behavior is in good agreement with recent experimental observations (see e.g. [34, 33]), and calculations [35, 36, 37], see Section 7 as well. In the case of electronic systems, Eq. (2.7) is valid at $x > x_{FC}$ when $r > 0$ [19, 38]. Such a behavior of the effective mass can be observed in the HF metals with a quite flat, narrow conduction band, corresponding to the large effective mass $M_L^*(x \simeq x_{FC})$, with strong electron correlations and the effective Fermi temperature $T_k \sim p_F^2/M_L^*(x)$ of the order of a few Kelvin or even lower (see e.g. [1]).

2.2   Fermion Condensation Quantum Phase Transition

As we have seen above at $T = 0$ when $r = (x - x_{FC}) \to 0$, the effective mass diverges, $M_L^*(r) \to \infty$, and eventually beyond the critical point x_{FC} the distance r becomes negative making the effective mass negative, as it follows from Eq. (2.7). To escape the possibility of being in unstable and in essence meaningless states with negative values of the effective mass, the system is to undergo a quantum phase transition at the critical point $x = x_{FC}$. Because the kinetic energy near the Fermi surface is proportional to the inverse effective mass, this phase transition is triggered by the frustrated kinetic energy and can be recognized as FCQPT [29, 39]. Therefore behind the critical point x_{FC} of this transition, the quasiparticle distribution represented by the step function does not minimize the Landau functional $E[n(\mathbf{p})]$. As a result, at $x < x_{FC}$ the quasiparticle distribution is determined by the standard equation that is used to search for the minimum of the energy functional [17]

$$\frac{\delta E[n(\mathbf{p})]}{\delta n(\mathbf{p}, T = 0)} = \varepsilon(\mathbf{p}) = \mu; \; p_i \leq p \leq p_f. \tag{2.8}$$

Equation (2.8) determines the quasiparticle distribution function $n_0(\mathbf{p})$, which minimizes the ground state energy E. Being determined by Eq. (2.8), $n_0(\mathbf{p})$ does not coincide with the step function in the region $(p_f - p_i)$, so that $0 < n_0(\mathbf{p}) < 1$, while outside the region it coincides with the step function. It follows from Eq. (2.8) that the single particle spectrum or the band is completely flat over the region. Such a state was called the state with FC because quasiparticles located in the region $(p_f - p_i)$ of momentum space are pinned to the chemical potential μ [19, 17, 22]. We note that this behavior was obtained for the band and quasiparticle distribution functions using some exactly solvable models [25, 24].

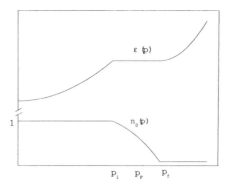

Figure 1: The quasiparticle distribution function $n_0(p)$ and energy $\varepsilon(p)$. Since $n_0(p)$ is the solution of Eq. (2.8) it implies $n_0(p < p_i) = 1$, $n_0(p_i < p < p_f) < 1$ and $n_0(p > p_f) = 0$, while $\varepsilon(p_i < p < p_f) = \mu$. The Fermi momentum p_F obeys the condition $p_1 < p_F < p_f$.

The possible solution $n_0(\mathbf{p})$ of Eq. (2.8) and the corresponding single particle spectrum $\varepsilon(\mathbf{p})$ are shown in Fig. 1.

As we shall see in Section 3 the relevant order parameter of the FC state, $\kappa(\mathbf{p}) = \sqrt{n_0(\mathbf{p})(1 - n_0(\mathbf{p}))}$, is the order parameter of the superconducting state with the infinitely small value of the superconducting gap. Therefore the entropy of this state is zero, $S(T = 0) = 0$ [19]. Thus, this state is of pure quantum nature and cannot exist at any finite temperatures. This quantum state with FC is driven by the density x: at $x > x_{FC}$ the system is on the disordered side of FCQPT; at $x = x_{FC}$, Eq. (2.8) has the non-trivial solutions $n_0(\mathbf{p})$ with $p_i = p_F = p_f$; at $x < x_{FC}$, the system is on the ordered side. We note that the solutions $n_0(\mathbf{p})$ of Eq. (2.8) can be viewed as new to the LFL theory. Indeed, at $T = 0$, the standard solution $n(\mathbf{p}, T \to 0) \to \theta(p_F - p)$ is not the only one possible. The "anomalous" solutions of Eq. (2.1) can exist because the logarithm on the right hand side of Eq. (2.1) is finite when p belongs to the region $(p_f - p_i)$ and $0 < n_0(\mathbf{p}) < 1$. Therefore in the region $T \ln[(1 - n_0(\mathbf{p}, T))/n_0(\mathbf{p}, T)]|_{T \to 0} \to 0$, and we again arrive at Eq. (2.8).

Let us assume that with the decrease of the density (or with the growth of the interaction strength) FC has just taken place. It means that $p_i \to p_f \to p_F$, and the deviation $\delta n(\mathbf{p}) = n_0(\mathbf{p}) - \theta(p_F - p)$ is small. Expanding the functional $E[n(\mathbf{p})]$ in Taylor's series with respect to $\delta n(\mathbf{p})$ and retaining the leading terms, one obtains from Eq. (2.8) the following relation

$$\mu = \varepsilon(\mathbf{p}) = \varepsilon_0(\mathbf{p}) + \int F(\mathbf{p}, \mathbf{p_1})\delta n(\mathbf{p_1})\frac{d\mathbf{p_1}}{(2\pi)^2} ; \quad p_i \leq p \leq p_f , \qquad (2.9)$$

where $F(\mathbf{p}, \mathbf{p_1}) = \delta^2 E/\delta n(\mathbf{p})\delta n(\mathbf{p_1})$ is the Landau amplitude. Both quantities, the interaction and the single-particle energy $\varepsilon_0(\mathbf{p})$ are calculated at $n(\mathbf{p}) = \theta(p_F - p)$. Equation

(2.9) acquires nontrivial solutions at some density $x = x_{FC}$. Thus, FCQPT takes place if the Landau amplitudes depending on the density are positive and sufficiently large, so that the potential energy is bigger than the kinetic energy. Then the transformation of the Fermi step function $n(\mathbf{p}) = \theta(p_F - p)$ into the smooth function defined by Eq. (2.9) becomes possible [17, 19]. The system in question can be considered as a strongly correlated Fermi liquid behind the point x_{FC}. It is seen from Eq. (2.9) that the FC quasiparticles form a collective state, since their energies are defined by the macroscopical number of quasiparticles within the momentum region $(p_f - p_i)$. The shape of the excitation spectra related to FC is not affected by the Landau interaction, which, generally speaking, depends on the system's properties, including the collective states, the irregularity of the composition, impurities, etc. The only thing determined by the interaction is the width of the FC region $(p_f - p_i)$ provided the interaction is sufficiently strong to produce the FC phase transition at all. Thus, we can conclude that the spectra related to FC are of a universal form, being dependent, as we shall see in Subsections 2.3 and 3.1, on the temperature and the superconducting gap. The existence of such spectra can be viewed as the characteristic feature of the "quantum protectorate" [40, 41].

2.3   The "Shadow" of the Fermion Condensate at Finite Temperatures

According to Eq. (2.1), the single-particle energy $\varepsilon(\mathbf{p}, T)$ within the interval $(p_f - p_i)$ at $T \ll T_f$ is linear in T [42]. At the Fermi level, one obtains by expanding $\ln(...)$ in terms of $n(\mathbf{p})$

$$\varepsilon(\mathbf{p}, T) - \mu(T) \; = \; T \ln \frac{1 - n(\mathbf{p})}{n(\mathbf{p})} \; \simeq \; T \frac{1 - 2n(\mathbf{p})}{n(\mathbf{p})} \bigg|_{p \simeq p_F} . \tag{2.10}$$

Here T_f is the temperature, above which FC effects become insignificant [20]

$$\frac{T_f}{\varepsilon_F} \; \sim \; \frac{p_f^2 - p_i^2}{2M\varepsilon_F} \; \sim \; \frac{\Omega_{FC}}{\Omega_F} . \tag{2.11}$$

In this formula Ω_{FC} is the FC volume, ε_F is the Fermi energy, and Ω_F is the volume of the Fermi sphere. We note that at $T \ll T_f$ the occupation numbers $n(\mathbf{p})$ are approximately independent of T, being given by Eq. (2.8). At finite temperatures according to Eq. (2.10), the dispersionless plateau $\varepsilon(\mathbf{p}) = \mu$ is slightly turned counter-clockwise about μ. As a result, the plateau is just a little tilted and rounded off at the end points. According to Eqs. (2.5) and (2.10), the effective mass M_{FC}^* related to FC is given by,

$$M_{FC}^* \; \simeq \; p_F \frac{p_f - p_i}{4T} . \tag{2.12}$$

To obtain Eq. (2.12) an approximation for the derivative $dn(p)/dp \simeq -1/(p_f - p_i)$ was used. It is seen from Eq. (2.12) that at $0 < T \ll T_f$, the heavy electron liquid behaves as if it were placed at QCP, in fact it is placed at the quantum critical line $x < x_{FC}$, that is the critical behavior is observed at $T \to 0$ for all $x \leq x_{FC}$.

Having in mind that $(p_f - p_i) \ll p_F$ and using Eqs. (2.11) and (2.12), the following estimates for the effective mass M_{FC}^* are obtained:

$$\frac{M_{FC}^*}{M} \; \sim \; \frac{N(0)}{N_0(0)} \; \sim \; \frac{T_f}{T} . \tag{2.13}$$

Eqs. (2.12) and (2.13) show the temperature dependence of M_{FC}^*. In Eq. (2.13) $N_0(0)$ denotes the density of states of noninteracting electron gas, and $N(0)$ is the density of states at the Fermi level. Multiplying both sides of Eq. (2.12) by $(p_f - p_i)$, we obtain the energy scale E_0 separating the slow dispersing low energy part related to the effective mass M_{FC}^* from the faster dispersing relatively high energy part defined by the effective mass M_L^* [29, 30, 43],

$$E_0 \simeq 4T. \tag{2.14}$$

It is seen from Eq. (2.14) that the scale E_0 does not depend on the condensate volume. The single particle excitations are defined according to Eq. (2.12) by the temperature and by $(p_f - p_i)$, given by Eq. (2.8). Thus, we conclude that the one-electron spectrum is of universal form and has the features of the "quantum protectorate".

It is pertinent to note that outside the FC region the single particle spectrum is not strongly affected by the temperature, being defined by M_L^*. Thus, we come to the conclusion that a system with FC is characterized by two effective masses: M_{FC}^* which is related to the single particle spectrum at lower energy scale and M_L^* describing the spectrum at higher energy scale. The existence of two effective masses is manifested by a break (or kink) in the quasiparticle dispersion, which can be approximated by two straight lines intersecting at the energy E_0. This break takes place at temperatures $T_c \leq T \ll T_f$, which is in accord with experimental data [44], and, as we will see, at $T \leq T_c$ which is also in accord with the experimental facts [44, 45]. Here T_c is the critical tempereture of the superconducting phase transition. The quasiparticle formalism is applicable to this problem since the width γ of single particle excitations is not large compared to their energy, being proportional to the temperature, $\gamma \sim T$ at $T > T_c$ [20]. The lineshape can be approximated by a simple Lorentzian [43]. This is consistent with experimental data obtained from scans at a constant binding energy [46] (see Sec. 4).

It is essential to have in mind, that the onset of the charge density wave instability in a many-electron system, such as an electron liquid, which takes place as soon as the effective inter-electron constant reaches its critical value $r_s = r_{cdw}$, is preceded by the unlimited growth of the effective mass, which is demonstrated Section 7. Here $r_s = r_0/a_B$ with r_0 being the average distance between electrons, while a_B is the Bohr radius. Therefore the FC occurs before the onset of the charge density wave. Hence, at $T = 0$, when r_s reaches its critical value r_{FC} corresponding to x_{FC}, $r_{FC} < r_{cdw}$, FCQPT already inevitably takes place [38]. It is pertinent to note that this growth of the effective mass with decreasing electron density was observed experimentally in a metallic 2D electron system in silicon at $r_s \simeq 7.5$ [34]. Therefore we can take this value as an estimate, $r_{FC} \sim 7.5$. On the other hand, there exist charge density waves or strong fluctuations of charge ordering in underdoped high-T_c superconductors [47, 48]. Thus, the formation of FC in high-T_c compounds can be thought more as a general property of low density electron liquid embedded in these solids rather than an unusual and anomalous solution of Eq. (2.8) [38]. Beyond the point of FCQPT, the condensate volume is proportional to $(r_s - r_{FC})$ as well as $T_f/\varepsilon_F \sim (r_s - r_{FC})/r_{FC}$ at least when $(r_s - r_{FC})/r_{FC} \ll 1$. Therefore we obtain

$$\frac{r_s - r_{FC}}{r_{FC}} \sim \frac{p_f - p_i}{p_F} \sim \frac{x_{FC} - x}{x_{FC}}. \tag{2.15}$$

FC serves as a stimulator that creates new phase transitions, which eliminates the degeneracy of the spectrum. For example FC can generate spin density waves or antiferromagnetic phase transition, thus leading to a whole variety of new properties of the system under consideration. Then, the onset of the charge density wave is preceded by FCQPT, and both of these phases can coexist at the sufficiently low density when $r_s \geq r_{cdw}$. The transition to superconductivity is strongly assisted by FC because both of the phases are characterized by the same order parameter. As a result, the superconductivity by removing the spectrum degeneracy "wins" the competition with other phase transitions up to the critical temperature T_c, see Section 3. We now turn to the consideration of the superconducting state and quasiparticle dispersions at $T \leq T_c$.

3 The Superconducting State with Fermion Condensate

In this Section we consider the superconducting state of 2D heavy electron liquid since the high-T_c superconductors are predominantly represented by 2D structures. On the other hand, our consideration can be easily adopted to the 3D case.

3.1 Superconducting State at $T = 0$

At $T = 0$, the ground state energy $E_{gs}[\kappa(\mathbf{p}), n(\mathbf{p})]$ of a 2D electron liquid is a functional of both the order parameter of the superconducting state $\kappa(\mathbf{p})$ and the quasiparticle occupation numbers $n(\mathbf{p})$. This energy is determined by the well-known equation of the Bardeen-Cooper-Schrieffer (BCS) weak-coupling theory of superconductivity, see e.g. [49, 50]

$$E_{gs} = E[n(\mathbf{p})] + \lambda_0 \int V(\mathbf{p}_1, \mathbf{p}_2) \kappa(\mathbf{p}_1) \kappa^*(\mathbf{p}_2) \frac{d\mathbf{p}_1 d\mathbf{p}_2}{(2\pi)^4} . \tag{3.1}$$

Here $E[n(\mathbf{p})]$ is the ground-state energy of a normal Fermi liquid, and

$$n(\mathbf{p}) = v^2(\mathbf{p}); \quad \kappa(\mathbf{p}) = v(\mathbf{p})u(\mathbf{p}) = \sqrt{n(\mathbf{p})(1 - n(\mathbf{p}))}, \tag{3.2}$$

Where $u(\mathbf{p})$ and $v(\mathbf{p})$ are the normalized coherence factors, $v^2(\mathbf{p}) + u^2(\mathbf{p}) = 1$. It is assumed that the pairing interaction $\lambda_0 V(\mathbf{p}_1, \mathbf{p}_2)$ is weak. We define the superconducting gap

$$\Delta(\mathbf{p}) = -\lambda_0 \int V(\mathbf{p}, \mathbf{p}_1) \kappa(\mathbf{p}_1) \frac{d\mathbf{p}_1}{4\pi^2}. \tag{3.3}$$

Minimizing E_{gs} with respect to $v(\mathbf{p})$ we obtain the equation connecting the single-particle energy $\varepsilon(\mathbf{p})$ to $\Delta(\mathbf{p})$,

$$\varepsilon(\mathbf{p}) - \mu = \Delta(\mathbf{p}) \frac{1 - 2v^2(\mathbf{p})}{2\kappa(\mathbf{p})}, \tag{3.4}$$

where the single-particle energy $\varepsilon(\mathbf{p})$ is determined by the Landau equation (2.2). With insertion of the value of $\kappa(\mathbf{p})$ into Eq. (3.3), we obtain the known equation for $\Delta(\mathbf{p})$

$$\Delta(\mathbf{p}) = -\frac{\lambda_0}{2} \int V(\mathbf{p}, \mathbf{p}_1) \frac{\Delta(\mathbf{p}_1)}{\sqrt{(\varepsilon(\mathbf{p}_1) - \mu)^2 + \Delta^2(\mathbf{p}_1)}} \frac{d\mathbf{p}_1}{4\pi^2}. \tag{3.5}$$

If $\lambda_0 \to 0$, then the maximum value Δ_1 of the superconducting gap $\Delta(\mathbf{p})$ tends to zero, and Eq. (3.4) reduces to Eq. (2.8)

$$\frac{\delta E[n(\mathbf{p})]}{\delta n(\mathbf{p})} = \varepsilon(\mathbf{p}) - \mu = 0, \text{ provided that } 0 < n(\mathbf{p}) < 1; \text{ or } \kappa(\mathbf{p}) \neq 0; \ p_i \leq p \leq p_f \,.$$

(3.6)

It is seen from Eq. (3.6) that at $x < x_{FC}$, the function $n(\mathbf{p})$ is determined by the standard equation to search the minimum of functional $E[n(\mathbf{p})]$ [20, 17]. Equation (3.6) determines the quasiparticle distribution function $n_0(\mathbf{p})$ which delivers the minimum value to the ground state energy $E_{gs}[\kappa(\mathbf{p}), n(\mathbf{p})]$ in the $\lambda_0 \to 0$ limit. Now we can study the relationships between the state defined by Eq. (3.6), or by Eq. (2.8), and the superconducting state. At $T = 0$, Eq. (3.6) defines the particular state of a Fermi-liquid with FC, for which the modulus of the order parameter $|\kappa(\mathbf{p})|$ has finite values in the range of momenta $p_i \leq p \leq p_f$, while $\Delta_1 \to 0$. Such a state can be considered as superconducting, with an infinitely small value of Δ_1, so that the entropy of this state is equal to zero. It is obvious that this state, being driven by the quantum phase transition, disappears at $T > 0$ [29, 30]. Any quantum phase transition, which takes place at temperature $T = 0$, is determined by a control parameter other than the temperature, for instance, by pressure, magnetic field, or the number density of mobile charge carriers x. In the case of FCQPT, as we have shown in Section 2, the control parameter is the density x of the system, which determines the strength of the Landau amplitudes. FCQPT occurs at the quantum critical point $x = x_{FC}$.

As any phase transition, FCQPT is related to the order parameter, which induces a broken symmetry. As it follows from Eq. (3.2), the order parameter of the state with FC is $\kappa(\mathbf{p})$. Thus, the solutions $n_0(\mathbf{p})$ of Eq. (3.6) represent a new class of the solutions of both BCS equations and LFL equations. In contrast to the conventional BCS solutions [49], the new ones are characterized by the infinitesimal value of the superconducting gap, $\Delta_1 \to 0$, while the order parameter $\kappa(\mathbf{p})$ is finite. At the same time, in contrast to the standard solutions of the LFL theory, the new ones are characterized by the superconducting order parameter $\kappa(\mathbf{p})$, that is at $T \to 0$, the quasiparticle distribution function does not tend to the step function $\theta(p_f - p)$, being the solution of Eq. (3.6). Thus, we can conclude that the solutions of Eq. (3.6) can be viewed as common solutions of both BCS and LFL equations, while Eq. (3.6) can be derived starting from either BCS or LFL theory. As we shall see, on the basis of the peculiarities of this new class of solutions, we can define the notion of the strongly correlated Fermi liquid and explain the main properties of the high-T_c superconductivity and HF metals. It is essential that the existence of state with FC can be revealed experimentally since the order parameter $\kappa(\mathbf{p})$ is suppressed by a magnetic field B. Destroying the state with FC, the magnetic field B converts the strongly correlated Fermi liquid into the normal LFL. In that case, the magnetic field plays the role of control parameter.

When $p_i \to p_F \to p_f$, Eq. (3.6) determines the critical density x_{FC} at which FCQPT takes place. When $x < x_{FC}$, the system becomes divided into two quasiparticle subsystems: the first subsystem is in the $(p_f - p_i)$ range and is characterized by the quasiparticles with the effective mass $M_{FC}^* \propto 1/\Delta_1$, as it follows from Eq. (3.4), while the second one is occupied by quasiparticles with finite mass M_L^* and momenta $p < p_i$. When $\lambda_0 \to 0$, the density of states near the Fermi level tends to infinity, $N(0) \sim M_{FC}^* \sim 1/\Delta_1$. The quasiparticles with M_{FC}^* occupy the same energy level and form pairs with binding energy of the order of Δ_1 and with average momentum p_0, $p_0/p_F \sim (p_f - p_i)/p_F \ll 1$. Therefore, this state strongly resembles the Bose-Einstein condensate, in which quasiparticles occupy the same energy level. But the FC quasiparticles have to be spread over the range $(p_f - p_i)$ in momentum space due to the exclusion principle. In contrast to the Bose-Einstein condensation, the fermion condensation temperature is $T_c = 0$. And in contrast to the ordinary superconductivity, FC is formed due to the Landau repulsive interaction $F(\mathbf{p}, \mathbf{p}_1)$ rather than by the relatively weak attractive quasiparticle-quasiparticle pairing interaction $\lambda_0 V(\mathbf{p}_1, \mathbf{p}_2)$.

If $\lambda_0 \neq 0$, Δ_1 becomes finite, leading to a finite value of the effective mass M_{FC}^*, which can be obtained from Eq. (3.4) upon differentiating both parts of this equation with respect to the momentum p and using Eq. (2.5) [29, 30, 43]

$$M_{FC}^* \simeq p_F \frac{p_f - p_i}{2\Delta_1} . \tag{3.7}$$

As to the energy scale, it is determined by the parameter E_0:

$$E_0 = \varepsilon(\mathbf{p}_f) - \varepsilon(\mathbf{p}_i) \simeq p_F \frac{(p_f - p_i)}{M_{FC}^*} \simeq 2\Delta_1 . \tag{3.8}$$

3.2   Superconducting State at Finite Temperatures

Let us assume that the range of FC is small, that is $(p_f - p_i)/p_F \ll 1$, and $2\Delta_1 \ll T_f$ so that the order parameter $\kappa(\mathbf{p})$ is governed mainly by FC [29, 30]. To solve Eq. (3.5) analytically, we take the BCS approximation for the interaction [49]: $\lambda_0 V(\mathbf{p}, \mathbf{p}_1) = -\lambda_0$ if $|\varepsilon(\mathbf{p}) - \mu| \leq \omega_D$, i.e. the interaction is zero outside this region, with ω_D being the characteristic energy, e.g. that of phonon. As a result, the gap becomes dependent only on the temperature, $\Delta(\mathbf{p}) = \Delta_1(T)$, being independent of the momentum, and Eq. (3.5) takes the form

$$1 = N_{FC}\lambda_0 \int_0^{E_0/2} \frac{d\xi}{\sqrt{\xi^2 + \Delta_1^2(0)}} + N_L\lambda_0 \int_{E_0/2}^{\omega_D} \frac{d\xi}{\sqrt{\xi^2 + \Delta_1^2(0)}} . \tag{3.9}$$

Here we set $\xi = \varepsilon(\mathbf{p}) - \mu$ and introduce the density of states N_{FC} in the $(p_f - p_i)$ range, or E_0 range. It follows from Eq. (3.7), that $N_{FC} = (p_f - p_F)p_F/2\pi\Delta_1(0)$. The density of states N_L in the range $(\omega_D - E_0/2)$ has the standard form $N_L = M_L^*/2\pi$. If the energy scale $E_0 \to 0$, Eq. (3.9) reduces to the BCS equation. On the other hand, assuming that $E_0 \leq 2\omega_D$ and omitting the second integral on the right hand side of Eq. (3.9), we obtain

$$\Delta_1(0) = \frac{\lambda_0 p_F(p_f - p_F)}{2\pi} \ln\left(1 + \sqrt{2}\right) = 2\beta\varepsilon_F \frac{p_f - p_F}{p_F} \ln\left(1 + \sqrt{2}\right), \tag{3.10}$$

where the Fermi energy $\varepsilon_F = p_F^2/2M_L^*$, and the dimensionless coupling constant β is given by the relation $\beta = \lambda_0 M_L^*/2\pi$. Taking the usual values of β for conventional superconductors as $\beta \simeq 0.3$, and assuming $(p_f - p_F)/p_F \simeq 0.2$, we get from Eq. (20) a large value of $\Delta_1(0) \sim 0.1\varepsilon_F$, while for conventional superconductors one has a much smaller gap, $\Delta_1(0) \sim 10^{-3}\varepsilon_F$. Taking into account the omitted above integral, we obtain

$$\Delta_1(0) \simeq 2\beta\varepsilon_F \frac{p_f - p_F}{p_F} \ln\left(1 + \sqrt{2}\right)\left(1 + \beta\ln\frac{2\omega_D}{E_0}\right). \tag{3.11}$$

It is seen from Eq. (3.11) that the correction due to the second integral is small, provided $E_0 \simeq 2\omega_D$. Below we shall show that $2T_c \simeq \Delta_1(0)$, which leads to the conclusion that the isotope effect is absent since Δ_1 is independent of ω_D. But this effect is restored as $E_0 \to 0$.

Assuming $E_0 \sim \omega_D$ but $E_0 > \omega_D$, we see that Eq. (3.11) has no standard solutions $\Delta(p) = \Delta_1(T = 0)$ because $\omega_D < \varepsilon(p \simeq p_f) - \mu$ and the interaction vanishes at these momenta. The only way to obtain solutions is to restore the condition $E_0 < \omega_D$. For instance, we can define such a momentum $p_D < p_f$ that

$$\Delta_1(0) = 2\beta\varepsilon_F \frac{p_D - p_F}{p_F} \ln\left(1 + \sqrt{2}\right) = \omega_D , \qquad (3.12)$$

while the other part in the $(p_f - p_i)$ range can be occupied by a gap Δ_2 of the different sign, $\Delta_1/\Delta_2 < 0$. It follows from Eq. (3.12) that the isotope effect is preserved, while both gaps can have s-wave symmetry.

At $T \simeq T_c$, Eqs. (3.7) and (3.8) are replaced by Eqs. (2.12) and (2.14), which is valid also at $T_c \leq T \ll T_f$

$$M_{FC}^* \simeq p_F \frac{p_f - p_i}{4T_c}, \quad E_0 \simeq 4T_c; \text{ if } T_c \leq T \text{ then}, M_{FC}^* \simeq p_F \frac{p_f - p_i}{4T}, \quad E_0 \simeq 4T .$$

$$(3.13)$$

Equation (3.9) is replaced by its conventional finite temperature generalization

$$1 = N_{FC}\lambda_0 \int_0^{E_0/2} \frac{d\xi}{\sqrt{\xi^2 + \Delta_1^2(T)}} \tanh \frac{\sqrt{\xi^2 + \Delta_1^2(T)}}{2T} +$$

$$+ N_L\lambda_0 \int_{E_0/2}^{\omega_D} \frac{d\xi}{\sqrt{\xi^2 + \Delta_1^2(T)}} \tanh \frac{\sqrt{\xi^2 + \Delta_1^2(T)}}{2T} . \qquad (3.14)$$

Putting $\Delta_1(T \to T_c) \to 0$, we obtain from Eq. (3.14)

$$2T_c \simeq \Delta_1(0) , \qquad (3.15)$$

with $\Delta_1(T = 0)$ being given by Eq. (3.11). Upon comparing Eqs. (3.7), (3.8) and (3.13), (3.15), we see that M_{FC}^* and E_0 are almost temperature independent at $T \leq T_c$.

3.3 Bogoliubov Quasiparticles

It is seen from Eq. (3.5) that the superconducting gap depends on the single-particle spectrum $\varepsilon(\mathbf{p})$. On the other hand, it follows from Eq. (3.4) that $\varepsilon(\mathbf{p})$ depends on $\Delta(\mathbf{p})$, provided that at $\lambda_0 \to 0$ Eq. (3.6) has the solution determining the existence of FC. Let us assume that λ_0 is small enough, so that the particle-particle interaction $\lambda_0 V(\mathbf{p}, \mathbf{p}_1)$ can only lead to a small perturbation of the order parameter $\kappa(\mathbf{p})$ determined by Eq. (3.6). It follows from Eq. (3.7) that the effective mass and the density of states $N(0) \propto M_{FC}^* \propto 1/\Delta_1$ are finite. As a result, we conclude that in contrast to the conventional theory of superconductivity the single-particle spectrum $\varepsilon(\mathbf{p})$ strongly depends on the superconducting gap and we have to solve Eqs. (2.2) and (3.5) in a self-consistent way. On the other hand, let us assume that Eqs. (2.2) and (3.5) are solved, and the effective mass M^*_{FC} is determined. Now one can fix the quasiparticle dispersion $\varepsilon(\mathbf{p})$ by choosing the effective mass M^* of the system in question equal to the obtained M^*_{FC} and then solve Eq. (3.5) as it is done in the case of the conventional BCS theory of superconductivity [49]. As a result, one observes that the superconducting state is characterized by the Bogoliubov quasiparticles (BQ) [51] with the dispersion

$$E(\mathbf{p}) = \sqrt{(\varepsilon(\mathbf{p}) - \mu)^2 + \Delta^2(\mathbf{p})},$$

and the normalization condition for the coherence factors $v(\mathbf{p})$ and $u(\mathbf{p})$ is held. We arrive to the conclusion that the observed features agree with the behavior of BQ predicted from the BCS theory. This observation suggests that the superconducting state with FC is BCS-like and implies the basic validity of the BCS formalism in describing the superconducting state [52]. Although the maximum value of the superconducting gap given by Eq. (3.11) and other exotic properties are determined by the presence of the fermion condensate. It is exactly the case that was observed experimentally in the high-T_c cuprate $Bi_2Sr_2Ca_2Cu_3O_{10+\delta}$ [53].

We have returned back to the LFL theory since the high energy degrees of freedom are eliminated and the quasiparticles are introduced. The only difference between the Landau Fermi-liquid, which serves as a basis when constructing the superconducting state, and Fermi liquid after FCQPT is that we have to expand the number of relevant low energy degrees of freedom by introducing a new type of quasiparticles with the effective mass M^*_{FC} given by Eq. (3.7) and the energy scale E_0 given by Eq. (3.8). Therefore, the dispersion $\varepsilon(\mathbf{p})$ is characterized by two effective masses M^*_L and M^*_{FC} and by the scale E_0, which define the low temperature properties including the line shape of quasiparticle excitations [29, 30, 54], while the dispersion of BQ has the standard form. We note that both the effective mass M^*_{FC} and the scale E_0 are temperature independent at $T < T_c$, where T_c is the critical temperature of the superconducting phase transition [54]. At $T > T_c$, the effective mass M^*_{FC} and the scale E_0 are given by Eqs. (2.12) and (2.14) respectively. Obviously, we cannot directly relate these new Landau Fermi-liquid quasiparticle excitations with the quasiparticle excitations of an ideal Fermi gas because the system in question has undergone FCQPT. Nonetheless, the main basis of the Landau Fermi liquid theory survives FCQPT: the notion of order parameter is preserved, and the low energy excitations of the strongly correlated liquid with FC are quasiparticles.

As it was shown above, properties of these new quasiparticles are closely related to the properties of the superconducting state. We may say that the quasiparticle system in the range $(p_f - p_i)$ becomes very "soft" and is to be considered as a strongly correlated liquid. On the other hand, the system's properties and dynamics are dominated by a strong collective effect having its origin in its proximity to FCQPT and determined by the macroscopic number of quasiparticles forming FC in the range $(p_f - p_i)$. Such a system cannot be perturbed by the scattering of individual quasiparticles and has features of a "quantum protectorate" and demonstrates the universal behavior [29, 30, 40, 41]. A few remarks related to the quantum protectorate and the universal behavior [40] are in order here. As the Landau theory of Fermi liquid, the theory of the high-temperature superconductivity based on FCQPT deals with the quasiparticles which are elementary excitations of low energy. As a result, this theory gives general qualitative description of both the superconducting and normal states. Of course, one can choose the phenomenological parameters and obtain the quantitative consideration of the superconductivity as it can be done in the framework of the Landau theory when describing a normal Fermi-liquid, say ^{3}He. Therefore, we conclude that any theory, which is capable to describe FC and compatible with the BCS theory, will give the qualitative picture of the superconducting state and the normal state that coincides with the picture based on FCQPT. Both of the approaches can be agreed at a numerical level provided the corresponding parameters are adjusted. For example, since the formation of the flat band is possible in the Hubbard model [25], generally speaking one can repeat the results of the theory based on FCQPT in the Hubbard model. It is appropriate to mention here that the corresponding numerical description limited to the case of $T = 0$ has been obtained within the Hubbard model [55, 56].

3.4   Pseudogap

Now let us discuss some special features of the superconducting state with FC [57, 58]. Consider two possible types of the superconducting gap $\Delta(\mathbf{p})$, namely that given by Eq. (3.5) and defined by the interaction $\lambda_0 V(\mathbf{p}, \mathbf{p}_1)$. If this interaction is dominated by an attractive phonon-mediated attraction, the solution of Eq. (3.5) with the s-wave, or the $s + d$ mixed waves will have the lowest energy. Provided the pairing interaction $\lambda_0 V(\mathbf{p}_1, \mathbf{p}_2)$ is the combination of both the attractive interaction and sufficiently strong repulsive interaction, the d-wave superconductivity can take place (see e.g. [59, 60]). But both the s-wave symmetry and the d-wave one lead to approximately the same value of the gap Δ_1 in Eq. (3.11) [54]. Therefore the non-universal pairing symmetries in high-T_c superconductivity are likely the result of the pairing interaction, while the d-wave pairing symmetry is not essential. This point of view is supported by the data [61, 62, 63, 64, 65].

We can define the critical temperature T^*, at which the superconductivity vanishes, as the temperature when $\Delta_1(T^*) \equiv 0$. At $T \geq T^*$, Eq. (3.14) has only the trivial solution $\Delta_1 \equiv 0$. On the other hand, the critical temperature T_c can be defined as a temperature, at which the superconductivity disappears while the gap occupies only part of the Fermi surface. Thus, as we shall see there are two different temperatures T_c and T^* in the case of the d-wave symmetry of the gap. It was shown [43, 57] that in the presence of FC there exist nontrivial solutions of Eq. (3.14) at $T_c \leq T \leq T^*$ when the BCS like interaction is replaced by the pairing interaction $\lambda_0 V(\mathbf{p}_1, \mathbf{p}_2)$, which includes the strong repulsive interaction leading to d-wave superconductivity. In that case, the gap $\Delta(\mathbf{p})$ as a function of angle ϕ, $\Delta(\mathbf{p}) = \Delta(p_F, \phi)$ possesses new nodes at $T > T_{node}$, as it is illustrated by Fig. 2 [43].

We show in Fig. 2 the ratio $\Delta(p_F, \phi)/T^*$ calculated at three different temperatures $0.9\,T_{node}$, T_{node}, and $1.2\,T_{node}$. An important difference between curve (a) and (b) and (c) is additional variations of the curves (b) and (c) marked by the two arrows. As seen from Fig. 2, the flattening occurs due to the appearance of the two new nodes. They appear at $T = T_{node}$ and move apart with increase of the temperature confining the area θ_c shown by the two arrows in Fig. 2. It is also seen from Fig. 2 that the gap Δ is extremely small over the range θ_c. It was recently shown in a number of papers (see, e.g., [66, 67]) that there exists an interplay between the magnetism and the superconductivity order parameters, leading to the damping of the magnetism order parameter below T_c. And vise versa,

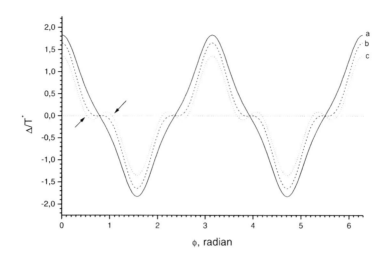

Figure 2: The gap $\Delta(p_F, \phi)$ as a function of ϕ calculated at three different temperatures expressed in terms of $T_{node} \simeq T_c$, while Δ is presented in terms of T^*. Curve (a), solid line, shows the gap calculated at temperature $0.9T_{node}$. Curve (b), dashed line, presents the same at $T = T_{node}$, and curve (c), dotted line, shows the gap calculated at temperature $1.2T_{node}$. Note the important difference in the curves (b) and (c) compared with curve (a) due to the flattening of the curves (b) and (c) about the nodes. The two arrows indicate the area θ_c emerging at T_{node}.

one can anticipate the damping of the superconductivity order parameter by the magnetism. Thus, we conclude that the gap in the range θ_c can be destroyed by strong antiferromagnetic correlations (or by spin density waves) existing in optimally doped and underdoped super-conductors. It is believed that impurities can easily destroy the gap $\Delta(\mathbf{p})$ in the considered area. As a result, the superconducting gap vanishes in the macroscopic area θ_c and causes the superconductivity to die out. We have to conclude that $T_c \simeq T_{node}$, with the exact value of T_c defined by the competition between the antiferromagnetic correlations (or spin density waves) and the superconducting correlations over the range θ_c. The behavior and the shape of the pseudogap is very similar to the ones of the superconducting gap as it is seen from Fig. 2. The main difference seen from Fig. 2 is that the pseudogap vanishes along segments θ_c of the Fermi surface, while the gap vanishes at the isolated d-wave nodes. Our calculations show that the function $\theta_c(x)$ increases very fast at small values of x, $\theta_c(x) \simeq \sqrt{x}$. Therefore T_c has to be close to T_{node}. This result is in accord with numerical calculations of the function $\theta_c([T - T_c]/T_c)$ plotted in Fig. 3.

Thus, we observe that the pseudogap state appears at temperatures $T \geq T_c \simeq T_{node}$ and vanishes at $T \geq T^*$ when Eq. (3.14) has only the trivial solution $\Delta_1 \equiv 0$. Quite naturally, one has to recognize that Δ_1 scales with T^*, while Eq. (3.15) takes the form

$$2T^* \simeq \Delta_1(0) . \tag{3.16}$$

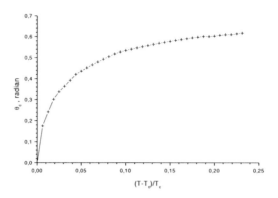

Figure 3: Calculated angle θ_c, pulling apart the two nodes, as a function of $(T - T_c)/T_c$.

It can then be concluded that the temperature T^* has the physical meaning of the BCS transition temperature between the state with the order parameter $\kappa \neq 0$ and the normal state.

At $T < T_c$ quasiparticle excitations are characterized by sharp peaks. When temperature becomes $T > T_c$ and $\Delta(\theta) \equiv 0$ in the range θ_c, there appear normal quasiparticle excitations with the width γ along the segments θ_c of the Fermi surface. There exists the pseudogap outside the segments θ_c, and the Fermi level is occupied by the BCS-type excitations. Both types of excitations have the width of the same order of magnitude transferring their energy and momentum to the normal quasiparticle excitations. We now estimate γ. For the entire Fermi level occupied by the normal state, the width is equal to $\gamma \approx N(0)^3 T^2/\varepsilon(T)^2$, with the density of states $N(0) \sim M^*(T) \sim 1/T$ (see Eq. (2.12)). The dielectric constant $\varepsilon(T) \sim N(0)$, and the width γ becomes $\gamma \sim T$ [20]. In our case, however, only a part of the Fermi surface within θ_c is occupied by the normal excitations. Therefore, the number of states allowed for quasiparticles and for quasiholes is proportional to θ_c, and the factor T^2 is replaced by $T^2 \theta_c^2$. Taking this into account, we obtain $\gamma \sim \theta_c^2 T \sim T(T - T_c)/T_c \sim (T - T_c)$. Here we have omitted the small contribution coming from the BCS-type excitations. That is why the width γ vanishes at $T = T_c$. Thus, the presented above analysis shows that in the pseudogap state at $T > T_c$, the superconducting gap smoothly transforms into the pseudogap. The excitations of that area of the Fermi surface that has the gap are of the same width $\gamma \sim (T - T_c)$. The region occupied by the pseudogap is shrinking with increasing temperature. It is worth noting that the normal state resistivity $\rho(T)$ behaves as $\rho(T) \propto T$ due to $\gamma \sim (T - T_c)$. Obviously at $T > T^*$ the behavior $\rho(T) \propto T$ remains valid up to temperatures $T \sim T_f$, and T_f can be as high as the Fermi energy, provided that FC occupies noticeable part of the Fermi volume, see Eq. (2.11).

The temperature T_{node} is determined mainly by the repulsive interaction being part of the pairing interaction $\lambda_0 V(\mathbf{p}_1, \mathbf{p}_2)$. In its turn, the repulsive interaction can depend on the properties of materials such as the composition, doping, etc. Since the superconductivity is destroyed at T_c, the ratio $2\Delta_1/T_c$ can vary in a wide range and strongly depends upon the material's properties, as it follows from considerations given above [43, 57, 58]. The ratio $2\Delta_1/T_c$ can reach very high values. For instance, in the case of $Bi_2Sr_2CaCu_2Q_{6+\delta}$ where the superconductivity and the pseudogap are considered to be of the common origin, $2\Delta_1/T_c$ is about 28, while the ratio $2\Delta_1/T^* \simeq 4$, which is in agreement with the experimental data for various cuprates [59]. Note that Eq. (3.16) gives also good description of the maximum gap Δ_1 in the case of the d-wave superconductivity, because the different regions with the maximum absolute value of Δ_1 and the maximal density of states can be considered as disconnected [60]. Therefore the gap in this region is formed by the attractive phonon interaction, which is approximately independent of the momenta. We can also conclude, that in the case of the s-wave pairing the pseudogap phenomenon is absent because there is no the sufficiently strong repulsive interaction. Thus, the transition from superconducting gap to pseudogap can take place only in the case of the d-wave pairing, so that the superconductivity is destroyed at $T_c \simeq T_{node}$, with the superconducting gap being smoothly transformed into the pseudogap, which closes at some temperature $T^* > T_c$ [43, 57, 58]. In the case of the s-wave pairing, we expect the absence of the pseudogap phenomenon in accordance with the observations (see e.g. [65] and references therein).

3.5   Dependence of T_c on the Doping

We now turn to consideration of the maximum value of the superconducting gap Δ_1 as a function of the number density x of the mobile charge carriers which is proportional to the doping. Using Eq. (2.15), we can rewrite Eq. (3.10) as follow

$$\frac{\Delta_1}{\varepsilon_F} \sim \beta \frac{(x_{FC} - x)x}{x_{FC}}. \tag{3.17}$$

Here we take into account that the Fermi level $\varepsilon_F \propto p_F^2$, the density $x \sim p_F^2/(2M_L^*)$, and thus, $\varepsilon_F \propto x$. We can reliably assume that $T_c \propto \Delta_1$ because the empirically obtained simple bell-shaped curve of $T_c(x)$ in the high temperature superconductors [2] should have only a smooth dependence upon x. Then, $T_c(x)$ in accordance with the data has the same bell-shaped form [68]

$$T_c(x) \propto \beta(x_{FC} - x)x. \tag{3.18}$$

As an example of the implementation of the previous analysis, let us consider the main features of a room-temperature superconductor. The superconductor has to be a quasi two-dimensional structure like cuprates. From Eq. (3.10) it follows, that $\Delta_1 \sim \beta\varepsilon_F \propto \beta/r_s^2$. Noting that FCQPT in 3D systems takes place at $r_s \sim 20$ and in 2D systems at $r_s \sim 8$ [38], we can expect that Δ_1 of 3D systems comprises 10% of the corresponding maximum value of 2D superconducting gap, reaching a value as high as 60 meV for underdoped crystals with $T_c = 70$ [69]. On the other hand, it is seen from Eq. (3.10), that Δ_1 can be even large, $\Delta_1 \sim 75$ meV, and one can expect $T_c \sim 300$ K in the case of the s-wave pairing as it follows from the simple relation $2T_c \simeq \Delta_1$. Indeed, we can safely take $\varepsilon_F \sim 300$ meV, $\beta \sim 0.5$ and $(p_f - p_i)/p_F \sim 0.5$. Thus, a possible room-temperature superconductor has to be the s-wave superconductor in order to get rid of the pseudogap phenomena, which can tremendously reduce the transition temperature T_c. The density x of the mobile charge carriers must satisfy the condition $x \leq x_{FC}$ and be adjustable to reach the optimal doping level $x_{opt} \simeq x_{FC}/2$.

3.6 The Gap and Specific Heat near T_c

Now we turn to the calculations of the gap and the specific heat at the temperatures $T \to T_c$. It is worth noting that this consideration is valid provided $T^* = T_c$. Otherwise the considered below discontinuity in the specific heat is smoothed out over the temperature range $T^* \div T_c$. For the sake of simplicity, we calculate the main contribution to the gap and the specific heat coming from FC. The function $\Delta_1(T \to T_c)$ is found from Eq. (3.14) by expanding the right hand side of the first integral in powers of Δ_1 and omitting the contribution from the second integral on the right hand side of Eq. (3.14). This procedure leads to the following equation [54]

$$\Delta_1(T) \simeq 3.4 T_c \sqrt{1 - \frac{T}{T_c}} . \tag{3.19}$$

Thus, the gap in the spectrum of the single-particle excitations has the usual behavior. To calculate the specific heat, the conventional expression for the entropy S [49] can be used

$$S = -2 \int [f(\mathbf{p}) \ln f(\mathbf{p}) + (1 - f(\mathbf{p})) \ln(1 - f(\mathbf{p}))] \frac{d\mathbf{p}}{(2\pi)^2} , \tag{3.20}$$

where

$$f(\mathbf{p}) = \frac{1}{1 + \exp[E(\mathbf{p})/T]} ; \quad E(\mathbf{p}) = \sqrt{(\varepsilon(\mathbf{p}) - \mu)^2 + \Delta_1^2(T)} . \tag{3.21}$$

The specific heat C is determined by the equation

$$C = T\frac{dS}{dT} \simeq 4\frac{N_{FC}}{T^2} \int_0^{E_0} f(E)(1 - f(E)) \left[E^2 + T\Delta_1(T)\frac{d\Delta_1(T)}{dT} \right] d\xi +$$

$$+ 4\frac{N_L}{T^2} \int_{E_0}^{\omega_D} f(E)(1 - f(E)) \left[E^2 + T\Delta_1(T)\frac{d\Delta_1(T)}{dT} \right] d\xi . \tag{3.22}$$

In deriving Eq. (3.22) we again used the variable ξ and the densities of states N_{FC} and N_L, just as before in connection with Eq. (3.9), and employed the notation $E = \sqrt{\xi^2 + \Delta_1^2(T)}$. Eq. (3.22) predicts the discontinuity $\delta C = C_s - C_n$ in the specific heat C at T_c because of the two last term in the square brackets on right hand side of Eq. (3.22). Here, C_s and C_n are the specific heat of the superconducting state and the normal one respectively. Using Eq. (3.19) to calculate the first term on the right hand side of Eq. (3.22), we obtain [54]

$$\delta C(T_c) \simeq \frac{3}{2\pi^2} (p_f - p_i) p_F^2. \tag{3.23}$$

This is in contrast to the conventional result where the discontinuity is a linear function of T_c. $\delta C(T_c)$ is independent of the critical temperature T_c because as seen from Eq. (3.13) the density of states varies inversely with T_c. Note that in deriving Eq. (3.23) we took into account the main contribution coming from FC. This term vanishes as soon as $E_0 \to 0$ and the second integral on the right hand side of Eq. (3.22) gives the conventional result.

A few remarks are in order here. As we shall demonstrate in Section 9, Eq. (9.4), the specific heat of systems with FC behaves as $C(T) \propto \sqrt{T/T_f}$. The specific heat discontinuity given by Eq. (3.23) is temperature independent. As a result, we obtain that

$$\frac{\delta C(T_c)}{C_n(T_c)} \sim \sqrt{\frac{T_f}{T_c}} \frac{(p_f - p_i)}{p_F}. \tag{3.24}$$

In contrast to the conventional case of normal superconductors, when $\delta C(T_c)/C_n(T_c) = 1.43$ [27], it is seen from Eq. (3.24) that the ratio $\delta C(T_c)/C_n(T_c)$ is not the constant and can be very large provided that $T_c/T_f \ll 1$.

4   The Dispersion and Lineshape of the Single-Particle Spectra

The newly discovered additional energy scale manifests itself as a break in the quasiparticle dispersion near $(50-70)$ meV, which results in a drastic change of the quasiparticle velocity [44, 45, 46]. Such a behavior is qualitatively different from what one could expect in a normal Fermi liquid. Moreover, this behavior can hardly be understood in the frames of either the Marginal Fermi Liquid theory or the quantum protectorate since there are no additional energy scales, or parameters, in these theories [41, 70, 71]. One could suggest that this observed strong self-energy effect, leading to the new energy scale, is due to the electron coupling with collective excitations. But in that case one has to give up the quantum protectorate idea, which would contradict observations [40, 41].

As we have seen in Sections 2 and 3, Eqs. (2.12) and (3.7), the system with FC is characterized by two effective masses: M_{FC}^* that is related to the single particle spectrum at lower energy scale, and M_L^* describing the spectrum at higher energy scale. These two effective masses manifest itself as a break in the quasiparticle dispersion, which can be approximated by two straight lines intersecting at the energy E_0, Eqs. (2.14) and (3.8). This break takes place at temperatures $T \ll T_f$ when the system both in its superconducting state and normal one. This beahvior is in good agreement with the experimental findings [44]. It is pertinent to note that at $T < T_c$, the effective mass M_{FC}^* does not depend on the momenta p_F, p_f and p_i as it follows from Eqs. (3.7) and (3.10),

$$M_{FC}^* \sim \frac{2\pi}{\lambda_0}. \tag{4.1}$$

Thus, it is seen from Eq. (4.1) that M_{FC}^* does not depend on x. This result is in good agreement with experimental facts [72, 73, 74]. The same is true for the dependence of the Fermi velocity $v_F = p_F/M_{FC}^*$ on x, because the Fermi momentum $p_F \sim \sqrt{n}$ only slightly depends on the electron density $n = n_0(1 - x)$ [72, 73]. Here n_0 is the single-particle electron density at the half-filling.

Since λ_0 is the coupling constant defining the pairing interaction, for example phonon-electron interaction, it could be reasonable to expect that the break in the quasiparticle dispersion comes from the phonon-electron interaction. The phonon scenario could explain the persistence of the break at $T > T_c$, since phonons are T-independent. On the other hand, it was shown that the quasiparticle dispersion tends to recover to the conventional one-electron dispersion when the energy is well above the typical phonon energies [75].

The experimental observations do not show that the recovery to the one-electron dispersion takes place [44].

The lineshape function $L(q, \omega)$ of the single-particle spectrum is a function of two variables. Measurements carried out at a fixed binding energy $\omega = \omega_0$, with ω_0 being the energy of a single-particle excitation, determine the lineshape $L(q, \omega = \omega_0)$ as a function of the momentum q. We have shown above that M_{FC}^* is finite and constant at $T \leq T_c$. Therefore, at excitation energies $\omega \leq E_0$, the system behaves like an ordinary superconducting Fermi liquid with the effective mass given by Eq. (3.7) [29, 30, 43]. At $T_c \leq T$ the low energy effective mass M_{FC}^* is finite and is given by Eq. (2.12). Once again, at the energies $\omega < E_0$, the system behaves as a Fermi liquid, the single-particle spectrum is well defined while the width of single-particle excitations is of the order of T [29, 30, 20]. This behavior was observed in experiments measuring the lineshape at a fixed energy [46, 76].

The lineshape can also be determined as a function of ω, $L(q = q_0, \omega)$, at a fixed $q = q_0$. At small ω, the lineshape resembles the considered above, and $L(q = q_0, \omega)$ has the characteristic maximum and width. At energies $\omega \geq E_0$, the contribution coming form quasiparticles with the mass M_L^* become important, leading to the increase of $L(q = q_0, \omega)$. As a result, the function $L(q = q_0, \omega)$ possesses the known peak-dip-hump structure [77] directly defined by the existence of the two effective masses M_{FC}^* and M_L^* [29, 30, 43]. We can conclude that in contrast to the Landau quasiparticles, these quasiparticles have a more complicated lineshape.

To develop deeper quantitative and analytical insight into the problem, we use the Kramers-Krönig transformation to construct the imaginary part $\text{Im}\Sigma(\mathbf{p}, \varepsilon)$ of the single-particle self-energy $\Sigma(\mathbf{p}, \varepsilon)$ starting with the real one $\text{Re}\Sigma(\mathbf{p}, \varepsilon)$, which defines the effective mass [78]

$$\frac{1}{M^*} = \left(\frac{1}{M} + \frac{1}{p_F} \frac{\partial \text{Re}\Sigma}{\partial p} \right) \bigg/ \left(1 - \frac{\partial \text{Re}\Sigma}{\partial \varepsilon} \right). \tag{4.2}$$

Here M is the bare mass, while the relevant momenta p and energies ε obey the following strong inequalities: $|p - p_F|/p_F \ll 1$, and $\varepsilon/\varepsilon_F \ll 1$. We take $\text{Re}\Sigma(\mathbf{p}, \varepsilon)$ in the simplest form which accounts for the change of the effective mass at the energy scale E_0:

$$\text{Re}\,\Sigma(\mathbf{p}, \varepsilon) = -\varepsilon \frac{M_{FC}^*}{M} + \left(\varepsilon - \frac{E_0}{2} \right) \frac{M_{FC}^* - M_L^*}{M} \left[\theta\left(\varepsilon - \frac{E_0}{2} \right) + \theta\left(-\varepsilon - \frac{E_0}{2} \right) \right]. \tag{4.3}$$

Here $\theta(\varepsilon)$ is the step function. Note that in order to ensure a smooth transition from the single-particle spectrum characterized by M_{FC}^* to the spectrum defined by M_L^* the step function is to be substituted by some smooth function. Upon inserting Eq. (4.3) into Eq. (4.2) we can check that inside the interval $(-E_0/2, E_0/2)$ the effective mass can be estimated as $M^* \simeq M_{FC}^*$, and outside the interval it is $M^* \simeq M_L^*$. By applying the Kramers-Krönig transformation to $\text{Re}\Sigma(\mathbf{p}, \varepsilon)$, we obtain the imaginary part of the self-energy [54]

$$\text{Im}\,\Sigma(\mathbf{p}, \varepsilon) \sim \varepsilon^2 \frac{M_{FC}^*}{\varepsilon_F M} + \frac{M_{FC}^* - M_L^*}{M} \left(\varepsilon \ln\left| \frac{\varepsilon + E_0/2}{\varepsilon - E_0/2} \right| + \frac{E_0}{2} \ln\left| \frac{\varepsilon^2 - E_0^2/4}{E_0^2/4} \right| \right). \tag{4.4}$$

We see from Eq. (4.4) that at $\varepsilon/E_0 \ll 1$ the imaginary part is proportional to ε^2, at $2\varepsilon/E_0 \simeq 1$ $\text{Im}\Sigma \sim \varepsilon$, and at $E_0/\varepsilon \ll 1$ the main contribution to the imaginary part is approximately constant. This is the behavior that gives rise to the known peak-dip-hump structure. It is seen from Eq. (4.4) that when $E_0 \to 0$ the second term on the right hand side tends to zero and the single-particle excitations become better defined, resembling the situation in a normal Fermi-liquid, and the peak-dip-hump structure eventually vanishes. On the other hand, the so called renormaliztaion constant, or the quasiparticle renormalization factor, $a(\mathbf{p})$ is given by [78]

$$\frac{1}{a(\mathbf{p})} = 1 - \frac{\partial \operatorname{Re} \Sigma(\mathbf{p}, \varepsilon)}{\partial \varepsilon}. \tag{4.5}$$

At $T \leq T_c$, as seen from Eqs. (4.3) and (4.5), the quasiparticle amplitude at the Fermi surface rises as the energy scale E_0 decreases. It follows from Eqs. (3.8) and (3.18) that $E_0 \sim (x_{FC} - x)/x_{FC}$. At $T > T_c$, it is seen from Eqs. (4.3) and (4.5) that the quasiparticle amplitude rises as the effective mass M^*_{FC} decreases. As seen from Eqs. (2.12) and (2.15), $M^*_{FC} \sim (p_f - p_i)/p_F \sim (x_{FC} - x)/x_{FC}$. As a result, we can conclude that the amplitude rises as the level of doping increases, while the peak-dip-hump structure vanishes and the single-particle excitations become better defined in highly overdoped samples. At $x > x_{FC}$, the energy scale $E_0 = 0$ and the quasiparticles are normal excitations of Landau Fermi liquid. It is worth noting that such a behavior was observed experimentally in highly overdoped Bi2212 where the gap size is about 10 meV [79]. Such a small size of the gap verifies that the region occupied by the FC is small since $E_0/2 \simeq \Delta_1$. Then, recent experimental data have shown that the Landau Fermi liquid does exist in heavily overdoped non-superconducting La$_{1.7}$Sr$_{0.3}$CuO$_4$ [80, 81].

5   Field-Induced LFL in Heavy Electron Liquid with FC

In this Section we consider the behavior of the heavy-electron liquid with FC in magnetic fields, assuming that the coupling constant is nonzero $\lambda_0 \neq 0$ but infinitely small. As we have seen in Section 3, at $T = 0$ the superconducting order parameter $\kappa(\mathbf{p})$ is finite in the FC range, while the maximum value of the superconducting gap $\Delta_1 \propto \lambda_0$ is infinitely small. Therefore, any small magnetic field $B \neq 0$ can be considered as a critical field and will destroy the coherence of $\kappa(\mathbf{p})$ and thus FC itself. To define the type of FC rearrangement, simple energy arguments are sufficient. On one hand, while the field is zero in the sample until then the state with FC is not destroyed. The energy gain ΔE_B due to removing the FC state is $\Delta E_B \propto B^2$ and tends to zero with $B \to 0$. On the other hand, occupying the finite range $(p_f - p_i)$ in the momentum space, the function $n_0(\mathbf{p})$ given by Eq. (2.8), or by Eq. (3.6), leads to a finite gain in the ground state energy [17]. Thus, a new ground state replacing FC should have almost the same energy as the former one. Such a state is given by the multiconnected Fermi spheres resembling an onion, where the smooth quasiparticle distribution function $n_0(\mathbf{p})$ in the $(p_f - p_i)$ range is replaced by a multiconnected distribution $\nu(\mathbf{p})$ [82, 83]

$$\nu(\mathbf{p}) = \sum_{k=1}^{n} \theta(p - p_{2k-1})\theta(p_{2k} - p). \tag{5.1}$$

Here the parameters $p_i \leq p_1 < p_2 < \ldots < p_{2n} \leq p_f$ are adjusted to obey the normalization condition:

$$\int_{p_{2k-1}}^{p_{2k+3}} \nu(\mathbf{p}) \frac{d\mathbf{p}}{(2\pi)^3} = \int_{p_{2k-1}}^{p_{2k+3}} n_0(\mathbf{p}) \frac{d\mathbf{p}}{(2\pi)^3}.$$

The corresponding multiconnected distribution is shown in Fig. 4.

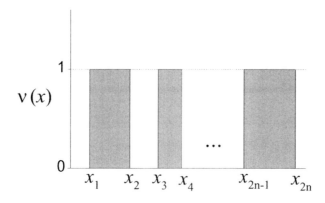

Figure 4: The function $\nu(\mathbf{p})$ for the multiconnected distribution replacing the function $n_0(\mathbf{p})$ in the range $(p_f - p_i)$ occupied by FC so that $p_i < p_F < p_f$.

For definiteness, let us consider the most interesting case of a 3D system, while the consideration of a 2D system also goes along the same line. We note that the idea of multiconnected Fermi spheres, with production of new, interior segments of the Fermi surface, has been considered some time ago [84, 85]. Let us assume that the thickness δp of each interior block is approximately the same $\delta p \simeq p_{2k+1} - p_{2k}$ and δp is defined by B. Using the Simpson's rule, we obtain that the minimum loss in the ground state energy due to formation of the blocks is about $(\delta p)^4$. This result can be understood by considering that the continuous FC function $n_0(\mathbf{p})$ delivers the minimum value to the energy functional $E[n(\mathbf{p})]$, while the approximation of $\nu(\mathbf{p})$ by the steps of size δp produces the minimum error of the order of $(\delta p)^4$. Taking into account that the gain due to the magnetic field is proportional to B^2 and equating both of the contribution we obtain

$$\delta p \propto \sqrt{B}. \tag{5.2}$$

Thus, at $T \to 0$ when $B \to 0$, the thickness δp goes to zero as well, $\delta p \to 0$, while the NFL behavior of the FC state is replaced by the behavior of LFL with the Fermi momentum p_f. It follows from Eq. (3.6) that $p_f > p_F$, while the number density x of itinerant electrons remains constant. As we shall see, this observation plays important role when considering the Hall coefficient $R_H(B)$ as a function of B at low temperatures in the HF metals with FC.

To calculate the effective mass $M^*(B)$ as function of the applied magnetic field B, we observe that at $T = 0$ the application of finite magnetic field B splits the FC state into the Landau levels and suppresses the superconducting order parameter $\kappa(\mathbf{p})$ thus destroying the FC state. Therefore the LFL behavior is expected to be restored [8, 86]. The Landau levels at the Fermi surface can be approximated by a single block whose thickness in momentum space is δp. Approximating the dispersion of quasiparticles within this block by $\varepsilon(p) \sim (p - p_F + \delta p)(p - p_F)/M$, we obtain that the effective mass $M^*(B) \sim M/(\delta p/p_F)$. The energy loss ΔE_{FC} due to rearrangement of the FC state related to this block can be estimated using the Landau formula [27]

$$\Delta E_{FC} = \int (\varepsilon(\mathbf{p}) - \mu) \delta n(\mathbf{p}) \frac{d\mathbf{p}^3}{(2\pi)^3}. \tag{5.3}$$

The region occupied by the variation $\delta n(\mathbf{p})$ has the thickness δp, while $(\varepsilon(\mathbf{p}) - \mu) \sim (p - p_F)p_F/M^*(B)$. As a result, we have $\Delta E_{FC} \sim \delta p^2/M^*(B)$. On the other hand, there is a gain $\Delta E_B \sim (B^2 \mu_B)^2 M^*(B) p_F$ due to the application of the magnetic field and coming from the Zeeman splitting. Equating ΔE_B to ΔE_{FC} and taking into account that in this case $M^*(B) \propto 1/\delta p$, we obtain the following relation

$$\frac{\delta p^2}{M^*(B)} \propto \frac{1}{(M^*(B))^3} \propto B^2 M^*(B). \tag{5.4}$$

It follows from Eq. (5.4) that the effective mass $M^*(B)$ diverges as

$$M^*(B) \propto \frac{1}{\sqrt{B - B_{c0}}}. \tag{5.5}$$

Here, B_{c0} is the critical magnetic field which drives both a HF metal to its magnetic field tuned QCP and the corresponding Néel temperature toward $T = 0$ [8]. We note that in some cases $B_{c0} = 0$, for example, the HF metal CeRu$_2$Si$_2$ shows neither evidence of the magnetic ordering, superconductivity down to the lowest temperatures nor the LFL behavior [87]. Equation (5.5) shows that by applying the magnetic field $B > B_{c0}$ the system can be driven back into LFL with the effective mass $M^*(B)$ depending on the magnetic field. This means that the following dependences are valid: for the coefficient $A(B) \propto (M^*(B))^2$ [88], for the specific heat, $C/T = \gamma_0(B) \propto M^*(B)$, and for the magnetic susceptibility $\chi_0(B) \propto M^*(B)$. The coefficient $A(B)$ determines the temperature dependent part of the resistivity, $\rho(T) = \rho_0 + \Delta\rho$, with ρ_0 being the residual resistivity and $\Delta\rho = A(B)T^2$. Since the coefficient is directly determined by the effective mass, we obtain from Eq. (5.5) that

$$A(B) \propto \frac{1}{B - B_{c0}}. \tag{5.6}$$

It is seen that the well-known empirical Kadowaki-Woods (KW) ratio [9], $K = A/\gamma_0^2 \simeq const$, is fulfilled. At this point, we stress that the value of K may be dependent on the degeneracy number of quasiparticles as it was recently shown. As a result, the grand-KW-relation produces good description of the data for whole the range of degenerate HF systems [89]. In the simplest case when the heavy electron liquid is formed by quasiparticles with spin $1/2$ and the degeneracy number is 2, K turns out to be close to the empirical value [88], called as the KW relation [9]. Therefore, we come to the conclusion that by applying magnetic fields the system is driven back into the LFL state where the constancy of the Kadowaki-Woods ratio is obeyed.

At finite temperatures, the system remains in the LFL state, but there exists a temperature $T^*(B)$, at which the NFL behavior is restored. To calculate the function $T^*(B)$, we observe that the effective mass M^* characterizing the single particle spectrum cannot be changed at $T^*(B)$ since there are no any phase transitions. In other words, at the crossover point, we have to compare the effective mass $M^*(T)$ defined by $T^*(B)$, Eq. (2.12), and that $M^*(B)$ defined by the magnetic field B, Eq. (5.5), $M^*(T) \sim M^*(B)$

$$\frac{1}{M^*(T)} \propto T^*(B) \propto \frac{1}{M^*(B)} \propto \sqrt{B - B_{c0}}. \tag{5.7}$$

As a result, we obtain

$$T^*(B) \propto \sqrt{B - B_{c0}}. \tag{5.8}$$

At temperatures $T \geq T^*(B)$, the system comes back to the NFL behavior with M^* defined by Eq. (2.12), and the LFL behavior disappear. We can conclude that Eq. (5.8) determines the line in the $B - T$ phase diagram that separates the region of the B dependent effective mass from the region of the T dependent effective mass. At the temperature $T^*(B)$, there occurs a crossover from the T^2 dependence of the resistivity to the T dependence. It follows from Eq. (5.8), that the heavy electron liquid at some temperature T can be driven back into the Landau Fermi-liquid by applying a strong enough magnetic field $(B - B_{c0}) \propto (T^*(B))^2$. We can also conclude, that at finite temperature $T < T^*(B)$, the heavy electron liquid shows a more pronounced metallic behavior at the elevated magnetic field B since the effective mass decreases (see Eq. (5.5)). The same behavior of the effective mass can be observed in the Shubnikov — de Haas oscillation measurements. We conclude that one obtains a unique possibility to control the essence of the strongly correlated liquid by magnetic fields which induce the change of the NFL behavior to the LFL liquid behavior.

Let us briefly consider the case when the system is very near FCQPT being on the ordered side and therefore $\delta p_{FC} = (p_f - p_i)/p_F \ll 1$. Since $\delta p \propto M^*(B)$, it follows from Eqs. (5.2) and (5.5) that

$$\frac{\delta p}{p_F} \sim a_c \sqrt{\frac{B - B_{c0}}{B_{c0}}}, \tag{5.9}$$

where a_c is a constant, which is expected to be of the order of a unit, $a_c \sim 1$. At bigger magnetic field B, the value of $\delta p/p_F$ becomes comparable with δp_{FC}, and the distribution function $\nu(\mathbf{p})$ vanishes being replaced by the conventional Zeeman splitting. As a result, we are dealing with the heavy electron liquid located on the disordered side of FCQPT. As we shall see in Section 9, the behavior of this system is quite different from that of the system with FC. It follows from Eq. (5.9) that relatively weak magnetic field B_{cr}

$$(\delta p_{FC})^2 \sim \frac{B_{cr} - B_{c0}}{B_{c0}}, \tag{5.10}$$

removes the system from the ordered side of the phase transition provided that $\delta p_{FC} \ll 1$.

6 Magnetic-Field Induced Landau Fermi Liquid in High-T_c Metals

The LFL theory has revealed that the low-energy elementary excitations of a Fermi liquid look like the spectrum of an ideal Fermi gas. These excitations are described in terms of quasiparticles with an effective mass M^*, charge e and spin $1/2$. The quasiparticles define the major part of the low-temperature properties of Fermi liquids. As we have shown in Section 5, at temperatures $T < T^*(B)$, the LFL behavior of the heavy electron liquid with FC is recovered by the application of magnetic field B larger than the critical field B_c suppressing the superconductivity. Thus, the heavy electron liquid with FC can be viewed as LFL induced by the magnetic field. In such a state, the Wiedemann-Franz (WF) law and the Korringa law are held and the elementary excitations are LFL quasiparticles. Our consideration is valid for relatively weak magnetic fields when the contributions coming from the magnetic field are proportional B^2 and the Zeeman splitting is much smaller then the Fermi momentum.

It was reported recently that in the normal state obtained by applying a magnetic field greater than the upper critical filed B_c, in a hole-doped cuprates at overdoped concentration $(Tl_2Ba_2CuO_{6+\delta})$ [11] and at optimal doping concentration $(Bi_2Sr_2CuO_{6+\delta})$ [90], there are no any sizable violations of the WF law. In the electron-doped copper oxide superconductor $Pr_{0.91}LaCe_{0.09}CuO_{4-y}$ (T_c=24 K) when superconductivity is eliminated by a magnetic field, it was found that the spin-lattice relaxation rate $1/T_1$ follows the $T_1T = constant$ relation, known as the Korringa law [91], down to temperature of $T = 0.2$ K [92]. At higher temperatures and applied magnetic fields of 15.3 T perpendicular to the CuO$_2$ plane, $1/T_1T$ as a function of T is a constant below $T = 55$ K. At 300 K $> T > 50$ K, $1/T_1T$ decreases with growing T [92]. Recent measurements for strongly overdoped non-superconducting $La_{1.7}Sr_{0.3}CuO_4$ have shown that the resistivity ρ exhibits T^2 behavior, and the WF law holds perfectly [80, 81]. Since the validity of the WF and the Korringa laws are the robust signature of LFL, these experimental facts demonstrate that the observed elementary excitations cannot be distinguished from the Landau quasiparticles. This imposes strong constraints for models describing the hole-doped and electron-doped high-temperature superconductors. For example, in the cases of a Luttinger liquid [93], spin-charge separation (see e.g. [94]), and in some solutions of $t - J$ model [95] a violation of the WF law was predicted.

As any phase transition, FCQPT is related to the order parameter, which induces a broken symmetry. It was shown in Section 3 that the relevant order parameter is the superconducting order parameter $\kappa(\mathbf{p})$, which is suppressed by the critical magnetic field B_c, when $B_c^2 \sim \Delta_1^2$. If the coupling constant $\lambda_0 \to 0$, the critical magnetic field $B_c \to 0$ will destroy the state with FC converting the strongly correlated Fermi liquid into LFL. The magnetic field plays the role of the control parameter determining the effective mass $M^*(B)$ as it follows from Eq. (5.5).

If λ_0 is finite, the critical field is also finite, and Eq. (5.5) is valid at $B > B_c$. In that case the system is driven back to LFL and has the LFL behavior induced by the magnetic field. Then, the low energy elementary excitations are characterized by $M^*(B)$ and cannot be distinguished from Landau quasiparticles. As a result, at $T \to 0$, the WF law is held in accordance with experimental facts [11, 90]. On the hand, in contrast to the LFL theory, the effective mass $M^*(B)$ depends on the magnetic field.

Equation (5.5) shows that by applying a magnetic field $B > B_c$ the system can be driven back into LFL with the effective mass $M^*(B)$ which is finite and independent of the temperature. This means that the low temperature properties depend on the effective mass in accordance with the LFL theory. In particular, the resistivity $\rho(T)$ as a function of the temperature behaves as $\rho(T) = \rho_0 + \Delta\rho(T)$ with $\Delta\rho(T) = AT^2$, and the factor $A \propto (M^*(B))^2$. Taking into account that in the case of the high-T_c superconductors B_{c0} is expected to be zero, we obtain from Eq. (5.5) that

$$\gamma_0 \sqrt{B} = const. \tag{6.1}$$

Here $\gamma_0 = C/T$ with C is the specific heat. Taking into account Eqs. (5.6) and Eq. (6.1), we obtain

$$\gamma_0 \sim A(B)\sqrt{B}. \tag{6.2}$$

At finite temperatures, the system remains LFL, but there is the crossover from the LFL behavior to the non-Fermi liquid behavior at temperature $T^*(B) \propto \sqrt{B}$. At $T > T^*(B)$, the effective mass starts to depend on the temperature $M^* \propto 1/T$, and the resistivity possesses the non-Fermi liquid behavior with a substantial linear term, $\Delta\rho(T) \propto T$ [29, 30, 83]. Such a behavior of the resistivity was observed in the cuprate superconductor $Tl_2Ba_2CuO_{6+\delta}$ ($T_c < 15$ K) [96]. At B=10 T, $\Delta\rho(T)$ is a linear function of the temperature between 120 mK and 1.2 K, whereas at $B = 18$ T, the temperature dependence of the resistivity is consistent with $\Delta\rho(T) = AT^2$ in the same temperature range [96].

In LFL, the nuclear spin-lattice relaxation rate $1/T_1$ is determined by the quasiparticles near the Fermi level whose population is proportional to M^*T, so that $1/T_1T \propto M^*$ is a constant [92, 91]. When the superconducting state is removed by the application of a magnetic field, the underlying ground state can be seen as the field induced LFL with effective mass depending on the magnetic field. As a result, the rate $1/T_1$ follows the $T_1T = constant$ relation, that is the Korringa law is held. Unlike the behavior of LFL, as it follows from Eq. (5.5), $1/T_1T \propto M^*(B)$ decreases with increasing the magnetic field at $T < T^*(B)$. At $T > T^*(B)$, we observe that $1/T_1T$ is a decreasing function of the temperature, $1/T_1T \propto M^* \propto 1/T$. These observations are in a good agreement with the experimental facts [92]. Since $T^*(B)$ is an increasing function of the magnetic field, see Eq. (5.8), the Korringa law retains its validity to higher temperatures at elevated magnetic fields. We conclude, that at temperature $T_0 \leq T^*(B_0)$ and bigger magnetic fields $B > B_0$ the system shows a more pronounced metallic behavior, since the effective mass decreases with increasing B (see Eq. (5.5)). Such a behavior of the effective mass can be observed in the de-Haas van Alphen-Shubnikov studies, $1/T_1T$ and the resistivity measurements. These experiments can shed light on the physics of high-T_c metals and reveal relationships between high-T_c metals and heavy-electron metals [97].

The existence of FCQPT can also be revealed experimentally because at densities $x > x_{FC}$, or beyond the FCQPT point, the system should be LFL at sufficiently low temperatures [86]. Recent experimental data have shown that this liquid exists in the heavily overdoped non-superconducting compound $La_{1.7}Sr_{0.3}CuO_4$ [80, 81]. It is remarkable that up to $T = 55$ K the resistivity exhibits the T^2 behavior with no additional linear term, and the WF law is verified to within the experimental resolution [80, 81]. While at elevated temperatures, a strong deviations from the LFL behavior are observed. We anticipate that in this case the system can be again driven back to the LFL behavior by the application of sufficiently strong magnetic fields.

Thus, the mentioned above striking measurements, which were used in studies of the nature of the high-T_c superconductivity, suggest that FCQPT and the emergence of the novel quasiparticles with effective mass strongly depending on the magnetic field and temperature and resembling the Landau quasiparticles are qualities intrinsic to the electronic system of the high-T_c superconductors.

7   Appearance of FCQPT in Different Fermi Liquids

It is widely believed that unusual properties of correlated liquids observed in the high-temperature superconductors, heavy-fermion metals, 2D ^{3}He and etc., are determined by quantum phase transitions. Therefore, immediate experimental studies of relevant quantum phase transitions and of their quantum critical points are of crucial importance for understanding the physics of the high-T_c superconductivity and HF systems. In case of the high-T_c superconductors, these studies are difficult to carry out, because all the corresponding area is occupied by the superconductivity. On the other hand, recent experimental data on different highly correlated Fermi liquids, when the system in question is approaching FCQPT from the disordered side, can help to illuminate both the nature of this point and the control parameter, by which this point is driven. We shall call Fermi liquids approaching FCQPT from the disordered side as highly correlated ones to distinguish them from strongly correlated liquids which have undergone FCQPT. Detailed explanations on this point are given in Section 8.

Experimental facts on high-density 2D ^{3}He [98, 99] show that the effective mass diverges when the density, at which 2D ^{3}He liquid begins to solidify, is approached [99]. Then, a sharp increase of the effective mass in a metallic 2D electron system was observed, when the density tends to the critical density of the metal-insulator transition point, which occurs at sufficiently low densities [34]. Note, that there is no ferromagnetic instability in both Fermi systems and the relevant Landau amplitude $F_0^a > -1$ [34, 99], in accordance with the almost localized fermion model [100].

Now let us consider the divergence of the effective mass in 2D and 3D highly correlated Fermi liquids at $T = 0$, when the density x approaches FCQPT from the side of normal Landau Fermi liquid, that is from the disordered phase. First, we calculate the divergence of M^* as a function of the difference $(x - x_{FC})$ in case of 2D Fermi liquid. For this purpose we use the equation for M^* obtained in [38], where the divergence of the effective mass M^* due to the onset of FC in different Fermi liquids (such as 2D and 3D electron and ^{3}He liquids) was predicted. At $x \to x_{FC}$, the effective mass M^* can be approximated as

$$\frac{1}{M^*} \simeq \frac{1}{M} + \frac{1}{4\pi^2} \int\limits_{-1}^{1} \int\limits_{0}^{g_0} \frac{v(q(y))}{[1 - R(q(y), \omega = 0, g)\chi_0(q(y), \omega = 0)]^2} \frac{y\,dy\,dg}{\sqrt{1 - y^2}}. \quad (7.1)$$

Here we adopt the notation $p_F\sqrt{2(1 - y)} = q(y)$ with $q(y)$ being the transferred momentum, M is the bare mass, ω is the frequency, $v(q)$ is the bare interaction, and the integral is taken over the coupling constant g from zero to its real value g_0. In Eq. (7.1), both $\chi_0(q, \omega)$ and $R(q, \omega)$ are the linear response function of a noninteracting Fermi liquid and the effective interaction respectively. They define the linear response function of the system in question

$$\chi(q, \omega, g) = \frac{\chi_0(q, \omega)}{1 - R(q, \omega, g)\chi_0(q, \omega)}. \quad (7.2)$$

In the vicinity of the charge density wave instability, occurring at the density x_{cdw}, the singular part of the function χ^{-1} on the disordered side is of the well-known form (see e.g. [2])

$$\chi^{-1}(q, \omega, g) \simeq a(x_{cdw} - x) + b(q - q_c)^2 + c(g_0 - g), \qquad (7.3)$$

where a, b and c are constants and $q_c \simeq 2p_F$ is the wavenumber of the charge density wave order. Upon substituting Eq. (7.3) into Eq. (7.1) and integrating, the equation for the effective mass M^* can be presented in the following form [31, 32]

$$\frac{1}{M^*} = \frac{1}{M} - \frac{C}{\sqrt{x - x_{cdw}}}, \qquad (7.4)$$

with C being some positive constant. The behavior of the effective mass as a function of the electron number density x in a silicon MOSFET is shown in Fig. 5. The fitting parameters are $x_{cdw} = 0.7 \times 10^{-11} \text{cm}^{-2}$, $C = 2.14 \times 10^{-6} \text{ cm}^{-1}$ and $x_{FC} = 0.9 \times 10^{-11} \text{cm}^{-2}$ [32]. It is seen from Fig. 5 that Eq. (7.4) describes rather good the data.

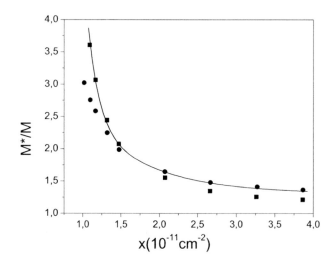

Figure 5: The ratio M^*/M in a silicon MOSFET versus the carrier number density x. The filled squares denote the Shubnikov – de Haas oscillations experimental data, the data obtained by the application of a parallel magnetic field are shown by the filled circles [34, 101].

It is seen from Eq. (7.4) that M^* diverges at some point x_{FC} referred to as the critical point, at which FCQPT occurs, as a function of the difference $(x - x_{FC})$.[31, 32]

$$\frac{M^*}{M} \simeq A + \frac{B}{x - x_{FC}}, \qquad (7.5)$$

where A and B are constants. It follows from the derivation of Eqs. (7.4) and (7.5) that their forms are independent of the bare interaction $v(q)$, which effects however A, B and x_{FC}

values. This result is in agreement with Eq. (2.7) which exhibits the same type of divergence independent of the interaction. Therefore both of these equations are also applicable to 2D ^{3}He liquid or to another Fermi liquid. It is also seen from Eqs. (7.4) and (7.5) that FCQPT precedes the formation of charge-density waves. As a consequence of this, the effective mass diverges at high densities in case of 2D ^{3}He, and at low densities in case of 2D electron systems, in accordance with experimental facts [34, 99]. Note, that in both cases the difference $(x - x_{FC})$ has to be positive, because x approaches x_{FC} when the system is on the disordered side of FCQPT with the effective mass $M^*(x) > 0$. Thus, in considering the 2D ^{3}He liquid we have to replace $(x - x_{FC})$ by $(x_{FC} - x)$ on the right hand side of Eq. (7.5). In case of a 3D system, at $x \to x_{FC}$, the effective mass is given by [38]

$$\frac{1}{M^*} \simeq \frac{1}{M} + \frac{p_F}{4\pi^2} \int_{-1}^{1} \int_{0}^{g_0} \frac{v(q(y))ydydg}{[1 - R(q(y), \omega = 0, g)\chi_0(q(y), \omega = 0)]^2}. \qquad (7.6)$$

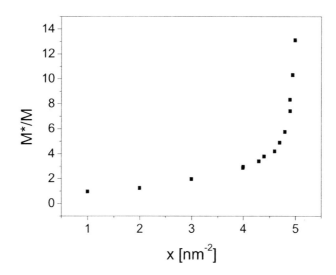

Figure 6: The ratio M^*/M in a 2D ^{3}He versus the fluid number density x inferred from the heat capacity measurements and the magnetization measurements [99]. The solid line represents $M^*/M = A + \frac{B}{(x_{FC}-x)}$, with A=1.09, B=1.68 nm^{-2} and $x_{FC} = 5.11$ nm^{-2}.

The comparison of Eq. (7.1) and Eq. (7.6) shows that there is no fundamental difference between these equations, and along the same way we again arrive at Eqs. (7.4) and (7.5). The only difference between 2D electron systems and 3D ones is that in the latter FCQPT occurs at densities which are well below those corresponding to 2D systems. For bulk ^{3}He, FCQPT cannot probably take place since it is absorbed by the first order solidification [99]. The apparent divergence of the effective mass $M^*(x)$ obtained in measurements on 2D ^{3}He [99] is shown in Fig. 6. This is in good agreement with the divergence given by Eqs. (7.5) and (2.7).

8   Heavy Fermion Metals with Highly Correlated Electron Liquid

In HF metals with strong electron correlations, quantum phase transitions at zero temperature may strongly influence the measurable quantities up to relatively high temperatures. These quantum phase transitions have recently attracted much attention because the behavior of HF metals is expected to follow universal patterns defined by the quantum mechanical nature of the fluctuations taking place at quantum critical points (see e.g. [3, 4]). Only recently, there appeared experimental facts which deliver experimental grounds to understand the nature of quantum phase transition producing the universal behavior of HF metals. It has been demonstrated that at low temperatures the main properties of HF metals such as the magnetoresistance, resistivity, specific heat, magnetization, susceptibility, volume thermal expansion, etc, strongly depend on temperature T and applied magnetic field B. As a result, these properties can be controlled by placing these metals at the special point of the field-temperature $B - T$ phase diagram. In the LFL theory, considered as the main instrument when investigating quantum many electron physics, the effective mass M^* of quasiparticle excitations determining the thermodynamic properties of electronic systems is practically independent of temperature and applied magnetic fields. Therefore, the observed anomalous behavior is uncommon and can hardly be understood within the framework of the conventional LFL theory based on the notion of quasiparticles. As a result, it is necessary to use theories that are based on the Landau concept of the order parameter which is introduced to classify phases of the state of matter. These theories connect the anomalous behavior with critical fluctuations of the magnetic order parameter. These fluctuations suppressing the quasiparticles are attributed to CQPT taking place when the system in question approaches its QCP. As it was noted in Introduction, the universal behavior is only observable if the electron system of HF metal is very near QCP, for example, when the correlation length is much larger than the microscopic length scales. In the case of CQPT, the physics is dominated by thermal and quantum fluctuations of the critical state, which is characterized by the absence of quasiparticles. It is believed that the absence of quasiparticle-like excitations is the main cause of the NFL behavior and other types of the critical behavior in the quantum critical region. However, theories based on CQPT fail to explain the experimental observations related to the divergence of the effective mass M^* at the magnetic field tuned QCP, the specific behavior of the spin susceptibility and its scaling properties, the thermal expansion behavior, etc, see e.g. [10, 6, 105, 87, 13, 102, 103, 104, 14, 108, 109, 106, 107, 110, 111].

The LFL theory rests on the notion of quasiparticles which represent elementary excitations of a Fermi liquid. Therefore these are appropriate excitations to describe the low temperature thermodynamic properties. The inability of the LFL theory to explain the experimental observations which point to the dependence of M^* upon the temperature T and applied magnetic field B may lead to the conclusion that the quasiparticles do not survive near QCP, and one might be further led to the conclusion that the heavy electron does not retain its integrity as a quasiparticle excitation (see e.g. [102, 106, 107, 112, 113]).

The mentioned above inability to explain the behavior of HF metals at QCP within the framework of theories based on CQPT may also lead to the conclusion that the other important Landau concept of the order parameter fails as well, see e.g. [106, 107, 112, 113]. Thus, we are left without the most fundamental principles of many body quantum physics while a great deal of interesting NFL phenomena related to the anomalous behavior and the experimental facts collected in measurements on the HF metals remain out of reasonable theoretical explanations.

On the other hand, it is the very nature of HF metals that suggests that their unusual properties are defined by a quantum phase transition related to the unlimited growth of the effective mass at its QCP. Moreover, a divergence to infinity of the effective electron mass was observed at a magnetic field-induced QCP [10, 105, 102]. In Section 2, we have demonstrated that such a quantum phase transition is to be FCQPT, an essential feature of which is the divergence of the effective mass M^* at its QCP, see Eq. (2.7).

8.1 Highly Correlated Heavy Electron Liquid

When a Fermi system approaches FCQPT from the disordered phase it remains the Landau Fermi liquid with the effective mass M^* strongly depending on the distance $r = (x - x_{FC})$, temperature T and a magnetic field B. This state of the system, with M^* essentially depending on T, r and B, resembles the strongly correlated liquid described in Section 2. However, in contrast to the strongly correlated liquid, there is no energy scale E_0 given by Eq. (2.14) and the system under consideration is the Landau Fermi liquid at sufficiently low temperatures with the effective mass $M^* \propto 1/r$ (see Eqs. (2.7) and (7.5)). Therefore this liquid can be called the *highly correlated liquid* which is obviously to have uncommon properties [26, 31]. Again, we use the heavy electron liquid model to study the universal behavior of the HF metals at low temperatures. As it was mentioned in Section 2, it is possible, since we consider processes related to the power-low divergence of the effective mass. This divergence is determined by small momenta transferred as compared to momenta of the order of the reciprocal lattice cell, and the contribution coming due to the lattice structure can be ignored. Thus, we may usefully ignore the complications related to the lattice and get rid of the specific peculiarities of a HF metal regarding the medium as homogeneous heavy electron isotropic liquid.

The effective mass M^* of quasiparticle excitations controlling the density of states determines the thermodynamic properties of electronic systems. To study the behavior of the effective mass $M^*(T, B)$ as a function of temperature and magnetic field, we use the Landau equation determining $M^*(T, B)$. In the case of a homogeneous liquid, at finite temperatures and low magnetic fields, this equations reads [27]

$$\frac{1}{M^*(T, B)} = \frac{1}{M} + \sum_{\sigma_1} \int \frac{\mathbf{p}_F \mathbf{p}_1}{p_F^3} F_{\sigma,\sigma_1}(\mathbf{p}_F, \mathbf{p}_1) \frac{\partial n_{\sigma_1}(\mathbf{p}_1, T, B)}{\partial p_1} \frac{d\mathbf{p}_1}{(2\pi)^3}. \quad (8.1)$$

Here $F_{\sigma,\sigma_1}(\mathbf{p}_F, \mathbf{p}_1)$ is the Landau amplitude depending on the momenta p and spins σ, p_F is the Fermi momentum, M is the bare mass of an electron and $n_\sigma(\mathbf{p}, T)$ is the quasiparticle distribution function. Since HF metals are predominantly three dimensional (3D) structures we treat the homogeneous heavy electron liquid as a 3D liquid also. For the sake of simplicity, we omit the spin dependence of the effective mass since in the case of a homogeneous liquid and weak magnetic fields, $M^*(T, B)$ does not noticeably depend on the spins. The quasiparticle distribution function is of the form

$$n_\sigma(\mathbf{p}, T) = \left\{ 1 + \exp \left[\frac{(\varepsilon(\mathbf{p}, T) - \mu_\sigma)}{T} \right] \right\}^{-1}, \quad (8.2)$$

where $\varepsilon(\mathbf{p}, T)$ is determined by Eq. (2.2). In our case, the single-particle spectrum does not noticeably depend on the spin, while the chemical potential may have a dependence due to the Zeeman splitting. We will show explicitly the spin dependence of a physical value when this dependence is of importance for understanding.

Replacing $n_\sigma(\mathbf{p}, T, B)$ by $n_\sigma(\mathbf{p}, T, B) \equiv \delta n_\sigma(\mathbf{p}, T, B) + n_\sigma(\mathbf{p}, T = 0, B = 0)$ where $\delta n_\sigma(\mathbf{p}, T, B) = n_\sigma(\mathbf{p}, T, B) - n_\sigma(\mathbf{p}, T = 0, B = 0)$, Eq. (1) takes the form

$$\frac{M}{M^*(T, B)} = \frac{M}{M^*(x)} + \frac{M}{p_F^2} \sum_{\sigma_1} \int \frac{\mathbf{p}_F \mathbf{p}_1}{p_F} F_{\sigma, \sigma_1}(\mathbf{p}_F, \mathbf{p}_1) \frac{\partial \delta n_{\sigma_1}(\mathbf{p}_1, T, B)}{\partial p_1} \frac{d\mathbf{p}_1}{(2\pi)^3}. \quad (8.3)$$

We assume that the heavy electron liquid is near FCQPT, therefore the distance r is small so that $M/M^*(x) \ll 1$, as it is seen from Eq. (2.7). In the case of normal metals with the effective mass of the order of a few bare electron masses, $M/M^*(x) \sim 1$, and up to temperatures $T \sim 100$ K, the second term on the right hand side of Eq. (8.3) is of the order of T^2/μ^2 and is much smaller than the first term. Thus, the system in question demonstrates the LFL behavior with the effective mass being practically independent of temperature, that is the corrections are proportional to T^2. One can check that the same is true when magnetic field up to $B \sim 30$ T is applied. Near the critical point x_{FC}, when $M/M^*(x \to x_{FC}) \to 0$, the behavior of the effective mass changes drastically, because the first term on the right hand side of Eq. (8.3) vanishes and the second term determines the effective mass itself rather than small corrections to $M^*(x)$ related to T and B. In that case, Eq. (8.3) becomes a homogeneous equation and determines the effective mass as a function of B and T. As we will see, Eq. (8.3) describes both the NFL behavior and the LFL one with the presence of quasiparticles. In contrast to the conventional Landau quasiparticles these are characterized by the effective mass that strongly depends on both the magnetic field and the temperature.

Let us turn to a qualitative analysis of solutions of Eq. (8.3) when $x \simeq x_{FC}$. We start with the case when $T = 0$ and B is finite. The application of magnetic field leads to the Zeeman splitting of the Fermi surface and the difference δp between the Fermi surfaces with "spin up" and "spin down" becomes $\delta p = p_F^\uparrow - p_F^\downarrow \sim \mu_B B M^*(B)/p_F$ with μ_B being the Bohr magneton. Upon taking this into account, we observe that the second term in Eq. (8.3) is proportional to $(\delta p)^2 \propto (\mu_B B M^*(B)/p_F)^2$, and Eq. (8.3) takes the form [8, 86, 114]

$$\frac{M}{M^*(B)} = \frac{M}{M^*(x)} + c \frac{(\mu_B B M^*(B))^2}{p_F^4}, \quad (8.4)$$

where c is a constant. Note that the effective mass $M^*(B)$ depends on x as well and this dependence disappears at $x = x_{FC}$. At the point $x = x_{FC}$, the term $M/M^*(x)$ vanishes, Eq. (8.4) becomes homogeneous and can be solved analytically [31, 86, 114]

$$M^*(B) \propto \frac{1}{(B - B_{c0})^{2/3}}. \quad (8.5)$$

Here B_{c0} is the critical magnetic field which drives both a HF metal to its magnetic field tuned QCP and the corresponding Néel temperature toward $T = 0$ [8]. We recall that in some cases $B_{c0} = 0$, for example, the HF metal CeRu$_2$Si$_2$ shows neither evidence of the magnetic ordering, superconductivity down to the lowest temperatures nor the LFL behavior [87].

Equation (8.5) shows the universal power low behavior of the effective mass which does not depend on the inter-particle interaction. We illustrate this behavior by calculations using a model functional [19, 18]

$$E[n(p)] = \int \frac{\mathbf{p}^2}{2M} \frac{d\mathbf{p}}{(2\pi)^3} + \frac{1}{2} \int V(\mathbf{p}_1 - \mathbf{p}_2) n(\mathbf{p}_1) n(\mathbf{p}_2) \frac{d\mathbf{p}_1 d\mathbf{p}_2}{(2\pi)^6}, \qquad (8.6)$$

with the inter-particle interaction

$$V(\mathbf{p}) = g_0 \frac{\exp(-\beta_0 |\mathbf{p}|)}{|\mathbf{p}|}. \qquad (8.7)$$

We normalized the effective mass by M, $M^* = M^*(B)/M$, temperature T_0 and magnetic field H by the Fermi energy ε_F^0, $T = T_0/\varepsilon_F^0$, $H = (\mu_B B)/\varepsilon_F^0$, and use the dimensionless coupling constant $g = (g_0 M)/(2/pi^2)$ and $\beta = \beta_0 p_F$. FCQPT takes place when the parameters reach their critical values, $\beta = b_c$ and $g = g_c$. On the other hand, FCQPT takes place when effective mass $M^* \to \infty$. This condition allows to relate b_c and g_c [19, 18]

$$\frac{g_c}{b_c^3}(1 + b_c) \exp(-b_c)[b_c \cosh(b_c) - \sinh(b_c)] = 1.$$

It follows from this equation that the critical point of FCQPT can be reached by changing g_0 if β_0 and p_F are fixed, or changing p_F if β_0 and g_0 are fixed, etc. For simplicity, we shall change g to reach FCQPT and investigate the properties of the system behind the critical point.

In Fig. 7 we present our calculations (triangles and squares) of the magnetic-field dependence of the effective mass at the critical point of FCQPT. At $\beta = b_c = 3$, FCQPT takes place when $g = g_c = 6.7167$. It is seen from Fig. 7 that the calculated power-low divergence of the effective mass is in accordance with Eq. (8.5).

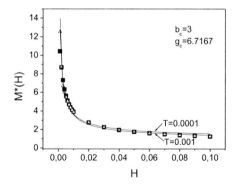

Figure 7: Calculated magnetic-field dependent effective mass M^* at fixed temperatures. Arrows indicate the temperatures. The vertical axis represents the normalized effective mass M^*. The horizontal axis is the normalized magnetic field H. Details of normalization are explained in the text. Solid lines represents $M^*(H) \propto H^{-2/3}$.

At densities $x > x_{FC}$, $M^*(x)$ is finite and we are dealing with the conventional Landau quasiparticles provided that the magnetic field is weak, so that $M^*(x)/M^*(B) \ll 1$ with $M^*(B)$ given by Eq. (8.5). In that case, the second term on the right hand side of Eq. (8.4) is proportional to $(BM^*(x))^2$ and represents small corrections. In the opposite case, when $M^*(x)/M^*(B) \gg 1$, the heavy electron liquid behaves as at QCP. Since in the LFL regime the main thermodynamic properties of the system is determined by the effective mass, it follows from Eq. (8.5) that we obtain a unique possibility to control the magnetoresistance, resistivity, specific heat, magnetization, volume thermal expansion, etc. At this point, we note that the large effective mass leads to the high density of states provoking a large number of states and phase transitions to emerge and compete with one another. Here we assume that these can be suppressed by the application of a magnetic field and concentrate on the thermodynamical properties.

To consider the qualitative behavior of $M^*(T)$ at elevated temperatures, we simplify Eq. (8.3) by omitting the variable B and simulating the influence of the applied magnetic field by the finite effective mass entering the denominator of the first term on the right hand side of Eq. (8.3). This effective mass becomes a function of the distance r, $M^*(r)$, which is determined by both B and $(x - x_{FC})$. If the magnetic field vanishes the distance is $r = (x - x_{FC})$. We integrate the second term over the angle variable, then over p_1 by parts and substitute the variable p_1 by z, $z = (\varepsilon(p_1) - \mu)/T$. In the case of the flat and narrow band, we use the approximation $(\varepsilon(p_1) - \mu) \simeq p_F(p_1 - p_F)/M^*(T)$ and finally obtain

$$\frac{M}{M^*(T)} = \frac{M}{M^*(r)} + \alpha \int_0^\infty F(p_F, p_F(1+\alpha z)) \frac{1}{1+e^z} dz - \alpha \int_0^{1/\alpha} F(p_F, p_F(1-\alpha z)) \frac{1}{1+e^z} dz. \tag{8.8}$$

Here the notations are used: $F \sim M d(F^1 p^2)/dp$, the factor $\alpha = TM^*(T)/p_F^2 = TM^*(T)/(T_k M^*(r))$, $T_k = p_F^2/M^*(r)$ The Fermi momentum is defined as $\varepsilon(p_F) = \mu$. We first assume that $\alpha \ll 1$. Then omitting terms of the order of $\exp(-1/\alpha)$, we expand the upper limit of the second integral on the right hand side of Eq. (8.8) to ∞ and observe that the sum of the second and third terms represents an even function of α. These are the typical expressions with Fermi-Dirac functions as integrands. They can be calculated using standard procedures (see e.g. [115]). Since we need only an estimation of the integrals, we represent Eq. (8.8) as

$$\frac{M}{M^*(T)} \simeq \frac{M}{M^*(r)} + a_1 \left(\frac{TM^*(T)}{T_k M^*(r)}\right)^2 + a_2 \left(\frac{TM^*(T)}{T_k M^*(r)}\right)^4 + ... \tag{8.9}$$

Here a_1 and a_2 are constants of the order of units. Equation (8.9) can be considered as a typical equation of the LFL theory with the only exception being the effective mass $M^*(r)$ which strongly depends on the distance $r = x - x_{FC} \geq 0$ and diverges at $r \to 0$. Nonetheless, it follows from Eq. (8.9) that when $T \to 0$ the corrections to $M^*(r)$ start with the T^2 terms provided that

$$\frac{M}{M^*(r)} \gg \left(\frac{TM^*(T)}{T_k M^*(r)}\right)^2 \simeq \frac{T^2}{T_k^2}, \tag{8.10}$$

and the system exhibits the LFL behavior. It is seen from Eq. (10) that when $r \to 0$, $M^*(r) \to \infty$, and the LFL behavior disappears. The free term on the right hand side of Eq.

(8.8) vanishes, $M/M^*(r) \to 0$, and Eq. (8.8) in itself becoming homogeneous determines the value and universal behavior of the effective mass.

At some temperature $T_1 \ll T_k$, the value of the sum on the right hand side of Eq. (8.9) is determined by the second term. Then Eq. (8.10) is not valid, and upon keeping only the second term in Eq. (8.9) this can be used to determine $M^*(T)$ in a transition region [114, 116]

$$M^*(T) \propto \frac{1}{T^{2/3}}. \tag{8.11}$$

The variation as $T^{-2/3}$ exponent with the temperature growth deserves a comment. Equation (8.11) is valid if the second term in Eq. (8.9) is much larger than the first one, that is

$$\frac{T^2}{T_k^2} \gg \frac{M}{M^*(r)}, \tag{8.12}$$

and this term is grater than the third one,

$$\frac{T}{T_k} \ll \frac{M^*(r)}{M^*(T)} \simeq 1. \tag{8.13}$$

Obviously, both Eqs. (8.12) and (8.13) can be simultaneously satisfied if $M/M^*(r) \ll 1$ and T is finite. The range of temperatures, over which Eq. (8.11) is valid, shrinks to zero as soon as $r \to 0$ because $T_k \to 0$. Thus, if the system is very near QCP, $x \to x_{FC}$, it is possible to observe the behavior of the effective mass given by Eq. (8.11) in a wide range of temperatures provided that the effective mass $M^*(r)$ is diminished by the application of high magnetic fields, that is, the distance r becomes larger under the action of B. When r is finite the $T^{-2/3}$ behavior can be observed at relatively high temperatures. To estimate the transition temperature $T_1(B)$, we observe that the effective mass is a continuous function of the temperature, thus $M^*(B) \sim M^*(T_1)$. Taking into account Eqs. (8.7) and (8.11), we obtain $T_1(B) \propto B$.

Then, at elevated temperatures, the system enters into a different regime. The coefficient α becomes $\alpha \sim 1$, the upper limit of the second integral in Eq. (8.8) cannot be expanded to ∞, and odd terms come into play. As a result, Eq. (8.9) is no longer valid, but the sum of both the first integral and the second one on the right hand side of Eq. (8.8) is proportional to $M^*(T)T$. Upon omitting the first term $M/M^*(r)$ and approximating the sum of the integrals by $M^*(T)T$, we solve Eq. (8.8) and obtain

$$M^*(T) \propto \frac{1}{\sqrt{T}}. \tag{8.14}$$

We illustrate the above consideration by numerical calculations (shown by filled squares in Fig. 8) based on the model functional (8.6). In Fig. 8, we show the evolution of the low temperature entropy from the transition region $T \sim T_1(B)$ to $S(T) \propto \sqrt{T}$ at higher temperatures upon applying magnetic field $H = 0.01$. The calculated behavior of $S(T)/T \propto M^*(T)$ is in accord with Eq. (8.14). Details of the normalization T and H are given at Fig. 7.

Thus, we can conclude that at higher temperatures when $x \simeq x_{FC}$ the system exhibits three types of regimes: the LFL behavior at $\alpha \ll 1$, when Eq. (8.10) is valid; the $M^*(T) \propto$

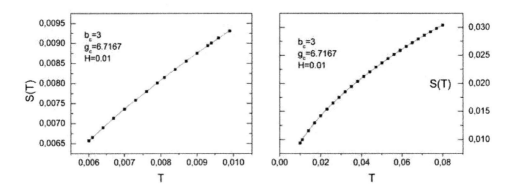

Figure 8: Calculated entropy $S(T)$ as a function of temperature at fixed magnetic field $H = 0.01$. Solid line represents $S(T)$ at the transition region $T = T_1(B)$ (left panel). At low temperatures $T < 0.007$, the system exhibits the LFL behavior, $S(T) \propto T$. The other solid (right panel) line represents $S(T) \propto T^{1/2}$.

$T^{-2/3}$ behavior and $S(T) \propto M^*(T)T \propto T^{1/3}$, when Eqs. (8.12) and (8.13) are valid; and the $1/\sqrt{T}$ behavior of the effective mass at $\alpha \sim 1$, while the entropy $S(T) \propto M^*(T)T \propto \sqrt{T}$, and the specific heat $C(T) = T(\partial S(T)/\partial T) \propto \sqrt{T}$.

   In the absence of magnetic field, calculated evolutions of $M^*(T)$, $S(T)$ and $C(T)$ based on the model functional (8.6) are shown by filled squares in Fig. 9, Fig. 10, and Fig. 11 respectively.

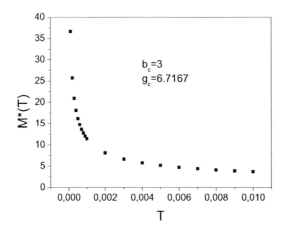

Figure 9: Calculated effective mass $M^*(T)$ as a function of temperature. Solid line represents $M^*(T) \propto 1/\sqrt{T}$.

Let us estimate the quasiparticles width $\gamma(T)$. Within the framework of the LFL theory

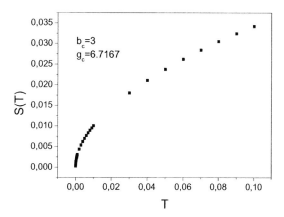

Figure 10: Calculated entropy $S(T)$ as a function of temperature. Solid line represents $S(T) \propto \sqrt{T}$.

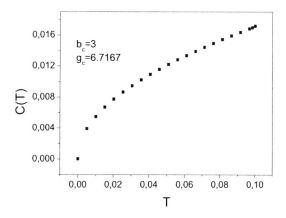

Figure 11: Calculated specific heat $C(T)$ as a function of temperature. Solid line represents $C(T) \propto \sqrt{T}$.

it is given by [27]

$$\gamma \sim |\Gamma|^2 (M^*)^3 T^2, \qquad (8.15)$$

where Γ is the particle-hole amplitude. In the case of a strongly correlated system with its large density of states related to the huge value of the effective mass, the amplitude Γ cannot be approximated by the bare particle interaction but can be estimated within the ladder approximation which gives $|\Gamma| \sim 1/(p_F M^*(T))$ [20]. As a result, we obtain that in the LFL regime $\gamma(T) \propto T^2$, in the $T^{-2/3}$ regime $\gamma(T) \propto T^{4/3}$, and in the $1/\sqrt{T}$ regime $\gamma(T) \propto T^{3/2}$. We observe that in all the cases the width is small as compared to the quasiparticle characteristic energy which is of the order of T, so the notion of a quasiparticle is meaningful. We can conclude that when the heavy electron liquid is near the QCP of FCQPT, being on the disordered side, its low energy excitations are quasiparticle excitations with the effective mass $M^*(B,T)$. At this point we note that at $x \to x_{FC}$, the quasiparticle renormalization factor $a(\mathbf{p})$ remains finite and approximately constant, and the divergence of the effective mass given by Eq. (2.7) is not related to vanishing $a(\mathbf{p})$ [117]. Thus the notion of the quasiparticles is preserved and these are the relevant excitations when considering the thermodynamical properties of the heavy electron liquid.

8.2 Resistivity of Heavy Fermion Metals

Since the resistivity, $\rho(T) = \rho_0 + \Delta\rho(T)$, is directly determined by the effective mass, because the coefficient $A(B,T) \propto (M^*(B,T))^2$ [88], the above mentioned temperature dependences can be observed in measurements of the resistivity of HF metals.

At $T \ll T_1$, the system in question demonstrates the LFL regime, the divergence of the effective mass at $x \to x_{FC}$ is described by Eq. (8.5) and the coefficient $A(B)$ diverges as

$$A(B) \propto \frac{1}{(B - B_{c0})^{4/3}}. \tag{8.16}$$

Thus, the resistivity, $\rho(B) \propto A(B)T^2$, as a function of magnetic field diverges as given by Eq. (8.16). As a function of temperature, the resistivity behaves as $\Delta\rho_1 = c_1 T^2/(B - B_{c0})^{4/3} \propto T^2$. The second is NFL regime which is determined by Eq. (8.11) and characterized by $\Delta\rho_2 = c_2 T^2/(T^{2/3})^2 \propto T^{2/3}$. At $T > T_1(B)$, the third NFL regime is given by Eq. (8.14) and represented by $\Delta\rho_3 = c_3 T^2/(\sqrt{T})^2 \propto T$. Here c_1, c_2, c_3 are constants. It is remarkable that all temperature dependences corresponding to these regimes were observed in measurements on the HF metals $CeCoIn_5$ and YbAgGe [104, 14, 105, 109]. If we consider the ratio $\Delta\rho_2/\Delta\rho_1 \propto ((B - B_{c0})/T)^{4/3}$, we come to a very interesting conclusion that the ratio is a function of only the variable $(B - B_{c0})/T$, thus representing the scaling behavior. This result is in excellent agreement with experimental facts [105].

8.3 Magnetic Susceptibility

The magnetic susceptibility is proportional to the effective mass, $\chi \propto M^*$, with M^* given by Eq. (8.5). Therefore, at $T \ll T_1$,

$$\chi(B) \propto M^*(B) \propto (B - B_{c0})^{-2/3}, \tag{8.17}$$

while the static magnetization $M_B(B)$ is given by

$$M_B(B) \propto BM^*(B) \propto (B - B_{c0})^{1/3}. \tag{8.18}$$

At $T \gg T_1$, as it follows from Eq. (8.14), Eq. (8.17) has to be rewritten as

$$\chi(T) \propto M^*(T) \propto \frac{1}{\sqrt{T}}. \tag{8.19}$$

The observed behavior of $\chi(B)$ and $M_B(B)$ and the behavior of $\chi(T)$ are in accord with results of measurements on $CeRu_2Si_2$ with the critical field $B_{c0} \to 0$ [87].

Consider the state of the system when $r \to 0$. Its properties are determined by magnetic fields B and temperature T because there are no other parameters to describe such a state. At the transition temperatures $T \simeq T_1(B)$, the effective mass depends on both T and B, while at $T \ll T_1(B)$, the system is LFL with the effective mass being given by Eq. (8.5), and at $T \geq T_1(B)$, the mass is defined by Eq. (8.11). Instead of solving Eq. (8.3), it is possible to construct a simple interpolation formula to describe the behavior of the effective mass over all regions,

$$M^*(B,T) = \frac{1}{c_1(B - B_{c0})^{2/3} + c_2 f(y)T^{2/3}}. \tag{8.20}$$

Here, $f(y)$ is a universal monotonic function of $y = (T/(B - B_{c0}))^{2/3}$ such that $f(y \sim 1) = 1$, and $f(y \ll 1) = 0$. It is seen from Eq. (8.20) that the behavior of the effective mass can be represented by a universal function of only one variable y if the temperature is measured in the units of the transition temperature $T_1(B)$, and the effective mass is measured in the units of $M^*(B)$ given by Eq. (8.5). This representation describes the scaling behavior of the effective mass. As seen from Eqs. (8.18) and (8.20), the scaling behavior of the magnetization can be represented in the same way, provided the magnetization is normalized by the saturated value at each field given by Eq. (8.5)

$$\frac{M_B(B,T)}{M_B(B)} \propto \frac{1}{1 + c_3 f(y)y}, \tag{8.21}$$

where c_3 is a constant. It is seen from Eq. (8.21), that magnetization is a monotonic function of y. Upon using the definition of susceptibility, $\chi = \partial M_B/\partial B$, we come to the conclusion that the susceptibility also exhibits the scaling behavior and can be presented as a universal function of only one variable y, provided it is normalized by the saturated value at each field given by Eq. (8.17)

$$\frac{\chi(B,T)}{\chi(B)} \propto \frac{1}{1 + c_3 f(y)y} + 2c_3 y\frac{f(y) + ydf(y)/dy}{(1 + c_3 f(y)y)^2}. \tag{8.22}$$

It is of importance to note that the susceptibility is not a monotonic function of y because the derivative is the sum of two contributions with different behavior. The second contribution on the right hand side of Eq. (8.22) gives the susceptibility a maximum [8, 116, 114]. As shown in Fig. 12, the above behaviors of the magnetization and susceptibility are in accord with numerical calculations of the susceptibility and magnetization [114] and with the facts observed in measurements on $CeRu_2Si_2$ [87].

It is seen from Fig. 12 that for finite B, the curve describing $\chi(B,T)$ acquires a maximum at some temperature T_P. This behavior is in good agreement with experimental facts [87] and associated with the suppression of the divergent NFL terms $\sim T^{-2/3}$ in $\chi(B,T)$ and recovery of the LFL behavior at static magnetic fields in which the Zeeman energy splitting $\mu_B B$ exceeds T [114]. Note that magnetization $M_B(B,T)$ does not exhibits a maximum. In Fig. 12, the temperature is normalized by T_P, the susceptibility is normalized by its peak height $\chi(B, T_P)$ and the magnetization by the saturated value at each field.

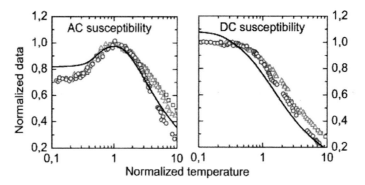

Figure 12: Normalized magnetic susceptibility $\chi(B,T)/\chi(B,T_P)$ (left panel) and normalized magnetization $M_B(B,T)/M_B(B,T_P)$ (right panel) for $CeRu_2Si_2$ in magnetic fields 0.20 mT (squares), 0.39 mT (triangles), and 0.94 mT (circles), plotted against normalized temperature T/T_P [87], where T_P is the temperature at peak susceptibility. The solid curves trace the calculated universal behavior [114].

8.4   Magnetoresistance

Using the results just presented, we consider the behavior of magnetoresistance (MR)

$$\rho_{mr}(B,T) = \frac{\rho(B,T) - \rho(0,T)}{\rho(0,T)}, \tag{8.23}$$

as a function of magnetic field B and T. Here the resistivity, $\rho(B,T) = \rho_0 + \Delta\rho(B,T) + \Delta\rho_{mr}(B)$, is measured at the magnetic field B and temperature T. We assume that the contribution $\Delta\rho_{mr}(B)$ coming from the magnetic field B can be treated within the low field approximation and given by the well-known Kohler's rule,

$$\frac{\Delta\rho_{mr}(B)}{\rho(0,T)} \simeq \lambda_\perp \left(\frac{B\rho(0,\Theta_D)}{B_0\rho(0,T)} \right)^2, \tag{8.24}$$

with Θ_D is the Debye temperature, B_0 is the characteristic field and $\lambda_\perp$ is a constant. Note, that the low field approximation implies that $\Delta\rho_{mr}(B) \ll \rho(0,T) \equiv \rho(T)$. We also assume that that temperature is not too low so that $\rho_0 \leq \Delta\rho(B=0,T)$, while $B \geq B_{c0}$. Substituting Eq. (8.24) into Eq. (8.23), we find that

$$\rho_{mr}(B,T) \sim \frac{c(M^*(B,T))^2T^2 + \Delta\rho_{mr}(B) - c(M^*(0,T))^2T^2}{\rho(0,T)}. \tag{8.25}$$

Here $M^*(B,T)$ denotes the effective mass which now depends on both the magnetic field and the temperature, and c is a constant determining the temperature dependent part of the resistivity, $c(M^*(B,T))^2T^2 = \Delta\rho(B,T)$.

Consider MR given by Eq. (8.25) as a function of B at some temperature $T = T_0$. At low magnetic fields when $T_0 > T_1(B) \propto B$, the main contribution to MR comes from $\Delta\rho_{mr}(B)$ since the effective mass depends mainly on temperature. Therefore, the ratio $|M^*(B,T) - M^*(0,T)|/M^*(0,T) \ll 1$, the main contribution is given by $\Delta\rho_{mr}(B)$, and MR is an increasing function of B. When B becomes so large that $T_1(B) \sim T_0$, the difference $(M^*(B,T) - M^*(0,T))$ becomes negative, and MR as the function of B reaches its maximum value at $T_1(B) \sim T_0$. We recall that $T_1(B)$ determines the crossover from T^2 dependence of the resistivity to the T dependence. At elevated B when $T_1(B) > T_0$, the ratio $(M^*(B,T) - M^*(0,T))/M^*(0,T) \sim -1$ and MR becomes negative being a decreasing function of B.

Consider now MR as a function of T at some B_0. At $T \leq T_1(B_0)$, we have LFL. At low temperatures $T \ll T_1(B_0)$, it follows from Eqs. (8.5) and (8.14) that $M^*(B_0)/M^*(T) \ll 1$, and MR is determined by the resistivity $\rho(0,T)$. Note, that B_0 has to be comparatively high to ensure the inequality, $M^*(B_0)/M^*(T) \ll 1$. As a result, $\rho_{mr}(B_0, T \to 0) \sim -1$, because $\Delta\rho_{mr}(B)/\rho(0,T) \ll 1$. Differentiating the function $\rho_{mr}(B_0,T)$ with respect to B_0 we can check that its slope becomes steeper as B_0 is decreased, being proportional $\propto (B_0 - B_{c0})^{-7/3}$. At $T \simeq T_1(B_0)$, MR possesses a node because at this point the effective mass $M^*(B_0) \simeq M^*(T)$, and $\rho(B_0,T) \simeq \rho(0,T)$. We can conclude that the crossover from the T^2 resistivity to the T resistivity, which occurs at $T \sim T_1(B_0)$, manifests itself in the transition from negative MR to positive MR. At $T \geq T_1(B_0)$, the main contribution to MR comes from $\Delta\rho_{mr}(B_0)$, and MR reaches its maximum value. Upon using Eqs. (8.14) and (8.25) and taking account that at this point $T \propto (B_0 - B_{c0})$, we obtain that the maximum value $\rho_{mr}^m(B_0)$ of MR is $\rho_{mr}^m(B_0) \propto 1/(B_0 - B_{c0})$. Thus, the maximum value is a decreasing function of B_0. At $T_1(B_0) \ll T$, MR is a decreasing function of the temperature. At these temperatures, MR becomes small comparatively to its maximum value $\rho_{mr}^m(B_0)$ because $|M^*(B,T) - M^*(0,T)|/M^*(0,T) \ll 1$ and $\Delta\rho_{mr}(B_0)/\rho(T) \ll 1$.

The recent paper [105] reports the measurements of the CeCoIn$_5$ resistivity in magnetic fields. The both transitions from negative MR to positive MR with increasing T and from positive MR to negative one with increasing B were observed [105]. Thus, the above observed behavior of MR is in good agreement with the experimental facts. We believe that an additional analysis of the data [105] can reveal that the crossover from T^2 dependence of the resistivity to the T dependence occurs at $T \propto (B - B_{c0}) \div (B - B_{c0})^{4/3}$, this analysis could reveal the above described supplementary peculiarities of MR as well [26].

9   HF Metals with Strongly Correlated Electronic Liquid

As we have seen in Section 2, at $T = 0$ when $r = (x - x_{FC}) \to 0$, the effective mass $M^*(r) \to \infty$ and eventually beyond the critical point $x = x_{FC}$ the distance r becomes negative making the effective mass negative as follows from Eq. (2.7). As it was shown in Section 2, the system is to undergo FCQPT. Therefore behind the critical point x_{FC} of this transition, the quasiparticle distribution function represented by the step function does not deliver the minimum to the Landau functional $E[n(\mathbf{p})]$. As a result, at $x < x_{FC}$ the quasiparticle distribution is determined by Eq. (2.8) to search the minimum of a functional, which determines the quasiparticle distribution function $n_0(\mathbf{p})$ delivering the lowest possible value to the ground state energy E. It was shown in Section 3 that the relevant order parameter $\kappa(\mathbf{p}) = \sqrt{n_0(\mathbf{p})(1 - n_0(\mathbf{p}))}$ is at the same time the order parameter of the superconducting state when the maximum value Δ_1 of the superconducting gap is infinitely small. Thus, this state cannot exist at any finite temperatures and driven by the parameter x:

at $x > x_{FC}$ the system is on the disordered side of FCQPT, while at $x < x_{FC}$, the system is on the ordered side. In Section 2 we have shown that this ordered state has a strong impact on the properties of the system at finite temperatures $T \ll T_f$: the effective mass $M^*(T)$ diverges as $1/T$, see Eq. (2.12), and the electronic system with FC is characterized by the energy scale E_0 given by Eq. (2.14). As a result, we can consider FCQPT as the phase transition, which separates the *highly correlated* heavy electron liquid from the *strongly correlated* one. It was shown in Section 8, that the highly correlated liquid at $T \to 0$ and $x > x_{FC}$ behaves as LFL, therefore, FCQPT separates the regions of LFL and strongly correlated liquids. Obviously, the strongly correlated electron liquid demonstrates the NFL behavior down to zero temperature.

At $0 < T \ll T_f$, the function $n_0(\mathbf{p})$ determines the entropy $S_{NFL}(T)$ of the heavy electron liquid in its NFL state. Inserting into Eq. (2.3) the function $n_0(\mathbf{p})$, one can check that behind the point of FCQPT there is a temperature independent contribution $S_0(r) \sim (p_f - p_i)/p_F \sim |r|/x_{FC}$, where $r = x - x_{FC}$. Another specific contribution is related to the spectrum $\varepsilon(\mathbf{p})$ which insures the connection between the dispersionless region $(p_f - p_i)$ occupied by FC and the normal quasiparticles located at $p < p_i$ and at $p > p_f$, and therefore it is of the form $\varepsilon(\mathbf{p}) \propto (p - p_f)^2 \sim (p_i - p)^2$. Such a form of the spectrum can be verified in exactly solvable models for systems with FC and leads to the contribution of this spectrum to the specific heat $C \sim \sqrt{T/T_f}$ [17]. Thus at $0 < T \ll T_f$, the entropy can be approximated as

$$S_{NFL}(T) \simeq S_0(r) + a\sqrt{\frac{T}{T_f}} + b\frac{T}{T_f}, \qquad (9.1)$$

with a and b are constants. The third term on the right hand side of Eq. (9.1) comes from the contribution of the temperature independent part of the spectrum $\varepsilon(\mathbf{p})$ and gives a relatively small contribution to the entropy.

The calculated evolution of $S(T)$, $C(T)$ and $M^*(T)$ based on the model functional (8.6) are shown by filled symbols in Fig. 13, Fig. 14, and Fig. 15 respectively. The calculations were carried out for $g = 7, 8, 12$ and $\beta = b_c = 3$, while the critical value $g = g_c = 6.7167$.

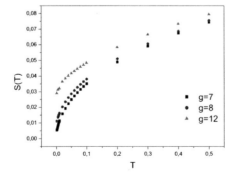

Figure 13: Calculated entropy $S(T)$ as a function of temperature. Solid lines represent $S(T)$ given by Eq. (9.1).

It is seen from Fig. 13 that $S_0(r)$ increases when the system being on the ordered side moves away from FCQPT.

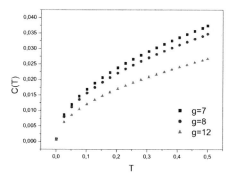

Figure 14: Calculated specific heat $C(T)$ as a function of temperature. Solid lines represent $C(T) \propto \sqrt{T}$.

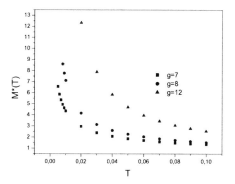

Figure 15: Calculated effective mass $M^*(T)$ as a function of temperature. Solid lines represent $M^*(T) \propto a_1/T + a_2\sqrt{T} + a_3$.

Obviously, the term $S_0(r)$ does not contribute to the specific heat. As a result the specific heat demonstrates the anomalous behavior $C(T) \propto \sqrt{T}$ as it is seen from Fig. 14. As to the effective mass $M^*(T) \propto S(T)/T$, it demonstrates the divergence $M^*(T) \propto 1/T$ in accordance with Eq. (2.5) and contains all terms defining the behavior of the entropy, as shown in Fig. 15.

9.1   The Grüneisen Ratio and Hall Coefficient in Heavy Fermion Metals; $T - B$ Phase Diagram

The temperature independent term $S_0(r)$ determines the specific NFL behavior of the system. For example, the existence of the temperature independent term $S_0(r)$ of the entropy can be illuminated by calculating the thermal expansion coefficient $\alpha(T)$ [7, 118], which is given by [27]

$$\alpha(T) = \frac{1}{3}\left(\frac{\partial(\log V)}{\partial T}\right)_P = -\frac{x}{3K}\left(\frac{\partial(S/x)}{\partial x}\right)_T. \tag{9.2}$$

Here, P is the pressure and V is the volume. The compressibility K is not expected to be singular at FCQPT and in systems with FC, because FC is attached to the Fermi level, and it moves along as $\mu(x)$ changes, while the compressibility $K = d\mu/d(Vx)$ is approximately constant [119]. Inserting Eq. (9.1) into Eq. (9.2), we find that

$$\alpha_{FC}(T) \simeq a_0 \sim \frac{M_{FC}^* T}{p_F^2 K}. \tag{9.3}$$

Here, a_0 is a number independent of temperature. When deriving Eq. (9.3) we keep only the main contribution coming from $S_0(r)$. On the other hand, the specific heat

$$C(T) = T\frac{\partial S(T)}{\partial T} \simeq \frac{a}{2}\sqrt{\frac{T}{T_f}}. \tag{9.4}$$

As a result, the Grüneisen ratio $\Gamma(T)$ diverges as

$$\Gamma(T) = \frac{\alpha(T)}{C(T)} \simeq 2\frac{a_0}{a}\sqrt{\frac{T_f}{T}}. \tag{9.5}$$

At this point, we consider how the behavior of the effective mass given by Eqs. (2.12) and (8.14) correspond to experimental observations. It was recently observed that the thermal expansion coefficient $\alpha(T)/T$ measured on $CeNi_2Ge_2$ shows a $1/\sqrt{T}$ divergence over two orders of magnitude in the temperature range from 6 K down to at least 50 mK, while measurements on $YbRh_2(Si_{0.95}Ge_{0.05})_2$ demonstrate that $\alpha/T \propto 1/T$ [6], contrary to the LFL theory which yields $\alpha(T)/T \propto M^* \simeq const$. Since the effective mass depends on T, we obtain that the $1/\sqrt{T}$ behavior, Eq. (8.14), is in excellent agreement with the result for the former system [7], and the $1/T$ behavior, Eq. (2.12), predicted in [118] corresponds to the latter HF metal.

We see that at $0 < T \ll T_f$, the heavy electron liquid with FC behaves as if it were placed at QCP. In fact it is placed at the quantum critical line $x < x_{FC}$, that is the critical behavior is observed at $T \to 0$ for all $x \le x_{FC}$. At $T \to 0$, the heavy electron liquid undergoes a first-order quantum phase transition because the entropy is not a continuous function: at finite temperatures the entropy is given by Eq. (9.1), while $S(T = 0) = 0$. Therefore, the entropy undergoes a sudden jump $\delta S = S_0(r)$ in the zero temperature limit. We make up a conclusion that due to the first order phase transition, the critical fluctuations are suppressed at the quantum critical line and the corresponding divergences, for example the divergence of $\Gamma(T)$, are determined by the quasiparticles rather than by the critical fluctuations as one could expect in the case of CQPT, see e.g. [4]. Note that according to the well known inequality, $\delta Q \le T\delta S$, the heat δQ of the transition from the ordered phase to the disordered one is equal to zero, because $\delta Q \le S_0(r)T \to 0$ at $T \to 0$.

To study the $B - T$ phase diagram of the heavy electron liquid with FC, we consider the case when the NFL behavior arises by the suppression of the antiferromagnetic (AF) phase upon applying a magnetic field B, for example, as it takes place in the HF metals YbRh$_2$Si$_2$ and YbRh$_2$(Si$_{0.95}$Ge$_{0.05}$)$_2$ [10, 6]. The AF phase is represented by the heavy electron LFL, with the entropy vanishing as $T \to 0$. For magnetic fields exceeding the critical value B_{c0} at which the Néel temperature $T_N(B \to B_{c0}) \to 0$ the weakly ordered AF phase transforms into weakly polarized heavy electron LFL. As it was discussed in Section 5, at $T = 0$ the application of the magnetic field B splits the FC state occupying the region $(p_f - p_i)$ into the Landau levels and suppresses the superconducting order parameter $\kappa(\mathbf{p})$ destroying the FC state. Such a state is given by the multiconnected Fermi sphere, where the smooth quasiparticle distribution function $n_0(\mathbf{p})$ in the $(p_f - p_i)$ range is replaced by a multiconnected distribution $\nu(\mathbf{p})$, see Fig. 4. Therefore the LFL behavior is restored being represented by the weakly polarized heavy electron LFL and characterized by quasiparticles with the effective mass $M^*(B)$ given by Eq. (5.5). At elevated temperatures $T > T^*(B - B_{c0}) \propto \sqrt{B - B_{c0}}$, the NFL state is restored and the entropy of the heavy electron liquid is given by Eq. (9.1). This behavior is displayed in the $T - B$ phase diagram shown in Fig. 16.

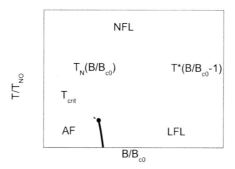

Figure 16: $T - B$ phase diagram of the heavy electron liquid. The $T_N(B/B_{c0})$ curve represents the field dependence of the Néel temperature. Line separating the antiferromagnetic (AF) and the non-Fermi liquid (NFL) state is a guide to the eye. The black dot at $T = T_{crit}$ marked by the arrow is the critical temperature, at which the second order AF phase transition becomes the first one. At $T < T_{crit}$, the thick solid line represents the field dependence of the Néel temperature when the AF phase transition is of the first order. The NFL state is characterized by the entropy S_{NFL} given by Eq. (9.1). The line separating the NFL state and the weakly polarized heavy electron Landau Fermi Liquid (LFL) is $T^*(B/B_{c0} - 1) \propto \sqrt{B/B_{c0} - 1}$.

In accordance with experimental facts we assume that at relatively high temperatures $T/T_{NO} \sim 1$ the AF phase transition is of the second order [10], where T_{NO} is the Néel temperature in the absence of the magnetic field. In that case, the entropy and the other thermodynamic functions are continuous functions at the transition temperature $T_N(B)$. This means that the entropy of the AF phase $S_{AF}(T)$ coincides with the entropy of the NFL state given by Eq. (9.1),

$$S_{AF}(T \to T_N(B)) = S_{NFL}(T \to T_N(B)). \tag{9.6}$$

Since the AF phase demonstrates the LFL behavior, that is $S_{AF}(T \to 0) \to 0$, Eq. (10) cannot be satisfied at sufficiently low temperatures $T \leq T_{crit}$ due to the temperature-independent term $S_0(r)$, see Eq. (9.1). Thus, the second order AF phase transition becomes the first order one at $T = T_{crit}$ as it is shown in Fig. 16. At $T = 0$, the critical field B_{c0}, at which the AF phase becomes the heavy LFL, is determined by the condition that the ground state energy of the AF phase coincides with the ground state energy $E[n_0(\mathbf{p})]$ of the heavy LFL, that is the ground state of the AF phase becomes degenerated at $B = B_{c0}$. Therefore, the Néel temperature $T_N(B \to B_{c0}) \to 0$, and the behavior of the effective mass $M^*(B \geq B_{c0})$ is given by Eq. (5.5), that is $M^*(B)$ diverges when $B \to B_{c0}$.

We note that the corresponding quantum and thermal critical fluctuations vanish at $T < T_{crit}$ because we are dealing with the first order AF phase transition. We can also reliably conclude that the critical behavior observed at $T \to 0$ and $B \to B_{c0}$ is determined by the corresponding quasiparticles rather than by the critical fluctuations accompanying second order phase transitions. When $r \to 0$ the heavy electron liquid approaches FCQPT from the ordered phase. Obviously, $T_{crit} \to 0$ at the point $r = 0$, and we are led to the conclusion that the Néel temperature vanishes at the point when the AF second order phase transition becomes the first order one. As a result, one can expect that the contributions coming from the corresponding critical fluctuations can only lead to the logarithmic corrections to the Landau theory of the phase transitions [115], and the power low critical behavior is again defined by the corresponding quasiparticles. Thus, we conclude that the Landau paradigm based on the notions of the quasiparticles and order parameter is applicable when considering the heavy electron liquid.

Now we are in position to consider the recently observed jump in the Hall coefficient at $B \to B_{c0}$ in the zero temperature limit [120]. At $T = 0$, the application of the critical magnetic field B_{c0} suppressing the AF phase (with the Fermi momentum $p_{AF} \simeq p_F$) restores the LFL with the Fermi momentum $p_f > p_F$. At $B < B_{c0}$, the ground state energy of the AF phase is lower then that of the heavy LFL, while at $B > B_{c0}$, we are dealing with the opposite case, and the heavy LFL wins the competition. At $B = B_{c0}$, both AF and LFL have the same ground state energy being degenerated. Thus, at $T = 0$ and $B = B_{c0}$, the infinitesimal change in the magnetic field B leads to the finite jump in the Fermi momentum because the distribution function becomes multiconnected, see Fig. 4, while the overall Fermi volume remains constant. That is, the number of itinerant electrons does not change. In response the Hall coefficient $R_H(B) \propto 1/x \propto 1/p_f^3$ undergoes the corresponding sudden jump. Here we have assumed that the low temperature $R_H(B)$ can be considered as a measure of the Fermi volume and, therefore, as a measure of the Fermi momentum [120]. As a result, we obtain

$$\frac{R_H(B = B_{c0} - \delta)}{R_H(B = B_{c0} + \delta)} \simeq 1 + 3\frac{p_f - p_F}{p_F} \simeq 1 + d\frac{S_0(r)}{x_{FC}}. \qquad (9.7)$$

Here δ is an infinitesimal magnetic field, $S_0(r)/x_{FC}$ is the entropy per one heavy electron, and d is a constant, $d \sim 5$. It follows from Eq. (9.1) that the abrupt change in the Hall coefficient tends to zero when $r \to 0$ and vanishes when the system in question is on the disordered side of FCQPT [121].

Now consider the magnetic susceptibility which is proportional to the effective mass given by Eq. (5.5). Therefore, at $T \ll T^*(B)$, the magnetic susceptibility given by Eq.

(5.8) is of the form

$$\chi(B) \propto M^*(B) \propto \frac{1}{\sqrt{B - B_{c0}}},$$ (9.8)

while the static magnetization $M(B)$ is given by [8]

$$M(B) \propto \sqrt{B - B_{c0}}.$$ (9.9)

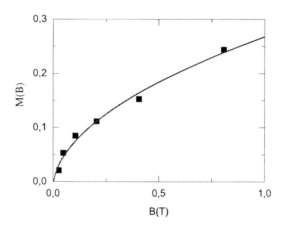

Figure 17: Magnetization $M(B)$ shown by filled squares as a function of magnetic field B [102]. The curve represents the field dependence of $M(B) = a_M \sqrt{B}$ given by Eq. (9.9) with a_M is a costnat.

As seen from Fig. 17, the field dependence of $M(B)$ given by Eq. (9.9) is in good agreement with the data obtained in measurements on YbRh$_2$(Si$_{0.95}$Ge$_{0.05}$)$_2$ [102]. We can also conclude that Eqs. (8.21) and (8.22) determining the scaling behavior of the effective mass, static magnetization and the susceptibility are also valid in the case of strongly correlated liquid, but the variable y is now given by $y = T/\sqrt{B - B_{c0}}$, while the function $f(y)$ can be dependent on $(p_f - p_i)/p_F$. This dependence comes from Eq. (2.12). As a result, we can obtain that at $T < T^*(B)$, the factor $d\rho/dT \propto A(B)T$ behaves as $A(B)T \propto T/(B - B_{c0})$, and at $T > T^*(B)$, it behaves as $A(B)T \propto 1/T$. These observations are in good agreement with the data obtained in measurements on YbRh$_2$(Si$_{0.95}$Ge$_{0.05}$)$_2$ [102].

We note that, as in the case of the highly correlated liquid, the susceptibility $\chi(B, T)$ of the strongly correlated liquid is not a monotonic function of y and possesses a maximum as a function of the temperature because the derivative $dM(B)/dB$ is the sum of two contributions. As it was shown in Section 5, the well-known empirical Kadowaki-Woods ratio [9], $K = A/\gamma_0^2 \simeq const$, is also obeyed in the case of the strongly correlated liquid. These results are in good agreement with facts [10, 122, 102].

As an application of the above consideration we study the $T - B$ phase diagram for the HF metal YbRh$_2$Si$_2$ [10] shown in Fig. 18. The LFL behavior is characterized by the effective mass $M^*(B)$ which diverges as $1/\sqrt{B - B_{c0}}$ [10]. We can conclude that Eq.

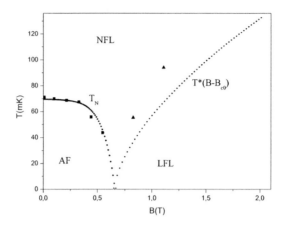

Figure 18: $T - B$ phase diagram for YbRh$_2$Si$_2$ [10, 102]. The T_N curve represents the field dependence of the Néel temperature. Line separating the antiferromagnetic (AF) and the non-Fermi liquid (NFL) state is a guide to the eye. The NFL state is characterized by the entropy S_{NFL} given by Eq. (9.1). Line separating the NFL state and LFL is $T^*(B - B_{c0}) = c\sqrt{B - B_{c0}}$, with c being an adjustable factor.

(5.5) gives good description of this experimental fact, and $M^*(B)$ diverges at the point $B \rightarrow B_{c0}$ with $T_N(B = B_{c0}) = 0$. It is seen from Fig. 18, that the line separating the LFL state and NFL can be approximated by the function $c\sqrt{B - B_{c0}}$ with c being a parameter. Taking into account that the behavior of YbRh$_2$Si$_2$ strongly resembles the behavior of YbRh$_2$(Si$_{0.95}$Ge$_{0.05}$)$_2$ [6, 102, 103, 122], we can conclude that in the NFL state the thermal expansion coefficient $\alpha(T)$ does not depend on T and the Grüneisen ratio as a function of temperature T diverges [6]. We are led to the conclusion that the entropy of the NFL state is given by Eq. (9.1). Taking into account that at relatively high temperatures the AF phase transition is of the second order [10], we predict that at lower temperatures this becomes the first order phase transition. Then, the described behavior of the Hall coefficient $R_H(B)$ is in good agreement with experimental facts [120].

Thus, we can conclude that the $T - B$ phase diagram of the heavy electron liquid with FC is in good agreement with the experimental $T - B$ phase diagram obtained in measurements on the HF metals YbRh$_2$Si$_2$ and YbRh$_2$(Si$_{0.95}$Ge$_{0.05}$)$_2$.

9.2   Heavy Fermion Metals Very near FCQPT on the Ordered Side

Let us consider the case when $\delta p_{FC} = (p_f - p_i)/p_F \ll 1$ and the electronic system of HF metal is very near FCQPT being on the ordered side. As we have seen in Section 5, Eq. (5.10), the application of magnetic field $(B - B_{c0})/B_{c0} \geq B_c$ removes the system from the ordered side placing it on the disordered one. As a result at $T \leq T_1(B)$, the effective mass $M^*(B)$ is given by Eqs. (8.5) and (8.11), while the corresponding resistivity is described in Section 8.2. In the absence of magnetic field or at $T_f \gg T > T_1(B)$, the system demonstrates the NFL behavior, the effective mass $M^*(T)$ is given by Eq. (2.12), and the entropy is given by Eq. (9.1). While the magnetic susceptibility $\chi(T) \propto M^*(T) \propto 1/T$, the thermal expansion coefficient $\alpha(T)$ is T-independent being determined by Eq. (9.3) and making the Grüneisen ratio divergent, as it follows from Eq. (9.5). Note, that the specific heat behaves as $C(T) \propto \sqrt{T}$ in both cases: when the electronic system is on the ordered side or on the disordered side of FCQPT, see Fig. 11 and Fig. 15. It follows from Eq. (2.12) that $\gamma(T) \propto T$. Therefore, the temperature dependent part $\Delta\rho(T)$ of the resistivity behaves as $\Delta\rho(T) \propto \gamma(T) \propto T$. Thus, the system demonstrates the NFL regime, $\Delta\rho(T) \propto T$, when it is either in the highly correlated or in the strongly correlated regimes.

At some temperature T_c, the system can undergo the superconducting phase transition. In contrast to the conventional superconductors where the discontinuity $\delta C(T_c)$ in the specific heat at T_c is a linear function of T_c, $\delta C(T_c)$ is independent of the critical temperature T_c. As seen from Eqs. (3.23) and (3.24), both the discontinuity $\delta C(T_c)$ and the ratio $\delta C(T_c)/C_n(T_c)$ can be very large as compared to the conventional case [54, 123].

Recent experiments show that the electronic system of the HF metal CeCoIn$_5$ can be considered as the system located near FCQPT and containing FC. Indeed, under the application of magnetic field it behaves as the highly correlated liquid (see Section 8) with the effective mass $A(B) \propto (B - B_{c0})^{-4/3}$ [105], as it is seen from Eq. (8.16). While in the NFL regime, $\alpha(T) \propto const$ and the Grüneisen ratio diverges [124], see Eqs. (9.3) and (9.5), respectively. Estimations of δp_{FC} based on the evaluation of the magnetic susceptibility show that $\delta p_{FC} \simeq 0.044$ [123]. The obtained value of δp_{FC} allows to explain the relatively big value of the discontinuity $\delta C(T_c)$ [123] observed at $T_c = 2.3$ K in measurements on CeCoIn$_5$ [125]. Thus, we can conclude that $B_{cr} \simeq 0.003$, as it follows from Eq. (5.10), and the HF metal CeCoIn$_5$ can be considered as placed near FCQPT.

10 Dissymmetrical Tunnelling in HF Metals and High-T_c Superconductors

Experiments on the HF metals explore mainly their thermodynamic properties. It would be desirable to probe the other properties of the heavy electron liquid such as the probabilities of quasiparticle occupations, which are not directly linked to the density of states or to the behavior of the effective mass M^* [126]. Scanning tunnelling microscopy (STM) being sensitive to both the density of states and the probabilities of quasiparticle occupations is an ideal technique for studying such effects at quantum level.

The tunnelling current I through the point contact between two ordinary metals is proportional to the driving voltage V and to the squared modulus of the quantum mechanical transition amplitude t multiplied by the difference $N_1(0)N_2(0)(n_1(p,T) - n_2(p,T))$ [127]. Here $n(p,T)$ is the quasiparticle distribution function and $N(0)$ is the density of states of the corresponding metal. On the other hand, the wave function calculated in the WKB approximation and defining t is proportional to $(N_1(0)N_2(0))^{-1/2}$. As a result, the density of states is dropped out and the tunnelling current is independent of $N_1(0)N_2(0)$. Upon taking into account that at $T \to 0$ the distribution $n(p, T \to 0) \to n_F(p)$, where $n_F(p)$ is the step function $\theta(p - p_F)$, one can check that within the LFL theory the differential tunnelling conductivity $\sigma_d(V) = dI/dV$ is a symmetric function of the voltage V.

In fact, the symmetry of $\sigma_d(V)$ holds provided that so called particle-hole symmetry is preserved as it is within the LFL theory. Therefore, the existence of the $\sigma_d(V)$ symmetry is quite obvious and common in the case of metal-to-metal contacts when these metals are ordinary metals and in their normal or superconducting states.

Now we turn to a consideration of the tunnelling current at low temperatures which in the case of ordinary metals is given by [127]

$$I(V) = 2|t|^2 \int [n_F(z - \mu) - n_F(z - \mu + V)] \, dz. \tag{10.1}$$

We use an atomic system of units: $e = m = \hbar = 1$, where e and m are electron charge and mass, respectively, and the energy z belongs to the interval E_0 given by Eq. (2.14)

$$\mu - 2T \leq z \leq \mu + 2T. \tag{10.2}$$

Since temperatures are low, we approximate the distribution function of ordinary metal by the step function n_F. It follows from Eq. (10.1) that quasiparticles with the energy z, $\mu - V \leq z \leq \mu$, contribute to the current, while $\sigma_d(V) \simeq 2|t|^2$ is a symmetrical function of V. In the case of the heavy electron liquid with FC, the tunnelling current is of the form [126]

$$I(V) = 2 \int [n_0(z - \mu) - n_F(z - \mu + V)] \, dz. \tag{10.3}$$

Here we have replaced the distribution function of ordinary metal by n_0 that is the solution of Eq. (2.8). We have also normalized the transition amplitude $|t|^2$ such that $|t|^2 = 1$. Assume that V satisfies the condition, $|V| \leq 2T$, while the current flows from the HF metal to the ordinary one. Quasiparticles of the energy z, $\mu - V \leq z$, contribute to $I(V)$, and the differential conductivity is givem by the relation $\sigma_d(V) \simeq 2n_0(z \simeq \mu - V)$. If the sign of the voltage is changed, the direction of the current is also changed. In that case, quasiparticles of the energy z, $\mu + V \geq z$, contribute to $I(V)$, and the differential conductivity $\sigma_d(-V) \simeq 2(1 - n_0(z \simeq \mu + V))$. The dissymmetrical part $\Delta\sigma_d(V) = (\sigma_d(-V) - \sigma_d(V))$ of the differential conductivity is of the form

$$\Delta\sigma_d(V) \simeq 2[1 - (n_0(z - \mu \simeq V) - n_0(z - \mu \simeq -V))]. \tag{10.4}$$

It is worth noting that according to Eq. (10.4) we have $\Delta\sigma_d(V) = 0$ if the considered HF metal is replaced by an ordinary metal. Indeed, the effective mass is finite at $T \to 0$, then $n_0(T \to 0) \to n_F$ being given by Eq. (2.4), and $1 - n(z - \mu \simeq V) = n(z - \mu \simeq -V)$. One might say that the dissymmetrical part vanishes due to the particle-hole symmetry. On the other hand, there are no reasons to expect that $(1 - n_0(z - \mu \simeq V) - n_0(z - \mu \simeq -V)) = 0$. Thus, we conclude that the differential conductivity becomes a dissymmetrical function of the voltage.

To estimate $\Delta\sigma_d(V)$, we observe that it is zero when $V = 0$, because $n_0(p = p_F) = 1/2$ as it should be and it follows from Eq. (2.1) as well. It is seen from Eq. (10.4) that $\Delta\sigma_d(V)$ is an even function of both $(z - \mu)$ and V. Therefore we can assume that at low values of the voltage V the dissymmetrical part behaves as $\Delta\sigma_d(V) \propto V^2$. Then, the natural scale to measure the voltage is $2T$, as it is seen from Eq. (10.2). In fact, the dissymmetrical part is to be proportional to $(p_f - p_i)/p_F$. As a result, we obtain

$$\Delta\sigma_d(V) \simeq c \left(\frac{V}{2T}\right)^2 \frac{p_f - p_i}{p_F} \simeq c \left(\frac{V}{2T}\right)^2 \frac{S_0(r)}{x_{FC}}. \tag{10.5}$$

Here, $S_0(r)$ is the temperature independent part of the entropy (see Eq. (9.1)), c is a constant, which is expected to be of the order of a unit. This constant can be evaluated by using analytical solvable models. For example, calculations of c within a simple model, when the Landau functional $E[n(p)]$ is of the form [19]

$$E[n(p)] = \int \frac{p^2}{2M} \frac{d\mathbf{p}}{(2\pi)^3} + V_1 \int n(p)n(p) \frac{d\mathbf{p}}{(2\pi)^3}, \qquad (10.6)$$

give $c \simeq 1/2$. It follows from Eq. (10.5), that when $V \simeq 2T$ and FC occupies a noticeable part of the Fermi volume, $(p_f - p_i)/p_F \simeq 1$, the dissymmetrical part becomes comparable with differential tunnelling conductivity, $\Delta\sigma_d(V) \sim V_d(V)$.

The dissymmetrical behavior of the tunnelling conductivity can be observed in measurements on both high-T_c metals in their normal state and the heavy fermion metals, for example, such as YbRh$_2$(Si$_{0.95}$Ge$_{0.05}$)$_2$ or YbRh$_2$Si$_2$ which are expected to have undergone FCQPT. In the case of HF metals, upon the application of magnetic field B the effective mass is to diverge as given by Eq. (5.5). Here B_{c0} is the critical magnetic field which drives the HF metal to its magnetic field tuned quantum critical point. The value of the critical exponent $\alpha = -1/2$ is in good agreement with experimental observations collected on these metals [10, 102]. The measurements of $\Delta\sigma_d(V)$ have to be carried out applying magnetic field B_{c0} at temperatures $T^*(B) < T \leq T_f$. In the case of these metals, T_f is of the order of few Kelvin. We note that at sufficiently low temperatures, the application of magnetic field $B > B_{c0}$ leads to the restoration of the LFL behavior with $M^*(B)$ given by Eq. (5.5). As a result, the dissymmetrical behavior of the tunnelling conductivity vanishes [126].

The dissymmetrical differential conductivity $\Delta\sigma_d(V)$ can also be observed when both the high-T_c metal and the HF metal in question go from normal to superconducting. The reason is that $n_0(p)$ is again responsible for the dissymmetrical part of $\sigma_d(V)$. As we have seen in Section 3, this $n_0(p)$ is not appreciably disturbed by the pairing interaction which is relatively weak as compared to the Landau interaction forming the distribution function $n_0(p)$. In the case of superconductivity, we have to take into account that the ratio,

$$\frac{N_s(E)}{N(0)} = \frac{|E|}{\sqrt{E^2 - \Delta^2}}, \qquad (10.7)$$

comes into the play because the density of states $N_s(E)$ of the superconducting metal is zero in the gap, that is when $|E| \leq |\Delta|$. Here E is the quasiparticle energy, while the normal state quasiparticle energy is $\varepsilon - \mu = \sqrt{E^2 - \Delta^2}$. Now we can adjust Eq. (10.4) for the case of superconducting HF metal, multiplying the right hand side of Eq. (10.4) by $N_s/N(0)$ and replacing the quasiparticle energy $z - \mu$ by $\sqrt{E^2 - \Delta^2}$ with E being represented by the voltage V. As a result, Eq. (10.5) can be presented in the following form

$$\Delta\sigma_d(V) \simeq \left|\frac{V}{\Delta}\right| \frac{\left(\sqrt{V^2 - \Delta^2}\right)^2}{|\Delta|\sqrt{V^2 - \Delta^2}} \frac{p_f - p_i}{p_F} \simeq \sqrt{1 - \left[\frac{\Delta}{V}\right]^2} \left[\frac{V}{\Delta}\right]^2 \frac{S_0(r)}{x_{FC}}. \qquad (10.8)$$

It is seen from Eqs. (10.5) and (10.8) that, as in the case described by Eq. (9.7) when the abrupt change in the Hall coefficient is defined by $S_0(r)$, the dissymmetrical part of the differential tunnelling conductivity is also proportional to the $S_0(r)$ vanishing at $r \to 0$.

Note that the scale $2T$ entering Eq. (10.5) is replaced by the scale Δ in Eq. (10.8). In the same way, as Eq. (10.5) is valid up to $V \sim 2T$, Eq. (10.8) is valid up to $V \sim 2|\Delta|$. It is seen from Eq. (10.8) that the dissymmetrical part of the differential tunnelling conductivity becomes as large as the differential tunnelling conductivity at $V \sim 2|\Delta|$ provided that FC occupies a large part of the Fermi volume, $(p_f - p_i)/p_F \simeq 1$. In the case of a d-wave gap, the right hand side of Eq. (10.8) has to be integrated over the gap distribution. As a result, $\Delta\sigma_d(V)$ is expected to be finite even at $V = \Delta_1$, where Δ_1 is the maximum value of the d-wave gap [126].

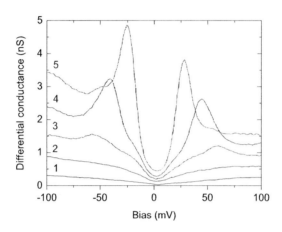

Figure 19: Spatial variation of the tunnelling differential conductance spectra measured on $Bi_2Sr_2CaCu_2O_{8+x}$. Curves 1 and 2 are taken at positions where the integrated LDOS is very small. The low differential conductance and the absence of a superconducting gap are indicative of insulating behavior. Curve 3 is for a large gap of 65 meV, with low coherence peaks. The integrated value of the LDOS at the position for curve 3 is small but larger than those in curves 1 and 2. Curve 4 is for a gap of 40 meV, which is close to the mean value of the gap distribution. Curve 5, taken at the position with the highest integrated LDOS, is for the smallest gap of 25 meV with two very sharp coherence peaks [128].

The presence of an electronic inhomogeneity in $Bi_2Sr_2CaCu_2O_{8+x}$ was recently discovered in observations using scanning tunnelling microscopy and spectroscopy [128]. This inhomogeneity is manifested as spatial variations in the local density of states (LDOS) spectrum, in the low-energy spectral weight, and in the magnitude of the superconducting energy gap. The inhomogeneity observed in the integrated LDOS is not induced by impurities, but rather is intrinsic in nature. The observations allowed to relate the magnitude of the integrated LDOS to the local oxygen doping concentration [128]. Spatial variation of the tunnelling differential conductance spectrum are shown in Fig. 19. The curves 1 and 2 can be considered as corresponding to the normal state of high-T_c superconductor. The other curves can be viewed as corresponding to high-T_c superconductors with different oxygen doping concentrations. It is seen from Fig. 19 that the tunnelling differential conductivity is strongly dissymmetrical in both the normal state and the superconducting state of the $Bi_2Sr_2CaCu_2O_{8+x}$ compound. Therefore, we can conclude that the dissymmetrical tunnelling described by Eqs. (10.5) and (10.8) is in good qualitative agreement with the experimental facts displayed in Fig. 19.

The dissymmetrical tunnelling conductivity given by Eqs. (10.5) and (10.8) can be observed in measurements on the heavy fermion metal $CeCoIn_5$ in its superconducting state and its NFL normal state. As it was discussed in Subsection 9.2, this metal is expected to have undergone FCQPT.

11   Summary and Conclusion

Through out this paper we have discussed the manifestations of the fermion condensation, which can be compared to the Bose-Einstein condensation. A number of experimental evidences have been presented that are supportive to the idea of the existence of FC in different natural and artificial substances. We have demonstrated also that numerous experimental facts collected in a whole variety of materials, belonging to the high-T_c superconductors, heavy fermion metals and strongly correlated 2D structures, can be explained within the framework of the theory based on FCQPT.

We have shown that FCQPT separates the regions of LFL and strongly correlated liquids. Beyond the FCQPT point the quasiparticle system is divided into two subsystems, one containing normal quasiparticles, while the other being occupied by FC localized at the Fermi level. In the superconducting state the quasiparticle dispersion $\varepsilon(\mathbf{p})$ in systems with FC can be approximated by two straight lines, characterized by effective masses M_{FC}^* and M_L^*, and intersecting near the binding energy E_0 which is of the order of the superconducting gap. The same quasiparticle picture and the energy scale E_0 persist in the normal state.

We have demonstrated that fermion systems with FC have features of "quantum protectorate" and shown that the theory of high-T_c superconductivity, based on FCQPT and on the conventional theory of superconductivity, permits the description of high values of T_c and of the maximum value of the gap Δ_1, which may be as big as $\Delta_1 \sim 0.1\varepsilon_F$ or even larger. We have also traced the transition from conventional superconductors to high-T_c ones. We have shown by a simple, although self-consistent analysis that both the pseudogap state and the general features of the shape of the critical temperature $T_c(x)$ as a function of the number density x of the mobile carriers in the high-T_c compounds can be understood within the framework of the theory.

We have also shown that striking experimental results on the magnetic-field induced LFL in high-T_c metals, which unveil the nature of the high-T_c superconductivity, suggest that FCQPT and the emergence of the novel quasiparticles with effective mass strongly depending on the magnetic field and temperature and resembling the Landau quasiparticles are qualities intrinsic to the electronic system of the high-T_c superconductors.

We have provided explanations of the experimental data on the divergence of the effective mass in 2D electron liquid and in 2D ^{3}He, as well as shown that above the critical point of FCQPT the system exhibits the LFL behavior. The behavior of the heavy electron liquid approaching FCQPT form the disordered phase can be viewed as the highly correlated one because the effective mass is very large and strongly depends on the density, temperature and magnetic fields.

At different temperatures, the behavior in magnetic fields of the highly correlated electron liquid approaching FCQPT from the disordered phase has been considered. We have shown that at sufficiently high temperatures $T_1(B) < T$, the effective mass starts to depend on T, $M^* \propto T^{-1/2}$. This $T^{-1/2}$ dependence of the effective mass at elevated temperatures leads to the non-Fermi liquid behavior of the resistivity, $\rho(T) \propto T$. The application of magnetic field B restores the common T^2 behavior of the resistivity. If the magnetic field $(B - B_{c0})$ decreases to zero, the effective mass M^* diverges as $M^* \propto (B - B_{c0})^{-2/3}$. At finite magnetic fields, the regime NFL is restored at some temperature $T_1(B) \propto (B - B_{c0})$ with $M^*(T) \propto T^{-2/3}$. We have demonstrated that this $B - T$ phase diagram has a strong impact on MR of the highly correlated electron liquid. At fixed B, MR as a function of the temperature exhibits a transition from the negative values of MR at $T \to 0$ to the positive values at $T \propto (B - B_{c0})$. While at low temperatures and elevated magnetic fields, MR goes from positive to negative. This behavior was observed in the heavy fermion metals.

We have demonstrated that the strongly correlated electron liquid with FC, which exhibits strong deviations from the LFL behavior down to lowest temperatures, can be driven into LFL by applying a magnetic field B. If the magnetic field $(B - B_{c0})$ decreases to zero, the effective mass M^* diverges as $M^* \propto 1/\sqrt{B - B_{c0}}$ and the Néel temperature T_N of the AF phase transition tends to zero, $T_N(B \to B_{c0}) \to 0$. The NFL regime is restored at some temperature $T^*(B) \propto \sqrt{B - B_{c0}}$. In that case and at $T \to 0$ and $B = B_{c0}$, the Grüneisen ratio as a function of temperature T diverges. While the entropy $S(T)$ possesses the specific low temperature behavior, $S(T) \propto S_0 + a\sqrt{T} + bT$ with S_0, a and b are temperature independent constants. We have shown that the obtained $T - B$ phase diagram is in good agreement with the experimental $T - B$ phase diagram obtained in measurements on the HF metals $YbRh_2Si_2$ and $YbRh_2(Si_{0.95}Ge_{0.05})_2$. We have also demonstrated that the abrupt jump in the Hall coefficient $R_H(B \to B_{c0}, T \to 0)$ is determined by the presence of FC. We have observed that the second order AF phase transition changes to the first order one below $T_{crit} < T_N(B = 0)$ making the corresponding quantum and thermal critical fluctuations vanish at the jump. Therefore, the abrupt jump and the divergence of the effective mass taking place at $T_N \to 0$ are defined by the behavior of quasiparticles rather than by the corresponding thermal and quantum critical fluctuations.

We have predicted that the differential tunnelling conductivity between a metallic point and an ordinary metal, which is commonly symmetric as a function of the voltage, becomes noticeably dissymmetrical when the ordinary metal is replaced by a HF metal, the electronic system of which has undergone FCQPT. This dissymmetry can be observed when the HF metal is both normal and superconducting. We have also discussed possible experiments to study the dissymmetry in measurements on the HF metals. In the case of the high-T_c superconductors, our consideration is in good agreement with available data.

In conclusion, we have shown that in contrast to covnetional quantum phase transitions, whose physics is dominated by thermal and quantum fluctuations and characterized by the absence of quasiparticles, the physics of Fermi systems and the heavy electron liquid near FCQPT or undergone FCQPT is determined by quasiparticles resembling the Landau quasiparticles. We have shown that the Landau paradigm based on the notion of quasiparticles and order parameters is still applicable when considering the low temperature properties of the heavy electron liquid, whose understanding has been problematic largely because of the absence of theoretical guidance. In contrast with the conventional Landau quasiparticles, the effective mass of the considered quasiparticles strongly depends on the temperature, applied magnetic fields, the number density x, pressure, etc. These quasiparticles and the order parameter are well defined and capable of describing both the LFL and the NFL behaviors of both the high-T_c superconductors and the HF metals and their universal thermodynamic properties down to the lowest temperatures. This system of quasiparticles determines the recovery of the LFL behavior under applied magnetic fields and preserves the Kadowaki-Woods ratio. Thus, we obtain a unique possibility to control the essence of HF systems and accordingly of the HF metals by magnetic fields in a wide range of temperatures.

Finally, our general consideration suggests that FCQPT and the emergence of novel quasiparticles at QCP and behind QCP and resembling the Landau quasiparticles are qualities intrinsic to strongly correlated substances, while FCQPT can be viewed as the universal cause of the non-Fermi liquid behavior observed in different metals and liquids.

Acknowledgments

We are grateful to P. Coleman, V.A. Khodel and M. Norman for valuable discussions.

This work was supported in part by the Russian Foundation for Basic Research, project No. 01-02-17189 and in part by INTAS, project No. 03-51-6170. The visit of VRS to Clark Atlanta University has been supported by NSF through a grant to CTSPS. VRS is grateful to the Racah Institute of Physics for the hospitality during his stay at the Hebrew University, Jerusalem. MYaA is grateful to the Binational Science Foundation, grant 2002064 and Israeli Science Foundation, grant 174/03. AZM is supported by US DOE, Division of Chemical Sciences, Office of Basic Energy Sciences, Office of Energy Research. PKG is supported by the grants of Russian Academy of Sciences.

References

[1] G.R. Stewart, *Rev. Mod. Phys.* **73**, 797 (2001).

[2] C.M. Varma, Z. Nussionov, and W. van Saarlos, *Phys. Rep.* **361**, 267 (2002).

[3] S. Sachdev, *Quantum Phase transitions* (Cambridge, Cambridge University Press, 1999).

[4] M. Vojta, *Rep. Prog. Phys.* **66**, 2069 (2003).

[5] D.V. Shopova and D.I. Uzunov, *Phys. Rep.* **379**, 1 (2003).

[6] R. Küchler, N. Oeschler, P. Gegenwart, T. Cichorek, K. Neumaier, O. Tegus, C. Geibel, J. A. Mydosh, F. Steglich, L. Zhu, and Q. Si, *Phys. Rev. Lett.* **91**, 066405 (2003).

[7] M.Ya. Amusia, A.Z. Msezane, and V.R. Shaginyan, *Phys. Lett. A* **320**, 459 (2004).

[8] V.R. Shaginyan, *JETP Lett.* **79**, 286 (2004).

[9] K. Kadowaki and S.B. Woods, *Solid State Commun.* **58**, 507 (1986).

[10] P. Gegenwart, J. Custers, C. Geibel, K. Neumaier, T. Tayama, K. Tenya, O. Trovarelli, and F. Steglich, *Phys. Rev. Lett.* **89**, 056402 (2002).

[11] C. Proust, E. Boaknin, R. W. Hill, L. Taillefer, and A. P. Mackenzie, *Phys. Rev. Lett.* **89**, 147003 (2002).

[12] A.J. Millis, A.J. Schofield, G.G. Lonzarich, and S.A. Grigera, *Phys. Rev. Lett.* **88**, 217204 (2002).

[13] A. Bianchi, R. Movshovich, I. Vekhter, P. G. Pagliuso, and J. L. Sarrao, *Phys. Rev. Lett.* **91**, 257001 (2003); F. Ronning, C. Capan, A. Bianchi, R. Movshovich, A. Lacerda, M. F. Hundley, J. D. Thompson, P. G. Pagliuso, and J. L. Sarrao, *Phys. Rev. B* **71**, 104528 (2005).

[14] J. Paglione, M.A. Tanatar, D.G. Hawthorn, E. Boaknin, F. Ronning, R.W. Hill, M. Sutherland, L. Taillefer, C. Petrovic, and P.C. Canfield, cond-mat/0405157.

[15] S.T. Belyaev, *Sov. Phys. JETP,* **7**, 289 (1958).

[16] S.T. Belyaev, *Sov. Phys. JETP,* **7**, 299 (1958).

[17] V.A. Khodel and V.R. Shaginyan, *JETP Lett.* **51**, 553 (1990).;

[18] V.A. Khodel and V.R. Shaginyan, *Nucl. Phys. A* **555**, 33 (1993).

[19] V.A. Khodel, V.R. Shaginyan, and V.V. Khodel, *Phys. Rep.* **249**, 1 (1994).

[20] J. Dukelsky, V.A. Khodel, P. Schuck, and V.R. Shaginyan, Z. Phys. **102**, 245 (1997); V.A. Khodel and V.R. Shaginyan, *Condensed Matter Theories,* **12**, 222 (1997).

[21] V.R. Shaginyan, Phys. Lett. A **249**, 237 (1998).

[22] G. E. Volovik, *JETP Lett.* **53**, 222 (1991).

[23] I.E. Dzyaloshinskii, *J. Phys. I* (France) **6**, 119 (1996).

[24] D. Lidsky, J. Shiraishi, Y. Hatsugai, and M. Kohmoto, *Phys. Rev. B* **57**, 1340 (1998).

[25] V.Yu. Irkhin, A.A. Katanin, and M.I. Katsnelson, *Phys. Rev. Lett.* **89**, 076401 (2002).

[26] V.R. Shaginyan, *JETP Lett.* **77**, 178 (2003).

[27] E.M. Lifshitz and L.P. Pitaevskii, *Statistical Physics* (Part 2, Butterworth-Heinemann, Oxford, 1999).

[28] L.Ja. Pomeranchuk, *JETP* **35**, 524 (1958).

[29] M.Ya. Amusia and V.R. Shaginyan, *JETP Lett.* **73**, 232 (2001);

[30] M.Ya. Amusia and V.R. Shaginyan, *Phys. Rev. B* **63**, 224507 (2001); V.R. Shaginyan, Physica B **312-313C**, 413 (2002).

[31] V.R. Shaginyan, *JETP Lett.* **77**, 99 (2003).

[32] V.M.Yakovenko and V.A. Khodel, *JETP Lett.* **78**, 398 (2003); cond-mat/0308380.

[33] A. Casey, H. Patel, J. Nye'ki, B. P. Cowan, and J. Saunders, *Phys. Rev. Lett.* **90**, 115301 (2003).

[34] A.A. Shashkin, S.V. Kravchenko, V.T. Dolgopolov, and T.M. Klapwijk, *Phys. Rev. B* **66**, 073303 (2002); A.A. Shashkin, M. Rahimi, S. Anissimova, S.V. Kravchenko V.T. Dolgopolov, and T. M. Klapwijk, *Phys. Rev. Lett.* **91**, 046403 (2003).

[35] J. Boronat, J. Casulleras, V. Grau, E. Krotscheck, and J. Springer, *Phys. Rev. Lett.* **91**, 085302 (2003).

[36] Y. Zhang, V. M. Yakovenko, and S. Das Sarma, *Phys. Rev. B* **71**, 115105 (2005).

[37] Y. Zhang and S. Das Sarma, *Phys. Rev. B* **70**, 035104 (2004).

[38] V.A. Khodel, V.R. Shaginyan, and M.V. Zverev, *JETP Lett.* **65**, 253 (1997).

[39] V.R. Shaginyan, J.G. Han, and J. Lee, *Phys. Lett. A* **329**, 108 (2004).

[40] R.B. Laughlin and D. Pines, *Proc. Natl. Acad. Sci. USA* **97**, 28 (2000).

[41] P.W. Anderson, cond-mat/0007185; cond-mat/0007287.

[42] V.A. Khodel, J.W. Clark, and V.R. Shaginyan, *Solid Stat. Comm.* **96**, 353 (1995).

[43] S.A. Artamonov and V.R. Shaginyan, *JETP* **92**, 287 (2001).

[44] P. V. Bogdanov, A. Lanzara, S. A. Kellar, X. J. Zhou, E. D. Lu, W. J. Zheng, G. Gu, J.-I. Shimoyama, K. Kishio, H. Ikeda, R. Yoshizaki, Z. Hussain, and Z. X. Shen, *Phys. Rev. Lett.* **85**, 2581 (2000).

[45] A. Kaminski, M. Randeria, J. C. Campuzano, M. R. Norman, H. Fretwell, J. Mesot, T. Sato, T. Takahashi, and K. Kadowaki, *Phys. Rev. Lett.* **86**, 1070 (2001).

[46] T. Valla, A. V. Fedorov, P. D. Johnson, B. O. Wells, S. L. Hulbert, Q. Li, G. D. Gu, and N. Koshizuka, *Science* **285**, 2110 (1999); T. Valla, A. V. Fedorov, P. D. Johnson, Q. Li, G. D. Gu, and N. Koshizuka, *Phys. Rev. Lett.* **85**, 828 (2000).

[47] G. Grüner, *Density Waves in Solids* (Addison-Wesley, Reading, MA, 1994).

[48] L. Tassini, F. Venturini, Q.-M. Zhang, R. Hackl, N. Kikugawa, and T. Fujita, *Phys. Rev. Lett.* **95**, 117002 (2005).

[49] J. Bardeen, L.N. Cooper, and J.R. Schrieffer, *Phys. Rev.* **108**, 1175 (1957).

[50] D.R. Tilley and J. Tilley, *Superfluidity and Superconductivity*, (Bristol, Hilger, 1985).

[51] N.N. Bogoliubov, *Nuovo Cimento* **7**, 794 (1958).

[52] M.Ya. Amusia and V.R. Shaginyan, *JETP Lett.* **77**, 671 (2003).

[53] H. Matsui, T. Sato, T. Takahashi, S.-C. Wang, H.-B. Yang, H. Ding, T. Fujii, T. Watanabe, and A. Matsuda, *Phys. Rev. Lett.* **90**, 217002 (2003).

[54] M.Ya. Amusia, S.A. Artamonov, and V.R. Shaginyan, *JETP Lett.* **74**, 435 (2001).

[55] A. Paramekanti, M. Randeria, and N. Trivedi, *Phys. Rev. Lett.* **87**, 217002 (2001); A. Paramekanti, M. Randeria, and N. Trivedi, cond-mat/0305611.

[56] P. W. Anderson, P. A. Lee, M. Randeria, T. M. Rice, N. Trivedi, and F. C. Zhang, *J Phys. Condens. Matter* **16**, R755 (2004).

[57] V.R. Shaginyan, *JETP Lett.* **68**, 527 (1998).

[58] M.Ya. Amusia and V.R. Shaginyan, *Phys. Lett. A* **298**, 193 (2002).

[59] M. Kugler, M. Fischer, Ch. Renner, S. Ono, and Y. Ando, *Phys. Rev. Lett.* **86**, 4911 (2001).

[60] A.A. Abrikosov, *Phys. Rev. B* **52**, R15738 (1995); A.A. Abrikosov, cond-mat/9912394.

[61] N.-C. Yeh, C.-T. Chen, G. Hammer, J. Mannhart, A. Schmeh, C. W. Schneider, R. R. Schulz, S. Tajima, K. Yoshida, D. Garrigus, and M. Strasik, *Phys. Rev. Lett.* **87**, 087003 (2001).

[62] A. Biswas, P. Fournier, M. M. Qazilbash, V. N. Smolyaninova, H. Balci, and R. L. Greene, *Phys. Rev. Lett.* **88**, 207004 (2002).

[63] J.A. Skinta, M.-S. Kim, T.R. Lemberger, T. Greibe, and M. Naito, *Phys. Rev. Lett.* **88**, 207005 (2002).

[64] J.A. Skinta, T.R. Lemberger, T. Greibe, and M. Naito, *Phys. Rev. Lett.* **88**, 207003 (2002).

[65] C.-T. Chen, P. Seneor, N.-C. Yeh, R.P. Vasquez, L.D. Bell, C.U. Jung, J.Y. Kim, M.-S. Park, H.-J. Kim, and S.-I. Lee, *Phys. Rev. Lett.* **88**, 227002 (2002).

[66] N. Metoki, Y. Haga, Y. Koike, and Y. Onuki, *Phys. Rev. Lett.* **80**, 5417 (1998).

[67] T. Honma, Y. Haga, and E. Yamamoto, *J. Phys. Soc. Jpn.* **68**, 338 (1999).

[68] M.Ya. Amusia and V.R. Shaginyan, *JETP Lett.* **76**, 651 (2002).

[69] N. Miyakawa, J. F. Zasadzinski, L. Ozyuzer, P. Guptasarma, D. G. Hinks, C. Kendziora, and K. E. Gray, *Phys. Rev. Lett.* **83**, 1018 (1999).

[70] C.M. Varma, P.B. Littlewood, S. Schmitt-Rink, E. Abrahams, and A.E. Ruckenstein, *Phys. Rev. Lett.* **63**, 1996 (1989).

[71] C.M. Varma, P.B. Littlewood, S. Schmitt-Rink, E. Abrahams, and A.E. Ruckenstein, *Phys. Rev. Lett.* **64**, 497 (1990).

[72] A. Ino, C. Kim, M. Nakamura, T. Yoshida, T. Mizokawa, A. Fujimori, Z.-X. Shen, T. Kakeshita, H. Eisaki, and S. Uchida, *Phys. Rev. B* **65**, 094504 (2002).

[73] X.J. Zhou, T. Yoshida, A. Lanzara, P.V. Bogdanov, S.A. Kellar, K.M. Shen, W. L. Yang, F. Ronning, T. Sasagawa, T. Kakeshita, T. Noda, H. Eisaki, S. Uchida, C. T. Lin, F. Zhou, J.W. Xiong, W.X. Ti, Z.X. Zhao, A. Fujimori, Z. Hussain, and Z.-X. Shen, *Nature* **423**, 398 (2003).

[74] W. J. Padilla, Y. S. Lee, M. Dumm, G. Blumberg, S. Ono, K. Segawa, S. Komiya, Y. Ando, D. N. Basov, *Phys. Rev. B* **72**, 060511 (2005).

[75] T. Valla, A.V. Fedorov, P.D. Johnson, and S.L. Hulbert, *Phys. Rev. Lett.* **83**, 2085 (1999).

[76] D.L. Feng, A. Damascelli, K.M. Shen, N. Motoyama, D.H. Lu, H. Eisaki, K. Shimizu, J.-I. Shimoyama, K. Kishio, N. Kaneko, M. Greven, G. D. Gu, X. J. Zhou, C. Kim, F. Ronning, N.P. Armitage, and Z.-X Shen, *Phys. Rev. Lett.* **88**, 107001 (2002).

[77] D.S. Dessau, B O. Wells, Z.X. Shen, W. E. Spicer, A.J. Arko, R.S. List, D. B. Mitzi, and A. Kapitulnik, *Phys. Rev. Lett.* **66**, 2160 (1991).

[78] A.B. Migdal, *Theory of Finite Fermi Systems and Applications to Atomic Nuclei* (Benjamin, Reading, MA, 1977).

[79] Z.M. Yusof, B.O. Wells, T. Valla, A.V. Fedorov, P.D. Johnson, Q. Li, C. Kendziora, S. Jian, and D. G. Hinks, *Phys. Rev. Lett.* **88**, 167006 (2002).

[80] S. Nakamae, K. Behnia, N. Mangkorntong, M. Nohara, H. Takagi, S. Yates, and N.E. Hussey, *Phys. Rev. B* **68**, 100502 (2003).

[81] N.E. Hussey, M. Abdel-Jawad, A. Carrington, A.P. Mackenzie, and L. Balicas, *Nature* **425**, 814 (2003).

[82] S. A. Artamonov, V.R. Shaginyan, and Yu.G. Pogorelov, *JETP Lett.* **68**, 942 (1998).

[83] Yu.G. Pogorelov and V.R. Shaginyan, *JETP Lett.* **76**, 532 (2002).

[84] M. de Llano and J. P. Vary, *Phys. Rev. C* **19**, 1083 (1979); M. de Llano, A. Plastino, and J.G. Zabolitsky, *Phys. Rev. C* **20**, 2418 (1979).

[85] M.V. Zverev and M. Baldo, *J. Phys. Condens. Matter* **11**, 2059 (1999).

[86] V.R. Shaginyan, M.Z. Msezane, and M.Ya. Amusia, Phys. Lett. A **338**, 393 (2005).

[87] D. Takahashi, S. Abe, H. Mizuno, D. A. Tayurskii, K. Matsumoto, H. Suzuki, and Y. Onuki, *Phys. Rev. B* **67**, 180407 (2003).

[88] V.A. Khodel and P. Schuck, *Z. Phys. B* **104**, 505 (1997).

[89] N. Tsujii, H. Kontani, and K. Yoshimura, *Phys. Rev. Lett.* **94**, 057201 (2005).

[90] R. Bel, K. Behnia, C. Proust, P. van der Linden, D.K. Maude, and S.I. Vedeneev, *Phys. Rev. Lett.* **92**, 17703 (2004).

[91] J. Korringa, *Physica (Utrecht)* **16**, 601 (1950).

[92] G.-q. Zheng, T. Sato, Y. Kitaoka, M. Fujita, and K. Yamada, *Phys. Rev. Lett.* **90**, 197005 (2003).

[93] C.L. Kane and M.P.A. Fisher, *Phys. Rev. Lett.* **76**, 3192 (1996).

[94] T. Senthil and M.P.A. Fisher, *Phys. Rev. B* **62**, 7850 (2000).

[95] A. Houghton, S. Lee, and J.P. Marston, *Phys. Rev B* **65**, 22503 (2002).

[96] A.P. Mackenzie, S.R. Julian, D.C. Sinclair, and C. T. Lin, *Phys. Rev. B* **53**, 5848 (1996).

[97] M.Ya. Amusia and V.R. Shaginyan, *Phys. Lett. A* **315**, 288 (2003).

[98] K.D. Morhard, C. Baüerle, J. Bossy, Yu. Bunkov, S.N. Fisher, and H. Godfrin, *Phys. Rev. B* **53**, 2658 (1996).

[99] A. Casey, H. Patel, J. Nyéki, B. P. Cowan, and J. Saunders, *Phys. Rev. Lett.* **90**, 115301 (2003).

[100] M. Pfitzner and P. Wölfe, *Phys. Rev. B* **33**, 2003 (1986).

[101] S.V. Kravchenko and M.P. Sarachik, *Rep. Prog. Phys.* **67**, 1 (2004).

[102] J. Custers, P. Gegenwart, H. Wilhelm, K. Neumaier, Y. Tokiwa, O. Trovarelli, C. Geibel, F. Steglich, C. Pépin, and P. Coleman, *Nature* **424**, 524 (2003).

[103] C. Pépin, *Phys. Rev. Lett.* **94**, 066402 (2005).

[104] S.L. Bud'ko, E. Morosan, and P.C. Canfield, *Phys. Rev.* **69**, 014415 (2004).

[105] J. Paglione, M.A. Tanatar, D.G. Hawthorn, E. Boaknin, R.W. Hill, F. Ronning, M. Sutherland, and L. Taillefer, *Phys. Rev. Lett.* **91**, 246405 (2003).

[106] T. Senthil, M. Vojta, and S. Sachdev, *Phys. Rev. B* **69**, 035111 (2004); T. Senthil, S. Sachdev, and M. Vojta, *Physica B* **359-361**, 9 (2005).

[107] T. Senthil, A. Vishwanath, L. Balents, S. Sachdev, and M.P.A. Fisher, *Science* **303**, 1490 (2004).

[108] R. Küchler, P. Gegenwart, K. Heuser, E.-W. Scheidt, G. R. Stewart, and F. Steglich, *Phys. Rev. Lett.* **93**, 096402 (2004).

[109] A. Bianchi, R. Movshovich, I. Vekhter, P.G. Pagliuso, and J.L. Sarrao, *Phys. Rev. Lett.* **91**, 257001 (2003).

[110] L. Zhu, M. Garst, A. Rosch, and Q. Si, *Phys. Rev. Lett.* **91**, 066404 (2003).

[111] K. Ishida, K. Okamoto, Y. Kawasaki, Y. Kitaoka, O. Trovarelli, C. Geibel, and F. Steglich, *Phys. Rev. Lett.* **89**, 107202 (2002).

[112] P. Coleman, *Lectures on the Physics of Highly Correlated Electron Systems VI* (Editor F. Mancini, American Institute of Physics, New York, 2002, p. 79).

[113] P. Coleman and A.J. Schofield, *Nature* **433**, 226 (2005).

[114] J.W. Clark, V.A. Khodel, and M.V. Zverev, *Phys. Rev B* **71**, 012401 (2005).

[115] E.M. Lifshitz and L.P. Pitaevskii, *Statistical Physics* (Part 1, Butterworth-Heinemann, 2000, p. 168).

[116] V.R. Shaginyan, *JETP Lett.* **80**, 263 (2004).

[117] J.W. Clark, V.A. Khodel, M.V. Zverev, and V.M. Yakovenko, *Phys. Rep.* **391**, 123 (2004).

[118] M.V. Zverev, V.A. Khodel, and V.R. Shaginyan, *JETP Lett.* **65**, 863 (1997).

[119] P. Nozières, *J. Phys. I* (France) **2**, 443 (1992).

[120] S. Paschen, T. Lühmann, S. Wirth, P. Gegenwart, O. Trovarelli, C. Geibel, F. Steglich, P. Coleman and Q. Si, *Nature* **432**, 881 (2004).

[121] V.R. Shaginyan, K.G. Popov, and S.A. Artamonov, *JETP Lett.* **82**, 234 (2005).

[122] P. Gegenwart, J. Custers, Y. Tokiwa, C. Geibel, and F. Steglich, *Phys. Rev. Lett.* **94**, 076402 (2005).

[123] V.A. Khodel, M.V. Zverev, and V.M. Yakovenko, cond-mat/0508275.

[124] N. Oeschler, P. Gegenwart, M. Lang, R. Movshovich, J.L. Sarrao, J.D. Thompson, and F. Steglich, *Phys. Rev. Lett.* **91**, 076402 (2003).

[125] E. D. Bauer, J.D. Thompson, J.L. Sarrao, L.A. Morales, F. Wastin, J. Rebizant, J.C. Griveau, P. Javorsky, P. Boulet, E. Colineau, G. H. Lander, and G. R. Stewart, *Phys. Rev. Lett.* **93**, 147005 (2004).

[126] V.R. Shaginyan, *JETP Lett.* **81**, 222 (2005).

[127] A.M. Zagoskin, *Quantum Theory of Many-Body Systems* (Springer-Verlag New York, Inc., 1998).

[128] S.H. Pan, J.P. O'Neal, R.L. Badzey, C. Chamon, H. Ding, J.R. Engelbrecht, Z. Wang, H. Eisaki, S. Uchida, A.K. Gupta, K.-W. Ng, E.W. Hudson, K.M. Lang, and J. C. Davis, *Nature* **413**, 282 (2001).

INDEX

D

E

F

G

Q

R

T

U

V

W

X

Z

Y